Farm Management

Principles and Strategies

Farm Management

Principles and Strategies

Kent D. Olson

 Iowa State Press
A Blackwell Publishing Company

Dr. Kent D. Olson has taught farm management (for 25 years) and agribusiness management on campus, in extension settings, and internationally. His research and outreach work emphasizes farm level management issues, especially financial management, the interaction between farming and the environment, sustainable agriculture, decision-making, alternative production technologies, and the impact of government policies. Dr. Olson has also been the state coordinator for two farm business management associations in Minnesota. Besides whole-farm analysis, his research has concentrated on crops (mainly maize, soybean, wheat, rice, cotton, alfalfa, and rangeland). He has worked in Sweden, Poland, Uganda, and Italy.

Professor Olson grew up on a farm in north-central Iowa and graduated from Iowa State University. He was an Extension Economist in the Department of Agricultural Economics at the University of California-Davis before joining the faculty at Minnesota. He and his siblings continue to own and manage the home farm.

© 2004 Iowa State Press
A Blackwell Publishing Company
All rights reserved

Iowa State Press
2121 State Avenue, Ames, Iowa 50014

Orders: 1-800-862-6657
Office: 1-515-292-0140
Fax: 1-515-292-3348
Web site: www.iowastatepress.com

Authorization to photocopy items for internal or personal use, or the internal or personal use of specific clients, is granted by Iowa State Press, provided that the base fee of $.10 per copy is paid directly to the Copyright Clearance Center, 222 Rosewood Drive, Danvers, MA 01923. For those organizations that have been granted a photocopy license by CCC, a separate system of payments has been arranged. The fee code for users of the Transactional Reporting Service is 0-8138-0418-3/2004 $.10.

♾ Printed on acid-free paper in the United States of America

International Standard Book Number: 0-8138-0418-3

Library of Congress Cataloging-in-Publication Data

Olson, Kent D.
   Farm management : principles and strategies / by Kent D. Olson—1st ed.
      p. cm.
   ISBN 0-8138-0418-3 (alk. paper)
   1. Farm management. I. Title.
   S561.O48 2003
   630′68—dc21   2003005501

The last digit is the print number: 9 8 7 6 5 4 3 2 1

To Linda, Erik, and Marie

Contents

Preface

I have written this text to help you, the reader, learn the principles and strategies of farm management. I have included both strategic and operations aspects to help you, a future or current manager, improve the efficiency, effectiveness, objectivity, and, ultimately, success of your decisions. The tools, ideas, and methods presented in this text are designed to help you make better decisions regardless of your definition of success. Profitability means success to some. Your definition of success may also include protection of the environment, a balance of work and leisure, business growth, and many other points. No matter what your goals are, I think you will benefit from learning the principles and strategies of farm management.

This book is for managers, not record keepers and not producers. Managers need to understand and use records and financial statements, but, in this competitive world, a manager who emphasizes record keeping over managing may not be in business long. Similarly, the person who just produces, because he or she "knows what needs to be done" but does not manage the farm, will have trouble surviving.

Throughout this text, I use the term "manager" to refer to whoever makes decisions whether that person is an owner, operator, partner, female, male, family, hired management, staff, supervisor, or someone else. In a multioperator farm, the management team is the "manager." The decision can be either small or large, frequent or infrequent.

This text covers topics that are essentially new to farm management, especially strategic management and quality management and some new techniques and tools in production and operations management. These ideas are widely used in general business but have not been discussed in traditional farm management. As farming becomes more and more complicated and close to economic pressures in the world, future farm managers will need to know these new management principles and strategies in addition to the traditional ones.

Topics include strategy, marketing, budgeting, production and operations, quality, finance, investment, risk, contracts, staffing, business organization,

and the future. We start in chapter 1 with an overview of the scope of farm
management including functions, decisions, and business plans. Chapter 2
introduces the principles of strategic management: developing a vision and a
mission, setting objectives, performing external and internal analyses, crafting
and implementing business strategy, and evaluating strategic performance.
Chapter 3 covers the basics of developing a marketing plan. Chapter 4
explains budgeting terms and the process for projecting costs and returns.
Chapter 5 explains production and operations management for both the enterprise and whole-farm levels. Chapter 6 introduces quality management for
farms. Chapters 7 and 8 cover financial analysis and financial management.
Chapter 9 shows how to analyze investments such as machinery and buildings.
Chapter 10 covers the elements of valuing land for purchase and evaluating the
terms of a rental agreement. Chapter 11 shows how to manage and reduce risk.
Chapter 12 explains the different types of production contracts and how to
evaluate contracts. Chapter 13 contains the essential elements of human
resource management and business organization. Finally, chapter 14 looks at
some ideas that farmers have for adjusting to change and how they are and will
be farming in the future.

Throughout this book I use examples based on real situations but with fictitious names. However, when I use a direct quotation without attributing a
name, the quotation and the person are real.

I encourage you to check the website for my farm management class. The
website includes study notes for each lesson, virtual field trips to real farms,
more examples, new analyses, background information, and other additions
that enhance this text. You can access the website through my homepage:
http://www.apec.umn.edu/faculty/kolson/.

Success comes in many designs and shapes. I hope the ideas, tools, and
techniques in this book help you achieve your definition of success. Good luck
and thanks for reading.

ACKNOWLEDGMENTS

I have not developed and written this text in isolation. Many people have
helped me learn and explore farm management. I thank them all, but I can only
remember the names of a few. Let me start with my late father, Norris Olson,
who gave me my first lesson in farm management when I was still a preteen.
As we stood by the feedlot on our family farm near Eagle Grove, Iowa, he
described how the whole farm could be thought of as a set of individual businesses that needed to be planned both separately and as part of the whole farm.
From that lesson, I went on to learn more about farm management from
R. V. Diggins, my first agriculture teacher in high school, and Earl Heady,

Ray Beneke, Sydney James, and Ron Winterboer at Iowa State University. Graduate school conversations with Bill Boggess were wonderful. Starting as a young professional at the University of California, Davis, and continuing at Minnesota, I benefited greatly from discussions with all my colleagues, especially Karen Klonsky, Lee Garoyan, Gordon Rowe, Hal Carter, Warren Johnston, Gordon King, Ben French, Erlin Weness, Dary Talley, Lorin Westman, Perry Fales, Jim Christensen, Bob Anderson, Jim Houck, Vernon Ruttan, and, most of all, Vernon Eidman.

Many farmers have taught me about farm management starting with our neighbors near Eagle Grove and others I knew through school, 4-H, FFA, and other contacts. To the farmers and producers throughout California and Minnesota who peppered me with questions and put up with mine, I say thank you.

Over the years, my students have kept me alert and not allowed me to stop learning. Marilyn Clement, my secretary, has been through many versions of this text and has made innumerable wise suggestions for improving my work. Thank you all very, very much.

To my family, Linda, Erik, and Marie, thank you for all the help, support, patience, and love you have given me through the years. Even Eddie, our dog, helped by always being by my side.

Farm Management

Principles and Strategies

1

The Scope of Farm Management

Managers do many things. They are responsible for many things. They are responsible for making sure the business or organization accomplishes what it is supposed to accomplish.

A business manager's job can be defined as the art and science that combines ideas, methods, and resources to produce and market a product profitably. This view limits the goal of the business to making a profit, but other goals are just as legitimate. Other goals include maximizing wealth, avoiding debt, reducing labor requirements, having family time, improving the environment, and contributing to the community. To include these and other goals, a manager's job can be described as the allocation, direction, and control of limited resources to achieve the goals of the firm efficiently. This definition also recognizes that money, land, labor, and other resources are always limiting.

In this general view, farmers are no different from other managers, so let us call them farm managers. They have a complicated job in a complicated world. The *Encyclopedia Britannica* says farm management is "making and implementing the decisions involved in organization and operating a farm for maximum production and profit." It also says "farmers manage the resources under their control and in ways to obtain as much satisfaction as possible from their decisions and actions." This, I believe, is a better fit for what farm managers and their families strive to do in today's environment.

A list of a manager's activities and duties provides a glimpse of the scope of management (fig. 1.1). A farm manager needs to integrate information from the biological, physical, and social sciences. A farm manager needs to acquire, direct, and control resources, including financial, physical, and human resources. Depending on the type of farm, a farm manager must understand a variety of subjects including soil structure, soil microbiology, livestock genetics, crop and animal nutrition and growth, weed and insect management, plant

3

Set goals and objectives.
Seek, sort, compile, integrate, and use information from many sources.
Consider and analyze alternative courses of action.
Forecast the future.
Anticipate change.
Make decisions.
Carry out decisions.
Take action.
Acquire resources.
Organize the use of resources.
Communicate with creditors, employees, government agencies, and others.
Train themselves, family members, and employees.
Buy and sell inputs and products.
Establish the timing of operations.
Monitor operations.
Recognize and identify problems and opportunities.
Respond and act when problems occur.
Mediate fights.
Evaluate results of decisions and actions.
Monitor financial conditions.
Monitor progress toward goals and objectives.
Accept responsibility.
Check on everything.

Figure 1.1. Examples of management activities. (Adapted from R. A. Milligan and B. F. Stanton, "What do farm managers do?" *1989 Yearbook of Agriculture: Farm Management*, United States Department of Agriculture, 3.)

and animal diseases, ecology, machinery management, economics, financial management, international food markets, leadership, human psychology, business organization, business law, communication, and strategic and operations management.

Many people and businesses have multiple goals. For farmers, multiple goals are due in part to the fact that farming is often both a production and consumption activity—life and work are closely intertwined. Some goals can be achieved together; some may conflict with one another. Goals can also be nested within other goals. For instance, a person may want to maximize income within the limits of time for family and protecting the environment. Goals may also be expressed as satisfaction not maximization. People may choose alternatives to satisfy income needs and maximize net worth, for example. Many people balance their income goals with their environmental concerns, community and family activities, and their ethics and morals in all their decisions. The balancing of goals may be done by never violating some basic ethical principles and then a personal trade-off between, say, income and community time. Of course, the different weights or importance put on the different goals and thus the trade-offs are the source of many conflicts within families and management teams. So conflict resolution skills are also needed.

Because the required knowledge and skills are so great, consultants and advisors are available so a manager does not need to be an expert in all these fields. In fact, a person does not need to own land to be a farm manager; many landowners hire professional farm managers to manage their farms (fig. 1.2). This text is an introduction to the skills needed to bring all this knowledge to bear on the management of one farm business.

The goal of many farm managers is to manage their own farm, but this is not the only career possible in farm management. Professional farm managers manage farms for land owners who choose to not be involved in the farm directly. The land owner may be a retired farmer, live far from the land, or simply view the land as an investment. Professional farm managers may work independently or be part of a local, regional, or national farm management company. Many are employed by banks and other institutions. These professional farm managers represent their clients in the management of the firm. Depending on the needs of their clients, the professional farm manager may take care of all aspects of farm management or selected parts of farm management. Besides the tasks listed in Figure 1.1, professional farm managers do many other tasks, including but not limited to the following:

Select the most capable farm operator available.
Estimate the benefits and costs of alternative leases.
Estimate the value of hiring a custom operator.
Negotiate lease terms.
Visit the farm to ensure good management is being practiced.
Consult with and make recommendations to the operator and owner.
Oversee construction of buildings and land improvements.
Supervise product marketing and, if necessary, input purchases.
Pay all proper expenses.
Check on and apply for insurance and government programs.
Prepare reports on production progress and results, financial statements, asset conditions, and other topics for the owner.
Project budgets for the next year.

Figure 1.2. Professional farm management.

THE FARMING ENVIRONMENT

Farm management is a complicated and demanding task. The farm business is surrounded by uncertainties, demands, and changing conditions in an unfriendly environment. This environment can be described as having four main components: resources, markets, institutions, and technology (fig. 1.3).

Resources

A farm's available *resources* define the farm. Resources include land, labor, capital (such as machinery, buildings, livestock, and supplies), and management skills as well as credit available and the farm's climate and weather. Information sources are also part of the farm's resources. Information sources include magazines, newsletters, marketing clubs, consultants, management services, extension staff and other advisors, and the World Wide Web. A farmer's own entrepreneurial ability is a critical resource. Perhaps the first resource or asset that should be listed is yourself and the skills and knowledge you have.

Markets

The *markets* are where products and inputs are bought and sold. The markets are where prices are discovered. Although agricultural markets are regulated

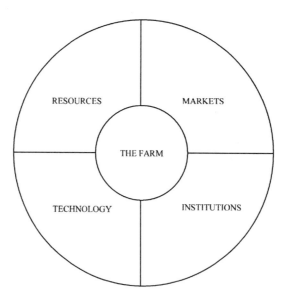

Figure 1.3. The economic environment of the farm.

to some extent depending on the commodity, prices are not set and known. Prices are discovered through producers, processors, consumers, buyers, and sellers talking together, sharing information, and settling on the price for that moment and place. Together these players determine the proximity, size, and price level of the market. Competition from neighbors, other countries, other food sources, and other products can affect farmers' profit margins and thus managerial choices.

Institutions

Many *institutions* affect a farmer's choices. The federal government with its policies and regulations is often seen as the biggest institution affecting farmers. However, state and local governments and their policies and regulations affecting credit, land use, contracts, and many other parts of business life also have a large impact on farmers. Environmental regulations are institutional constraints. Banks and other credit institutions have a large impact. Social institutions, such as community groups, churches, and even coffee shops affect the alternatives considered and ultimately chosen.

Technology

Technology is more than machines. It includes physical, mechanical, and biological processes and techniques and the information and management knowledge required to use them. Many scientists—engineers, plant breeders, soil scientists, animal scientists, geneticists, plant pathologists, system analysts, economists, management scientists—give us technology. Past, current, and future technology choices affect the productivity and, thus the viability of the farm.

A farm manager has to be concerned with and aware of the entire environment. At times, one area may be emphasized, say, technology at the time of a building program, but over time a balance is needed. The manager cannot look at one aspect only and ignore the others. Changes can happen that create problems or opportunities. If these changes are missed, the problems may become larger or the missed opportunities prove costly in terms of what is not gained.

THE FUNCTIONS OF MANAGEMENT

Management has four major functions: planning, organizing, directing, and controlling. They are separate yet intertwined. In any well-managed business, these four functions occur simultaneously with some variations in the attention paid to each function. When the business is in full swing, organizing and directing can dominate, but planning and controlling can be just as critical to the success of the business.

These management functions should not be confused with business functions. Business functions include production, marketing, finance, and personnel management. They are found in different parts of the business. Management functions are not isolated in different parts of the business but are carried out in each of the business functions. Crop production, for example, needs planning to decide which crops to grow, organizing to gather the inputs, directing to produce and market the crop, and control to decide whether an insecticide application is needed. Likewise, financial management needs planning to decide how much credit is needed, organizing to obtain credit, directing to use the credit when needed, and control to check that the correct credit terms have been used.

In general business terms, a farm manager is a general manager, that is, a manager who is responsible for many, if not all, parts of a business. As a general manager, a farm manager is responsible for all management functions and all business functions although the manager may not personally or physically do all these functions.

Planning

Planning is the determination of the intended strategy and course of action (that is, the plan) for the business. The basic process of planning generally involves the following steps:

1. Appraising the goals and objectives of the business owners and operators
2. Assessing the industry in which the farm will operate
3. Preparing an inventory of the farm resources available and analyzing the farm's situation within the industry
4. Selecting alternatives to be analyzed
5. Determining physical inputs and probable production for each alternative
6. Selecting prices to apply to the input and production data
7. Calculating expected costs and returns for each alternative
8. Estimating the potential range of costs and returns due to variability of prices, costs, and yields
9. Analyzing the probable results of the alternatives
10. Developing an operating plan based on probable results and the goals and objectives of the business, manager, and (or) owner

Part of planning also is the establishment of control standards—the benchmarks by which the manager wants to evaluate how well the plan is being carried out. Another part of planning is the response to past control results, that is, feedback control. Because of feedback, plans may need to be modified because prior results have not provided the desired results.

Tools of the planning process include budgets, economic analysis methods, and information on the physical needs and responses of plants and animals. Larger farms, more complex farms, and more people will increase the need to use these tools formally with written results.

Budgets report the quantified estimates of expected results due to carrying out a specific plan or set of actions. There are four main types:

- Enterprise budgets: Expected costs and returns for specific enterprises
- Whole-farm budgets: Expected costs and returns for a specific combination of enterprises
- Cash flow budgets: A record of cash transfers into and out of the business listed by specific period
- Partial budgets: A listing of only those costs and returns expected to change due to a proposed change in the business

Economic tools include microeconomic and macroeconomic principles, investment analysis and other financial tools, price forecasting, operations scheduling, risk management, and other techniques explained in this text.

Plans will vary depending on length of planning horizon, problem, or part of the business involved. Usually, a plan includes some general policies, procedures, or strategies to be carried out, and specific methods to be followed. Effective plans have several common elements:

1. They are based on attainable goals and objectives.
2. They reflect the manager's knowledge, skills, and interests.
3. They consider the quantity and quality of land, labor, and capital resources available to the manager.
4. They consider information about comparable operations and the past performance of the ranch or farm, but they are unique to a specific farm or ranch rather than being copies of other operations or models.
5. They are based on expectations for the future, not mere repetition of past practices.
6. They include short, intermediate, and long-term components.
7. They include strategies for managing production, marketing, finances, and personnel.
8. They provide methods for monitoring and evaluating situations and results (that is, the "control" function of management).
9. They are kept current, and are designed so they can be modified when objectives, resources, knowledge, expectations, or conditions change.

Policy and procedure statements are also the result of planning. They are guidelines to operational or short-run decision making. They provide established routine and thus operational efficiency for a business. Policy and procedure statements communicate decisions between individuals and through time. Although they need to be reevaluated as conditions and knowledge change, policy and procedure statements help avoid "reinventing the wheel" each time a decision is faced. The chosen feeding program on a livestock farm is an example of a policy and procedure statement. By deciding beforehand how and what to feed animals at different weights and stages, the daily feeding routine is quicker and more efficient than evaluating prices and nutritional needs every day. Whether these policy and procedure statements are formally written or informally discussed depends on the size and complexity of the business.

Organizing

Organizing involves the acquisition and organization of the necessary land, labor, machinery, livestock, capital, and management resources. It includes the negotiation of the terms necessary for resource acquisition. Organizing involves deciding who is responsible for what in the business and who reports to whom; that is, organizing involves determining the organization chart. Organizing also involves choosing whether the business will be a sole proprietorship, partnership, corporation, or other legal form.

Directing

Directing the chosen plan involves the coordination of the land, labor, machinery, livestock, capital, and management resources. It involves directing of physical activities; scheduling production and marketing; and recruiting, selecting, motivating, and directing labor resources—that is, personnel management. An effective director uses leadership, supervision, delegation, communication, motivation, and personnel development as primary management tools.

Controlling

Controlling the farm's performance consists of two parts: (1) comparing the results of implementing the chosen plan with the farm's initial goals and objectives and (2) taking corrective actions, if needed. Evaluation is usually done with budgets, records, and financial statements. "Keeping the books," although important, is only part of control management. Many farmers are interested in keeping the books only as a scorecard, to find out how much money they made or because they are required to file income tax forms. They realize it is too late to influence the results, so keeping the books is, for most, a boring activity. So an important part of control management is the development of alternative measures or clues that can predict good or poor results while it is still possible to change plans.

Once evaluation has pointed out problem areas, the manager needs to determine possible causes for these problems. Causes may be improper planning or implementation, improper setting of norms and standards, or changes in external forces, such as the economy, weather, or government policy. Finally, the manager must decide what actions will regain control of the situation. Even if the problems are not correctable, the manager must decide what actions are necessary to regain control. A control plan will have the corrective actions specified for situations in which the measured variable falls outside the "in-control" range.

The control function completes a circle that began with planning. A farm manager may deal with all three functions on any given day if the business is small. In large businesses, these functions may be divided between several employees. In either case, communication between the functions is essential.

THE DECISION PROCESS

A standard feature in farm management texts is a list of decision-making steps (fig. 1.4). Since these steps are usually presented as a linear list , they are often thought of as a set of steps that one follows in a sequence from beginning to end. However, this is not always the case. One group of researchers describe decision making as a "groping, cyclical process" (Mintzberg, Raisingham, and Théorêt 1976). The process usually has many loops and feedback steps as new information is obtained and consequences are estimated and considered.

---

1. Determining values and setting goals
2. Problem detection
3. Problem definition
4. Observation
5. Analysis
6. Development of intention
7. Implementation
8. Responsibility bearing

Figure 1.4. A linear model of decision making.

Matrix Model

Based on interviews with farmers, my colleagues and I have revised this traditional model of decision making (Öhlmér, Olson, and Brehmer 1998). Instead of eight functions, our decision process has a combination of four phases and four subprocesses. The phases are problem detection, problem definition, analysis and choice, and implementation. The subprocesses are information search and paying attention; planning; estimating consequences, evaluation, and choice; and bearing responsibility. This model is best viewed as a matrix, not as a list (table 1.1). A matrix better reflects the nonlinear process of making decisions.

In this matrix model, a farmer's values and goals should be developed before any decision process is started. Observation (included in the subprocess of information search and paying attention) is part of every phase. Development of intention is part of the subprocess of bearing responsibility and checking the choice. Bearing responsibility is also seen in all phases. A farmer knows that he or she is responsible for meeting values and goals; this concern is what starts the process. Bearing responsibility is the driving force behind searching for problems and opportunities, defining the problem and solution alternatives, analyzing and choosing the best alternative, and implementing the decision.

Besides revising the traditional model, we saw five characteristics in farmers' decision making. First, farmers continually update their problem perceptions, ideas of options, plans and expectations when new information is obtained. Second, farmers often use a qualitative approach to forming expec-

Table 1.1.   A Revised Conceptual Model of the Decision-Making Process

Phase	Subprocess			
	Searching and paying attention	Planning	Evaluating and choosing	Bearing responsibility
Problem detection	Information scanning, Paying attention		Consequence evaluation, Does a problem exit?	Checking the choice
Problem definition	Information search, Finding options		Consequence evaluation, Choose options to study	Checking the choice
Analysis and choice	Information search	Planning	Consequence evaluation, Choice of option	Checking the choice
Implementation	Information search, Clues to outcomes		Consequence evaluation, Choice of corrective action(s)	Bearing responsibility for final outcome, Feed forward information

Source: Öhlmér, Olson, and Brehmer 1998.

tations and estimating consequences expressed in directions from the current condition. Third, farmers tend to prefer a "quick and simple" decision approach over a detailed, elaborate approach. Fourth, farmers prefer to collect information and avoid risk through small tests and incremental implementation. Fifth, during implementation, farmers continually check clues to form their evaluation of long-run actions in a feed forward and compensation approach, rather than a postimplementation evaluation.

Another Approach to Decision Making

Another approach to decision making is to define the type of problem (or opportunity) as analysis, management, or design (table 1.2). Knowing the type of problem can help a manager focus on what needs to be done. These three types are defined based on which of three components (inputs, outputs, and structure) are not known. Inputs are, for example, land, labor, fertilizer, new equipment, animals, money. Outputs are the products: for example, corn, milk, calves, vegetables, custom work. The structure is the people, processes, rules, and current equipment that interact to create the outputs from the inputs.

In *analysis problems*, one must predict output based on given inputs and structure. Examples of analysis problems are the common "what-if" questions, building budgets, and forecasting prices and yields. In *management problems*, one must identify a feasible set of controllable inputs that will produce the desired output given the structure. Examples of management problems are cropping plans, production schedules, and marketing plans. In *design problems*, the task is to choose a structure that will produce the desired output from the available inputs. Examples of design problems are facility and building designs, incentive pay systems, and organizational design.

THE FARM BUSINESS LIFE CYCLE

Most businesses change as the age of the people and the business changes. Goals, activities, and involvement change. These changes can be understood

Table 1.2.   Three Kinds of Problems: Analysis, Management, and Design

System component	Problem Type		
	Analysis	Management	Design
Inputs	given	unknown	given
Outputs	unknown	given	given
Structure	given	given	unknown

by considering the three stages of the life cycle: (1) entry or establishment, (2) growth and survival (sometimes called expansion and consolidation), and (3) exit or disinvestment.

These stages are readily apparent with family farms—especially sole proprietorships. They are less apparent but still evident in corporations and partnerships.

Entry or Establishment Stage

The prospective farmer evaluates farming as a career versus other career opportunities. If farming is chosen, a person must acquire a minimum set of capital resources and managerial skills necessary to establish an economically viable business. Historically, the entry stage has been characterized by a career progression known as the "agricultural career ladder." The traditional ladder consisted of the following steps: begin as a hired worker or as family labor, gradually acquire a basic set of machinery or livestock, rent land or buildings, establish part ownership of farmed land, and become the principal operator of the farm business.

A common variation of the ladder involves, first, part-time farming with an off-farm job for additional income and, then, through growth, farming as a full-time occupation. Sometimes part-time farming is the goal. This can be the rational choice of many due to personal and family goals and the situation in which the farm and family find themselves.

Other methods of entry may be necessary as the money needed to start farming continues to increase. This increase is due to machinery and equipment being substituted for labor and rapid increases in the costs of machinery, buildings, land, and borrowing. Some other methods are:

- Become a partner of an established farmer who may or may not be a relative—expansion is a key at this stage
- Obtain financial backing from a third party to purchase capital and land
- Become a laborer/manager in an existing business and gradually invest in the business
- Marry into a farming family

The keys to entry are management skills, land tenure, and capital. Possessing strong management skills is critical to gaining tenure and financial resources.

Corporations and partnerships will also have an entry and establishment phase. It may be shorter than for a sole proprietorship because in farming, corporations and partnerships are often starting from an established farm or farms. The establishment phase may be the learning phase for the owners and operators. They are learning how to manage together and communicate with each other.

Often a corporation or partnership is established to allow for growth of the farm or for transfer of ownership. In these cases, the business not only is in the establishment stage but also is in the growth or disinvestment stage.

In the entry and establishment phase, goals such as net worth accumulation and resource productivity usually have a higher priority than income and consumption. The basic goal is to establish both the business and a performance record on which to build the future. Persons in this stage may be willing to take riskier actions for two reasons. First, since they likely have less money invested in the business, they have less to lose. Second, if the business is not successful in this stage, it will not be around for growth. Also, since this stage often involves younger people, they are more willing to take risks because they have more time to try something else if this venture does not work.

The Growth and Survival Stage

This stage involves strategies to improve efficiency and expand the resource base of the business. The manager recognizes resource bottlenecks and inefficiencies and makes needed corrections. Financial considerations are important. The survival of the farm requires risk management, that is, taking measures to avoid the adverse effects of the market, weather, and disease. Depending on the persons and the business situation, goals such as income, community support, and free time may increase in importance in this stage.

The Exit or Disinvestment Stage

In this stage, the farmer plans for and reaches retirement. Resources and managerial responsibilities are transferred to the next generation or other parties. The farmer may sell other assets but retain ownership of the land. Most people in this stage place a high priority on avoiding risk and providing for their retirement years—perhaps balanced by the goal to transfer the farm to the next generation.

A corporation or partnership may or may not go through this stage. If it were set up to transfer the farm from one generation to another, a corporation or partnership might have owners going through exit and disinvestment. However, the next generation having already started will probably keep the business in the growth and survival stage.

A BUSINESS PLAN

Volatile prices for farm inputs and products create adjustment pressures for farm managers that only add to the uncertainty caused by weather and biological production. Managers must continually monitor and adjust their strategic and operating plans to control costs, maintain profits, and accomplish other goals. With high interest rates and small profit margins, managers are forced to keep a close eye on cash-flow requirements. Traditional rules of thumb cannot be trusted as the economic environment changes. In times of economic stability, this year's plan could be as simple as repeating what was done last year. In times of change and uncertainty, it does not work that way. Today, the safest prediction is that the future will not be a continuation of the past.

In an uncertain environment, a farm business needs a structured approach to planning. A written business plan provides that structured approach. As shown in figure 1.1, a farmer must choose and do many things. Rather than making decisions "on the fly" or "by the seat of their pants," successful farmers spend considerable time preparing a plan for the future. The plan needs to be for the short term (that is, today or this year) and for the long term (say, the next five to ten years or longer).

Few managers can afford large speculative agricultural and economic experiments. Most people would prefer to assess the economic consequences and risks of alternative actions, on paper, before implementation than in grim reality afterward. Yet knowing what to put on paper to produce these assessments is no easy task.

The common method of assessment of alternatives and choices is the preparation of a business plan. Making a business plan not only forces a manager to make and defend choices, but also provides a document to show to creditors, investors, and customers. A standard business plan includes an analysis of the farm's industry, the farm's position within that industry, and the farm's plans for marketing, production, financing, organizing, and staffing. A sample outline is shown in figure 1.5. An individual farm may not need to complete every part of this outline. However, by considering every section, the farmer will have considered all the necessary points. The Center for Farm Financial Management has developed software to help farmers prepare a business plan. It is available at http://www.cffm.umn.edu/Software/BusinessPlan/.

Every farm should have a written business plan. The act of writing will improve thinking, decisions, communication, and memory. The size and complexity of the business will determine how detailed the plan is. A farm with two or more operators, partners, or owners will benefit from the communication required to prepare a written plan. A major expansion obviously requires a major effort to prepare a plan that will stand up to intense scrutiny by potential creditors or partners. However, a small or part-time farm can benefit from a written plan as a record of why decisions were made.

I. Executive summary
II. General description of the farm
       A. Type of business
       B. Products and services
       C. Market description
       D. Location(s), legal descriptions
       E. History of the farm and operators
       F. Owners, partners, operators
III. Strategic plan
       A. Vision, mission, goals, and objectives
       B. External analysis
       C. Internal analysis
       D. Chosen business strategy
       E. Strategy evaluation and control
IV. Marketing plan
       A. The target market
       B. Pricing strategy
       C. Product quality management
       D. Inventory and delivery timetables
       E. Market risk and control management
V. Production and operations plan
       A. Production process
           1. Product choice
           2. Product and process design
           3. Technology choice
           4. Environmental considerations
       B. Raw materials, facilities, and equipment
       C. Location of production
       D. Management of process quality
       E. Production risk and control management
       F. Production and operations schedule
VI. Financial plan
       A. Financial statements: historical and projected
           1. Balance sheets
           2. Income statements
           3. Cash flow statements
           4. Ratio analysis
       B. Capital needed
       C. Investment analysis
       D. Financial risk and control management
VII. Organization and staffing plan
       A. Personnel needs
       B. Sources of personnel
           1. Owner and other family labor
           2. Hired employees
           3. Consultants
       C. Structure and responsibilities
           1. Business organization
           2. Brief job descriptions
       D. Basic personnel policies
           1. Compensation
           2. Evaluation
           3. Training
       E. Workforce risk and control management

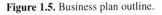

Figure 1.5. Business plan outline.

If the actual results do not occur as planned, a written record can help any manager evaluate what went wrong, whether better decisions could have been made, and how the business could have protected itself better. Without a written record, incorrect memories may cause incorrect changes.

The Plan

The business plan starts with the *executive summary*. The executive summary is for the reader who does not read the entire plan. The executive summary should be a concise discussion of the strategic direction of the farm, how this strategic plan will be implemented, and what resources are needed. The rest of the plan is required to provide details and rationale for direction and resource requirements.

The *general description* of the farm is a short summary of where the farm is economically, physically, and historically. Any creditor will want to know the names of the owners, partners, and operators to provide both a human side to the management and the reputation of these people.

The *strategic plan* lays out the farm's vision and direction. It describes the farm's view of its chosen industry and how the farm is seen fitting into that industry. This section shows how and why the strategy was chosen and how it will be evaluated.

The *marketing plan* shows how the farm's products will be marketed physically and the initial plan for pricing. The plans for monitoring market conditions and for changing the marketing plan should be spelled out. For instance, if the plan is to use options to price corn, what indicators will be monitored to make buy-and-sell decisions? What are the contingency plans for sudden market shifts? The marketing plan should also include some discussion on the purchasing plan for inputs such as fertilizer and feed. Are advance purchases needed? Do feed requirements need to be hedged on the futures market?

The *production and operations plan* will probably take the most time to prepare and, if prepared well, will save the most time during organization and directing. In this section, the reasons for choosing certain production processes are listed. The amount and timing of raw materials such as equipment, fertilizer, seed, feed, and livestock are listed. The plans for controlling quality are described. Contingency plans for bad weather or equipment breakdowns are spelled out.

In the *financial plan*, historical and projected financial statements are presented. Capital requirements and timing are shown. The plans and steps taken to protect the business financially need to be listed. Although it is usually seen as extremely painful and time-consuming, the work put into the marketing plan and the production plan should make the preparation of financial statements a matter of assembling numbers in the right places. Then interpretation of the statements can take precedence over preparation.

The final section describes the *plan for organizing and staffing* the business. Again, the size and detail of this section will depend on the size and complexity of the farm business. However, even a sole proprietorship needs to consider whether additional labor and additional knowledge (such as consultants) are needed. Writing and discussing job descriptions can be very helpful for employees and bosses and even for family members and longtime partners. Just saying (or worse, thinking), "we'll do it," can be detrimental to obtaining credit, the long-term success of the business, and family harmony.

The rest of this book shows how to complete a rudimentary business plan. This is not the complete and final word on farm management, but it introduces all the pieces of management needed to plan, organize, direct, and control the farm business.

SUMMARY POINTS

- Farm management is allocation, direction, and control of limited resources to achieve the goals of the farm and farm family.
- Farmers are general managers not just production managers and responsible for many activities on a farm.
- People have many goals, which means trade-offs may be needed between goals.
- The business environment of a farm has four main components: resources, markets, institutions, and technology.
- The four major functions of management are planning, organizing, directing, and controlling. Management functions differ from the business functions of production, marketing, financing, and personnel management.
- Planning is the determination of the intended strategy and course of action.
- Organizing is the acquisition and organization of the necessary land, labor, machinery, livestock, capital, and management resources.
- Directing is the coordination of resources to produce the chosen products.
- Controlling consists of comparing actual results to expected results and taking corrective actions when needed.
- Making decisions is not a linear process but involves many loops, feedbacks, and iterative steps.
- Four phases of the decision process are problem detection, problem definition, analysis and choice, and implementation.
- Four subprocesses of decision making are information search and paying attention; planning; estimating consequences, evaluation, and choice; and bearing responsibility.
- The business life cycle has three stages: entry or establishment, growth and survival, and exit or disinvestment.
- Preparing a business plan provides a structured approach for planning and helps ensure that important points are considered.

REVIEW QUESTIONS

1. What do you think are the most important things farm managers do? Use the list in figure 1.1 and develop your own ideas.
2. Describe farm management in your own words.
3. Picturing a farm you know (or one you can imagine), describe the environment in which it operates. Include some items from each of the four areas mentioned: resources, markets, institutions, and technology. Do you think this farm is keeping a proper balance with these four areas?
4. Is profit maximization the only goal of farmers? If not, what are five other, common goals that farmers may have?
5. Describe and contrast the four functions of management: planning, organizing, directing, and control.
6. How would you describe the decision-making process?
7. One psychologist developed a theory of decision making called the "garbage can theory" because problems and solutions seemed to meet by chance. How does that description compare with the decision-making process described in this chapter?
8. Do a person's goals change over time? Why can two generations have different goals?
9. Why is a business plan needed? What size of farm needs to prepare a written business plan?
10. What are the major sections in a typical business plan outline? Describe what is included in each section.

2

Strategic Management

Strategy is the pattern of actions used by a farmer (or any business or organization) to accomplish goals and objectives. Strategy consists of the moves and approaches crafted to strengthen the farm's position, satisfy customers, achieve performance targets, and accomplish long-term goals. Having a strategy helps a farmer make good choices among alternative courses of action in an uncertain world.

Strategy consists of the answers to four questions. The questions look simple, but answering them well requires considerable thought, work, and communication with the people involved in the farm.

1. What do we want to do?
2. What do we bring to the table?
3. Where should we put our efforts and why?
4. What do we need to do to compete, survive, and meet our goals?

Answering these questions should not be seen as a one-time event. Strategy is not a big, complicated plan that never changes, rather a farmer needs to keep watching and answering these questions anew as the world changes. A farmer's actual strategy is made up of both planned actions and relatively quick reactions to changing circumstances.

Strategic management consists of planning a strategy, implementing it, controlling the outcomes of the implementation, and adjusting the chosen strategy over time as conditions change (fig. 2.1). *Strategic planning* involves selecting the vision, mission, values, and objectives of the farm; analyzing the external and internal environments; and crafting the best strategy for the farm in its current environment. *Strategy implementation* involves designing organizational structures and procedures needed for the strategy and obtaining and

21

Strategic Planning
    Identification of stakeholders
    Development of vision, mission, objectives
    External analysis
    Internal analysis
    Crafting strategy

Strategy Implementation
    Obtain and organize the farm resources
    Direct the resources

Strategic Control
    Measure and evaluate farm performance
    Monitor external events
    Take corrective actions as needed

Figure 2.1. The elements of strategic management.

directing the resources needed to put the strategy into action. *Strategic control* involves designing control systems, comparing actual results to goals and objectives, monitoring the business environment, and modifying the organizational structure, implementation plan, or even the strategy to meet goals and objectives. These three parts of strategic management are examined in this chapter.

GAME PLAN

The sports metaphors of choosing a "game plan" or "positioning the business" are often used when describing strategic management. Good team players read the playing field before deciding what to do during the game. They know their own strengths and weaknesses. They see where their own team members are and know what their strengths and weaknesses are. They see where the competitors are and know what their strengths and weaknesses are. They see where the ball is relative to the goal. They see the opportunities and threats and move to the best position to help the team accomplish its goal. Wayne Gretzky, a former player with the Los Angeles Kings and a member of the hockey hall of fame, described his strategy for hockey: "I don't skate to where the puck is. I skate to where the puck will be."

Using this sports metaphor, a farmer crafts a strategy by understanding the business environment, seeing where and what is happening, looking for strengths and weaknesses in both his or her own farm and in the competition, and then the farmer moves the farm to the best position to take advantage of opportunities, to protect it from threats, and to help accomplish goals and objectives. Successful strategic management will position the farm, not where the profits were or are, but where they will be in the future.

CRAFTING A STRATEGY

Crafting a strategy can help keep a farmer, as manager, focused on what is truly important when making decisions that will affect the success and survival of the business. Short-term opportunities (a good deal on machinery, for example) or threats ("sign now or lose this chance," for example) may create distractions and possibly decisions that do not fit the chosen strategy or contribute to long-run goals. When a long-run strategy has been developed, it can be used to evaluate potential opportunities and threats for their impact on the strategic goals of the farm.

However, let us return to the idea that crafting a strategy is not a one-time event. Short-term opportunities and threats should not be ignored completely. They may signal that the business environment has changed so that a strategy needs to change. Taking advantage of land unexpectedly becoming available in an economic downturn may signal the need to reevaluate a strategy that does not include growth. Thus, paying attention to short-term events is part of scanning the environment, as described later in this chapter.

The need for strategic management is as vital for a one-person or one-family farm business as it is for a multipartner, multiemployee farm. The sole farmer's focus can be diverted from long-term goals just as easily as a multiperson farm management team. If the sole proprietor has taken the time to develop a strategy, he or she can use that strategy to guide his or her own day-to-day decisions.

A disciplined strategic management process has many advantages over the potential results of following decisions made on the basis of freewheeling improvisation, gut feeling, good deals, and drifting along. A good strategic management process will:

1. Stimulate thinking about the future
2. Make farmers more alert to the winds of change, new opportunities, and threatening developments
3. Provide clear statements of the farm's goals for both a single owner-operator farm and a farm with many employees
4. Help coordinate the numerous strategy-related decisions by managers and partners across the organization or as the sole proprietor performs different functions for the farm

5. Create a more proactive management posture and counteract tendencies for decisions to be reactive, defensive, or spur of the moment
6. Encourage the allocation of resources, capital, and staff to areas that will support the chosen strategy and help attain strategic and financial objectives
7. Allow for the constant adjustment of the business model to produce sustained success in a changing environment
8. Help farmers become better decision makers

While every farmer may not see every advantage, all farmers can benefit from strategic thinking and management. Large farms can benefit from focusing and communicating. Smaller farms can benefit because their owners are doing so many tasks that focusing on strategy for some duties can give direction and focus to other duties.

Following the steps of strategic management will not guarantee success. But farmers who follow a strategic management process do have a higher probability of success in meeting both strategic and financial objectives. Farmers who have good intentions, work hard, and hope but do not look into the future or outside the farm do not have as high a probability of success. The difference between these two types of farmers lies in the advantages listed above.

STRATEGIC PLANNING

Strategic planning has five elements that will provide the answers to the questions posed at the beginning of this chapter.

1. Identification of stakeholders and their values, philosophy, and ethics
2. Development of the farm's vision and mission, clarifying values, and setting objectives
3. External analysis of the farm's competitive environment including the general economy and the industry to which the farm belongs (e.g., dairy or corn)
4. Internal analysis of the farm's operating environment
5. Crafting the strategy for the whole farm and a complementary strategy for each part of the farm (corn production, hogs, custom work, for example) and each management function (marketing and finance, for example)

Even though these elements of strategic planning are listed in a linear fashion, they are done with many loops backward and forward to other elements of the process. Each element needs to be done (and many times redone) before the final strategy is determined for the farm in its current situation.

Identification of Stakeholders

A farm involves more than one person. Many people are interested in the farm itself, the outcomes (e.g., profit, products, runoff), and the processes of farming. The farmer, the farm family, the partners, the creditors, the community all hold an interest—a stake—in the farm. If these stakeholders are ignored in strategy development, conflict will likely occur. As a result the farm may not operate as efficiently as possible and may not obtain the goals envisioned by the farmer and the farm family. Thus, identifying the stakeholders in the farm is a critical step in strategic planning.

Stakeholder groups are people, businesses, and institutions that have some claim on the farm. Stakeholders can be divided into internal and external stakeholders. Internal stakeholders include the farmer, the farm family, partners, employees, other owners or stockholders, and creditors. External stakeholders may include the farm's customers, suppliers, governments, unions, competitors, local communities, and the general public.

The farm owner/operator is an obvious internal stakeholder. On most farms, members of the farm family are also included as internal stakeholders. The family is very concerned and interested in what the farm does, how it is done, and the results. Because they feel so close to the farm, family members may want, and even demand, to be part of strategic planning even if they neither work on nor own part of the farm. Older and younger generations of a farm family may be, or consider themselves to be, stakeholders also. For some families and circumstances, the yet-to-be-born generation may also be considered as stakeholders in the farm.

If these potential stakeholders are not recognized and their connections are not dealt with in some manner, the farm may have trouble formulating and implementing a plan for the future. At the least, social friction in the family may result if the actual and perceived connections are not dealt with explicitly.

Values

Sharing the same values within a management team is also critical to the success of a farm. By identifying the internal stakeholders, they can discuss and identify their shared values, their philosophy of business, and their operating ethics for the farm. These values, philosophy, and ethics can be communicated to employees, suppliers, creditors, and others doing business with the farm to help the farm accomplish its objectives.

Values, philosophies, and ethics can directly affect both what a farm produces, what production methods it chooses, and how it relates to its employees, customers, suppliers, and the community. These values, philosophies, and ethics come from personal beliefs, social mores, religion, humanitarian relationships, environmental concerns, and so on. They also include beliefs in how hard a person should work, what time a person should start work, how many hours a day a person should work, how much leisure time is needed, the value of manual work compared to desk work, and the need to be fully honest with others. These seemingly obvious values to some may be viewed differently by others. Differences over even small things like the time to start work in the morning can create problems or be the manifestation of other problems within the management team.

Concern for the environment and farming's impact on the environment is another example of the role of values. Stakeholders on one farm may hold the value that using any chemically processed herbicides, insecticides, or fertilizers is not good for the environment. These values will likely push this farm into using organic production methods and selling in the organic market. Another set of stakeholders on another farm may have a multiple set of values: to produce food as inexpensively as possible without harming the environment. This farm may use some chemicals but not others. It may also evaluate the topography of the land and not farm land on steep slopes and close to water. The public debate over the use of chemicals and the adoption of biotechnology can be viewed as a debate over society's values about the environment and even society's beliefs of what does or does not harm the environment.

The importance of shared values, philosophies, and ethics cannot be overestimated. If the internal stakeholders share the same values, philosophies, and ethics, the rest of strategic planning and even day-to-day management is greatly simplified, and the probability of success is greatly enhanced. A unified team does not guarantee success, however. Goals and objectives may still not be accomplished due to events outside the farm. If differences are not discussed and reconciled within the internal stakeholders, there is no question that conflict will occur and goals will not be accomplished.

The reason to discuss and develop a set of ethics for any business is to give people the tools for dealing with moral complexity and the ability to identify and consider the moral implications of their decisions. For example, knowing that all internal stakeholders believe in complete truthfulness will keep a farmer or family member from making decisions that would violate that ethic even though higher profits are given up in a low-income year.

Some people argue that businesses (including farms) should also incorporate social responsibility or social criteria and goals into their decision-making process. Part of this group has a very strict environmental ethic that cannot be violated regardless of the profit potential. Another part of this group says that social responsibility is a sound financial investment for the business.

The opposing argument is that a business' only social responsibility is to use its resources wisely to increase profits as long as it obeys laws and engages in open and free competition without deception or fraud. The differing strategies that may result due to differing ethics can easily be seen. One group may choose to consider only organic production methods. Another group may be very careful in its use of processed inputs. Another group may be willing to use any chemical input that is labeled for the crops and livestock on the farm. Each person and farm needs to choose which ethic is to be followed.

Vision, Mission, and Objectives

Vision and mission are almost interchangeable terms. A business mission is usually seen as dealing more with the present and near future while a vision deals with the long term. A mission statement defines a farm's current business directions and goals and indicates what a farm is trying to do for its customers. In contrast, a strategic vision is the picture of what the stakeholders want the farm to look like in the future, say ten years or more into the future.

Together, the vision and mission statements should have three main elements:

- A statement of the overall vision and mission of the farm
- A statement of management's key philosophical values
- Key objectives that management is trying to accomplish

In today's world, a farm's mission statement needs to define its business using a customer orientation rather than a product orientation. That is, a farm needs to view its business in terms of what customers it is producing for, not what products it is producing. Abell (1980) provides three questions to help develop a customer orientation for a farm.

Which consumer or customer groups are being satisfied? A farmer needs to answer this for three kinds of customers. The first group is the final consumers of products produced on the farm or produced from products produced on the farm. These are the people who consume food, wear cotton or wool clothes, and use agricultural products in construction. For example, a dairy farmer needs to decide whether the consumer groups being satisfied are all consumers or only those who want to avoid rBST, the recombinant bovine growth hormone. Similarly, a soybean producer needs to decide whether he or she is aiming to satisfy livestock producers who want a protein source, consumers who want soy products for human consumption, organic customers, or specialty soybean customers.

The second group of consumers comprises those that buy farm products and transform them into the products used by the final consumer (or the next step in the supply chain). These consumers can be processors such as cereal producers, meat processors, dairies, sugar processors, and cotton millers. This group of consumers also includes other farmers who buy grain and soybean meal to feed animals. A rice exporter will want different varieties depending on their international customers. Meat processors want a certain size of animal because their equipment is designed to handle that size and that size will provide the size of meat cuts their customers want.

The third group of consumers is made up of those who consume a more intangible set of products produced by farmers. These intangible products include, but are not limited to, environmental quality, hunting access, and rural landscapes. Except for the possibility of paying for hunting access, these consumers most likely do not pay farmers directly for these products. But they may stimulate the imposition of alternative policies and regulations that affect how a farmer can operate and what potential payments they may receive.

What needs are being satisfied? For the chosen consumers, do they want low cost products, a certain set of product characteristics, or something else such as dependable delivery of a consistent quality product, certified organic products, or goods produced locally by family farmers? Cereal producers, such as General Mills, need specific wheat varieties that produce the cereal characteristics that their consumers need. A soybean crusher may need a certain type of soybean that produces an oil with the baking characteristics needed by bakers who need to make products with certain characteristics that the retailer and ultimately the final consumer needs. A hunter wants to have success hunting. The public wants a clean, safe water supply. The list is endless and varies with consumers, geographical location, and time.

How are customer needs being satisfied? How can the farmer produce to be sure that customer needs are being met? This often involves knowing how the entire food chain operates from the farm's input supplier through the local buyers, shippers, processors, wholesalers, retailers, and ultimately the consumer. Identity preservation is becoming increasingly important for meeting the needs of many customers and consumers who want specific characteristics. For example, if a dairy farmer is aiming at those consumers who want rBST-free milk, several things need to happen to meet the needs of those consumers. The dairyman cannot use rBST of course, but then the milk processor needs to keep that milk separate from milk that may have been produced using rBST and the retail label has to report that the milk comes from cows not treated with rBST. Finally, the milk has to be sold in stores where these consumers will shop.

The second element of a vision and mission statement should be a statement of the core values of the farm. The number of core values is usually small, no more than five, to ensure that they are indeed the core values. These core values drive or guide other values and decisions; they are not changed easily. Core values are kept even if the market changes and penalizes the company for holding those values. If the penalties are large enough, the stakeholders will change the farm mission and operation (perhaps even selling the farm) to alleviate the penalties, but they will not change their core values. Although the exact definitions are debated, examples of the values held by most farmers include land stewardship, honesty, family, commitment to the local community, and hard work.

The mission statement is the first step in communicating the stakeholders' shared values to a larger audience. Communicating their values can help solidify the commonality of values among the internal stakeholders and let external stakeholders (such as creditors) know how management plans to operate the farm.

Key objectives need to be identified and communicated as the third element of vision and mission statements. The purpose of setting objectives is to convert managerial statements of business mission and company direction into specific, measurable performance targets. Measurable objectives help track performance and progress toward the vision. Without measurement, the organization does not know how well it is performing and whether its long-term vision will be accomplished with the current strategy.

Key objectives need to be both financial and strategic. If only one type is developed, the other type may not receive proper management attention and the vision may not be realized.

Financial objectives focus on specific measures of financial performance: profitability, liquidity, solvency, financial efficiency. If a farm does not have acceptable financial performance, it may not receive the resources (capital and other types) that it needs to meet other objectives and the farmer's vision for the farm.

Strategic objectives focus on activities that affect competitive position: entry into a new market, being a low-cost producer, recognition for quality. Strategic objectives are needed to encourage managerial efforts to strengthen a farm's overall business and competitive position.

A farm also needs long-term and short-term objectives. While long-term objectives are needed to track progress toward the long-term vision, a business also needs to monitor its progress in smaller steps. Performance targets and objectives five or more years ahead cause managers to take actions now to meet those long-term objectives and to consider the impact of today's decisions on longer-term performance. Short-term objectives should be steps toward long-term objectives. They provide the ability to check progress so the strategy can be adjusted, if needed, to achieve the long-term objectives and vision.

Some commonly expressed goals make poor visions and missions, but they can be converted into good objectives. For example, stating the farm's mission as "maximizing profit" does not describe the business of the farm and probably places too much emphasis on money. However, obtaining a certain income level and its related standard of living is an obvious financial objective for many farmers. Similarly, reducing estimated phosphorus runoff to a certain level and achieving a specific corn yield are good examples of strategic objectives. However, zero erosion or maximizing crop yields do not describe a farm's business or mission.

Objectives need to be challenging but achievable. Thus, several points need to be considered when setting objectives:

- Objectives need to be focused on important issues.
- Objectives need to be precise and measurable.
- Objectives need to specify a time period in which they are to be accomplished.
- Desired performance levels (e.g., profitability, productivity, efficiency) have to reflect what the industry and competitive conditions realistically allow. Performance objectives have to meet industry minimums if the farm is to remain a viable business and, at the same time, not surpass realistic maximums for the industry.
- Objectives need to be set to ensure the farm will be successful in terms of its overall vision and mission.
- Objectives need to reflect the potential capability of the farm.
- To improve future performance, objectives need to require stretch and disciplined effort on the part of the farm and those working on the farm.
- Having a challenge of trying to close the gap between actual and desired performance levels can help a farm improve its operations, be more inventive, feel some urgency in improving both its financial performance and its business position, and be more intentional and focused in its actions.

External Analysis of Competitive Environment

An external analysis of the farm's competitive environment includes understanding the forces operating within both the general economy and the industry in which the farm belongs (e.g., dairy or corn). External analysis evaluates the environment in which the farm operates, that is, outside of the farm itself. The process of external analysis can be seen as answering eight questions (fig. 2.2).

1. What are the conditions and trends in the macro environment?
2. What are the industry's dominant economic traits?
3. What is competition like and how strong are each of the competitive forces?
4. What is causing the industry's structure and business environment to change?
5. Which farms are in the strongest/weakest competitive positions?
6. What strategic moves are others likely to make next?
7. What are the key factors for competitive success?
8. Is the industry attractive and what are the prospects for above-average profitability?

Figure 2.2. Eight key questions for external analysis (Adapted from Thompson and Strickland 2003).

The Macro Environment

The farm and its industry operate within a larger, or macro, business environment. Changes in that environment may be important in crafting strategy for the farm. Five dimensions of the macro environment need to be considered: macroeconomic, technological, social, demographic, and political and legal.

The macroeconomic environment. Changes in the growth rate of the economy, interest rates, currency exchange rates, and inflation rates are major determinants of the overall level of demand for an industry's products as well as the supply of inputs. By understanding how the macroeconomy is changing, a farmer can better understand how these forces will affect the farm and what strategic changes may be needed.

The technological environment. Technological change can easily make established products and processes obsolete but at the same time create new products and processes. Continuing to use the same production methods even though new technology has made them physically or economically obsolete will keep a farm from being competitive. Monitoring, evaluating, and updating technology appropriately will help keep a farm competitive.

The social environment. What is socially acceptable and desirable changes over time. This creates both threats and opportunities for an industry. The changing social environment can be seen in even a very minimal list of current debates within and about agriculture and farming: globalization, industrialization, organic versus chemicals, animal welfare, farm size and structure, etc. These could be seen as threats to farmers' freedom to choose production methods. Alternatively, they could be seen as opportunities to serve new and perhaps more profitable markets.

The demographic environment. Changing proportions of the population by age and ethnic segments can have major impacts on the industry. New opportunities can develop with these changes as well as threats of decreased demand for current products. Some demographic changes are slow trends (an aging population, for example); however, they can affect the long-term profitability of major investments. Other trends, such as changes in the farm labor force, can be rapid enough to affect short-term investments and, perhaps, encourage a farmer to learn a second language.

The political and legal environment. Changes occurring within the political and legal environment can impact the demand for products, the supply of inputs, and the availability of production processes. In general, the trend toward more open borders and, hence, freer trade (NAFTA, for example) should increase demand for farm products and the supply of farm labor. Political changes that would alter that trend could cause changes in product demand and input supply. Greater concern over the environment may translate into stronger environmental legislation, which could result in decreased availability of some inputs, decreased potential to use some production methods, and, perhaps, larger penalties for not following recommended or regulated production methods. Changes in the U.S. farm bill could also affect the potential income of farms and the regulations on what can be produced. Individual state differences can affect the competitiveness of farms within each state. These are a few examples of the changes that need to be monitored. Each industry will have a different list and place different priorities on each element.

Dominant Economic Traits

The second step in external analysis draws our attention to the chosen industry. The industry considered here is not just agriculture or food but the subindustry within agriculture or farming, say, hogs, grains, fruit, and so on. To better understand an industry, many factors need to be considered to describe its economic traits. By considering the following list even for a familiar industry, a farmer will be in a better position to craft a strategy for his or her own farm.

Market size. What is the total amount sold of the product? In a more detailed view, what is the amount sold of fluid milk, cheese, organic milk, high oil corn, buffalo meat, wheat, barley, processed carrots, and so on? Small markets do not tend to attract big or new competitors. Large markets, however, often draw the interest of farms interested in expanding.

Scope of competitive rivalry. Is it a local, regional, national, international or global industry? Where are the producers located geographically? How important (in quantitative terms) is each production area? Which areas are growing? Which are shrinking?

Market growth rate and where the industry is in the growth cycle. What stage is the industry in: early development, rapid growth and takeoff, early maturity and saturation, stagnant and aging, decline and decay? Fast growth breeds new entry; growth slowdown spawns increased rivalry and a shake-out of weak competitors.

A corn grower may view the #2 yellow market as stagnant and aging due to slow growth while the organic, blue corn market could be viewed as early development or moving into a rapid growth stage. The search for new consumer products by commodity groups is part of the response to stagnant markets. The intensity of competition within the hog industry could be viewed as due to its being a stagnant and aging industry.

Number of rivals and their relative sizes. Is the industry fragmented with many small farms or concentrated and dominated by a few large farms? For agricultural commodities, most industries consist of many relatively small farms. This is true even if only farms with sales over $100,000 are counted. The poultry, hog, some vegetables, and now the dairy industries have experienced considerable concentration and have a few large producers relative to the size of the total market.

Number of buyers and their relative sizes. The number of product buyers has a very large impact on how the market works for farmers as producers. More buyers mean more competition and presumably fairer prices for farmers. Fewer buyers mean less competition and a higher chance of less than fair prices for farmers due to a shift in market power. The number of buyers needed to ensure a competitive market is a central question in the concentration debate whether it is at a local, national, or international level.

Prevalence of backward and forward integration. Integration usually raises capital requirements and often creates competitive differences and cost differences among fully versus partially integrated firms. If an industry is highly integrated through either ownership or contractual relationships, the possibility of entry by new firms may be limited.

Ease of entry and exit. High barriers protect positions and profits of existing firms. Low barriers make existing firms vulnerable to entry.

Pace of technological change. The pace of change in both production processes and products can affect an industry. Rapid technological change increases risk. Investments in technology, facilities, and equipment may become obsolete before they wear out. Rapid product innovation shortens product life cycle and increases risk because of opportunities for leapfrogging.

Level of differentiation of the rival product(s)/service(s). With standardized products, buyers have more power because of their ease in switching from seller to seller. With product(s)/service(s) that are highly differentiated, or even weakly differentiated, the seller has more power.

Presence of economies of scale. Can companies realize scale economies in purchasing, manufacturing, transportation, marketing, or advertising?

Economies of scale increase volume and market share needed to be cost competitive.

Capacity utilization levels. High rates of capacity utilization are crucial to achieving low-cost production efficiency. Surplus capacity pushes prices and profit margins down; shortages pull them up.

Impact of learning and experience. If the industry has a strong learning or experience curve, average unit cost declines as cumulative output builds up (because the experience of "learning by doing" builds up).

Capital requirements. Large capital requirements make investment decisions critical, create a barrier to entry and exit, and make timing of either investment or entry and exit important.

Industry profitability. Whether the industry profitability is above or below par will affect the pressure from new entrants or alleviate the pressure due to exits. High-profit industries attract new entrants; depressed conditions encourage exits.

Competitive Forces

Farming is often described as an industry with perfect competition. That is, with so many farmers, none of them can control the price they receive for their products. However, in today's marketplace, this view of a farmer facing a perfect market is not correct and will lead a farmer to make incorrect decisions. Today's market places a farmer much closer to the consumer, which means a farmer needs to understand the industry and the competitive forces in which he or she operates. Farmers also need to understand the forces affecting the processors to which they sell their products and the forces affecting the suppliers of their inputs.

Michael **Porter's five forces model** provides an excellent framework for analyzing industry competition. In his framework, Porter (1980) describes competition in an industry as a composite of five forces.

Force #1: Risk of entry by potential competitors. For farmers, potential rivals and competitors may be just down the road, in another part of the nation, or across the ocean. Each of these three groups needs to be considered when evaluating the risk of entry by competitors.

If the risk of entry is low, farmers will have more bargaining power when negotiating both input and product prices. If the risk of entry is high, farmers will have less bargaining power. This competitive pressure may show up at the processor level also and affect farmers indirectly. If the risk of entry by potential competitors is high for a processor, that processor's negotiating power is not as strong with its customers so it can not obtain as favorable a price for its products, which translates into a lower price that the processor can afford to pay the farmer for the raw product.

The level of the competitive pressures coming from the risk of entry by potential competitors depends on the height of barriers to entry. These barriers to entry are determined by several factors:

a. The extent to which established companies have brand loyalty. Brand loyalty builds barriers to entry by the unwillingness of consumers to switch from a brand they know and want to a new brand. While brand loyalty may not have a strong, direct effect on farms, the connection is stronger for those farms who are closer to the consumer (direct marketers, contractees, members of cooperatives who sell directly, for example) compared to farms who produce commodities such as #2 yellow corn or hogs for the open market.

b. The extent to which established companies enjoy an absolute cost advantage over potential entrants. An absolute cost advantage may occur due to a depreciated plant, established dealers, and/or being farther down the learning curve.

c. The extent to which established companies have scale economies, that is, they are large enough to be able to operate at lower costs than a new entrant who would likely be smaller.

d. The extent to which government regulation restricts entry. The most obvious government regulations creating barriers to entry are those that govern utilities. Because they have this entry protection, private utilities are usually monitored and controlled by the government. For farmers, government regulations affect how farms operate and thus create advantages to current farms due to knowing and understanding the bureaucratic system of rules, regulations, forms, and potential payments.

Force #2: Rivalry among established farms. Competitive rivalry is generated by the competitive forces created by jockeying for better market position and competitive advantage within an industry. The extent of this rivalry will affect how a farmer operates and how well a farmer can expect to achieve his or her financial and strategic goals.

The extent of rivalry among established farmers depends on several factors.

a. The industry and its competitive structure. Industry structures are characterized as perfect competition, monopolistic competition, oligopoly, or monopoly. Each type has its own set of market pressures, conditions, and rivalries. Farming is close to but not exactly perfect competition. Market-level oligopolies and monopolies do not occur in farming, but, in some smaller markets, monopolistic competition may be a better description of the economic environment than perfect competition.

b. Demand conditions. Is demand growing with new customers, growing with existing customers, stagnant, or perhaps declining? When demand is growing with new customers (for example, new buyers of high-oil corn a few years ago), farmers have a much better chance of finding a market or contract for selling the product. However, if demand is increasing but with existing customers, farmers may find it harder to find a buyer because existing buyers and farmers have established relationships that are operating well. If demand is stagnant, farmers will have more trouble finding buyers as the buyers start to look for better and/or lower-cost producers. The extent of rivalry, and thus competitive pressure, is the greatest when demand is declining, because established farmers (those who have had contracts to produce, for example) vie for a shrinking supply of the production contracts from the processor.

c. Exit barriers. Higher exit barriers increase rivalry and competitive pressures. Exit barriers can keep the established farms from quitting. Thus, they compete to remain or increase profitability and achieve other goals. These exit barriers can take several forms: investments in specialized assets (e.g., a farrowing barn), high fixed costs of exit (e.g., removing a building or cleaning waste facilities), emotional attachments to an industry, relationships between businesses (e.g., production contracts), and dependence on an industry.

In the farming industry, the question "Who are a farmer's competitors?" can have a complicated answer. Farmers do not always see their immediate neighbors as competitors. In a commodity market, immediate neighbors are not a farmer's direct competitors because they do not compete for market share (unless the local elevator is almost full during harvest). However, neighbors can be competitors in other markets. For example, farmers know that other farmers want a chance to produce (that is, have a share of the market for) processing vegetables or high-oil corn. This knowledge or perception of how many farmers want the chance or contract to produce and how much they want the contract will affect how hard a farmer negotiates a contract with the buyer. More competition will mean less ability to negotiate a more favorable contract; less competition means the farmer has more negotiating power. Farmers in other parts of the country and the world can easily be seen as competitors when they visibly try to take market share away from the local group of farmers.

Force #3: Bargaining power of buyers. Buyers can be viewed as a competitive threat when they force down prices or when they demand higher quality and better services. With buyers, the competitive pressure grows out of their ability to exercise bargaining power and leverage when negotiating or setting prices for farmers' products. This ability depends on the buyer's size and power relative to that of the farmer(s). The recent mergers of buyers of agricultural products have increased the level of concern about the decline of bargaining power of farmers. The increasing use of contracts is another aspect of the buyer gaining bargaining power by controlling the amount of price information in the marketplace.

Force #4: Bargaining power of suppliers. Suppliers can be viewed as a threat when they are able to force up the price a farmer must pay for inputs or to reduce the quality of goods supplied. With sellers, the competitive pressure grows out of their ability to exercise bargaining power and leverage when negotiating or setting prices for farmers' inputs. The recent mergers of input suppliers and consolidation of present and future input technologies (e.g., seed and pesticide as a single input) will increase the supplier's bargaining power relative to the farmer and allow the suppliers to retain the profit potential of the inputs.

Force #5: Substitute products. Substitute products limit the price that farmers can seek or ask for without losing customers to the substitute products. With substitute products, competitive pressure comes from the market attempts of outsiders to win buyers over to their products. The advertising campaigns of the pork, beef, and poultry industries are an obvious example of the competitive pressures due to substitute products: Each industry feels forced to spend money advertising and also, cannot charge as much as they would like without pushing their customers into buying the other products.

Using the five forces model to understand the competitive environment of an industry does have some limitations. It does present a static picture of competition. The five forces model is more useful when an industry is fairly stable. It may be of limited use during periods of turbulence resulting from rapid innovation or some other discontinuity. However, even during times of rapid change, there can be value in seeing a series of "snap shots" that repeated analysis of the five forces model would provide. The five forces model also underemphasizes the importance of differences among farms. There are wide differences in profit rates of individual farms. Individual resources and capabilities are far more important determinants of the farm's profitability than the characteristics of the industry or subindustry. Farms that are more innovative in pioneering new products, processes or strategies can often earn much better profits than the industry average. However, even with these limitations, the five forces model can help provide an understanding of an industry and how it may be changing.

Change in Structure and Environment

The forces of structural change can be divided into major and minor forces. Usually, only three or four forces qualify as major driving forces of change. The most common driving forces and some agricultural examples are listed below.

1. Changes in the long-term industry growth rate. Is the market expanding or declining?
   - Chicken and turkey consumption is increasing in both total and per capita measures.
   - The per capita consumption of beef and pork has been more stable.
   - The consumption of organic food products is the fastest increasing category in percentage terms, but it is still a small portion of total food sales.
2. Changes in who buys the product and how they use it
   - Male or female; old or young; ethnic preferences
   - At home, away, or take-home; fast food or slow
3. Product innovation
   - Convenience foods
   - Irradiated foods
4. Technological change
   - Biotechnology
   - Changes in inputs
   - New and better products
   - Management information: gathering and processing
   - Changes in capital requirements, plant sizes, management size
5. Marketing innovation
   - Direct buying by processors
   - More specifications coming from grocery stores
6. Entry or exit of major firms (and countries)
   - In international grain trade: is China in or out? Russia? The European Union?
   - Mergers and acquisitions (Northrup King by BASF, IBP by Tyson)
   - Foreign companies have entered the U.S. market
   - U.S. companies have entered foreign markets
7. Diffusion of technical knowledge
   - To other companies, countries
   - Large hog facilities being built in more states and by more companies
8. Increasing globalization of the industry
   - Transportation improvements allowing more rapid and cheaper movement of products and thus closer alignment of local markets with world markets
   - NAFTA, WTO, and other agreements and institutions
9. Changes in cost and efficiency
   - Improvements in technology, varieties, machines, communications
10. Emerging buyer preferences for differentiated products instead of a commodity product (or for a more standardized product instead of strongly differentiated products)

- For many farm products, differentiation is not a powerful force. For example, beef is beef when it is fresh; it is not until the processor makes a branded sausage or canned beef stew that consumers can differentiate between products. Similarly, wheat is wheat until it is made into branded products. Recently, however, efforts are being made to brand fresh meat.
11. Regulatory influences and government policy changes
    - New U.S. farm policies
    - New environmental regulations
    - Changes in local zoning rules
12. Changing societal concerns, attitudes, and lifestyles
    - Health concerns for less fat and less pesticide use
    - Environmental concerns caused by chemical use, erosion
    - Leisure versus work time
13. Changes in uncertainty and business risk

Strong and Weak Competitive Positions

What are the competitive characteristics that differentiate farms? What are the characteristics of the strongest? The weakest? Characteristics that are important to consider include these six points: size, location, production methods, age of equipment and/or workforce, specialization, diversification, and vertical integration. A good evaluation of these positions requires good data on the financial performance of farms over time as well as an understanding of how the forces and changes identified in other parts of external analysis will affect the different types of farms.

The Next Strategic Moves

Based on their competitive positions, what moves are competitors likely to make? Which companies, regions, or countries may change? What are the makers of substitute products doing or what might they be changing? Will their potential changes make a large impact? This may seem like a game of chess in some ways but a serious discussion of the strengths and weaknesses of each player in the context of the future industry is critical to deciding how an individual farm should react and perform in the future.

Key Success Factors

Farmers could worry about many measures and conditions and spend valuable time and resources to improve them. But within each agricultural industry there are key success factors (KSFs) that farms must identify and keep at performance levels required by the industry. If this is done, other measures of success will follow. If the KSFs are not met, the farm will not perform adequately and its viability is threatened.

A KSF can be identified as the word or phrase that would complete this sentence for a farmer: "If we _____, we will be successful." A dairy farmer found and worded one of his KSFs this way: "After visiting several new facilities, I knew we had to milk at least 120 cows per hour per worker in order for our expansion to be successful." Once they are identified, each farm needs to develop plans and procedures to monitor and improve KSFs.

KSFs can be grouped in several ways:

Technology-related: Science, production process, expertise in marketing, use of computers internally and externally

Manufacturing-related: Scale economies, quality in production, labor skills and costs, flexibility versus costs

Distribution-related: Local versus regional markets; the networks of roads, railroads, barges; costs of distribution

Marketing-related: Substitute products—competition for shelf space, customer desires and needs ("fat free," for example), direct marketing—meeting a customer's needs in quantity and quality

Skills-related: Labor abilities, quality (getting it done correctly), expertise, management

Organizational capability: Ability to respond to changes, management, information systems

Other factors: Reputation, location, access to capital

Attractiveness and Profitability

This is an overall assessment of the industry's attractiveness or unattractiveness, special issues and problems, and profit outlook. Attractiveness is a very subjective term. So each person or set of stakeholders needs to evaluate all the information obtained in the external analysis and decide whether he or she would like to work in the industry. Is it a good industry to be in or is it nasty? Information about the profit outlook can be found in reports available from the Economic Research Service of the USDA, the annual reports of the farm record associations available in many states, Federal Reserve reports, and many other sources of analyses and commentary on national and local economic trends.

In this section, we examined the procedures for external analysis. These procedures help a firm and manager evaluate and understand the whole industry in which a farm operates (or wants to operate). Understanding the whole industry is critical to developing strategies that will allow an individual farm to achieve its objectives and ultimately fulfill the stakeholders' vision of the farm in the future.

In the next section, we look inward and evaluate the condition and situation within the farm itself. This internal analysis is also critical for developing the correct strategies for an individual farm and its objectives and vision.

Internal Analysis of the Farm's Operating Environment

A full internal analysis includes consideration of the three main internal factors that shape an individual farm's strategy: (1) shared values and culture; (2) personal ambitions, philosophies, and ethical principles of the managers; and (3) the farm's strengths, weaknesses, and competitive capabilities. The importance of the stakeholders developing a set of shared values was discussed earlier in this chapter. Beyond shared values, the managers themselves shape the strategies of the farm by their own ambitions, philosophies, and principles. The work culture created by the shared values and individual managers creates a culture within the farm that affects which strategies are chosen, how they are carried out, and whether they will be successful within the industry as understood through the external analysis described in the previous section. How these first two internal factors are developed and woven together has a tremendous impact on the ability of the farm to take advantage of its strengths and capabilities and overcome its weaknesses identified in the third phase of internal analysis.

The third phase of internal analysis is the evaluation of the farm's strengths, weaknesses, and competitive capabilities. A farm's strategy needs to be grounded in what it is good at doing, its strengths and competitive capabilities. Until they are defended against or overcome, weaknesses cannot be the basis for success. The third phase of internal analysis centers on five key questions (fig. 2.3). The first question analyzes the farm's financial condition and past performance and evaluates how well it is meeting the current vision, mission, and objectives. The next questions identify the farm's strengths, weaknesses, opportunities, and threats and compare the farm to other farms in the industry. The last question develops the list of strategic issues that need to be addressed when crafting strategy, the last step in strategic planning.

How Well Is the Present Strategy Working?

This question can also be asked this way: How well is the farm achieving its financial and strategic objectives and progressing toward the stakeholder's

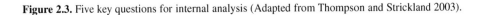

1. How well is the present strategy working?
2. What are the farm's strengths, weaknesses, opportunities, and threats?
3. Are the farm's costs competitive with rivals?
4. How strong is the farm's competitive position?
5. What strategic issues need to be addressed?

Figure 2.3. Five key questions for internal analysis (Adapted from Thompson and Strickland 2003).

vision for the farm? We start our answer by assessing the farm's current financial condition and performance. Starting with financial data allows us to look at the "concrete" results of past strategic and operational decisions. As explained in chapter 7, we are interested in the farm's profitability, solvency, liquidity, repayment capacity, and efficiency. While profitability is often of foremost interest, a major part of this first question of internal analysis is our evaluation of the farm's success at balancing profit objectives with other financial and strategic objectives.

Another reason for starting with financial analysis is to find out whether the farm has a sustained competitive advantage, that is, its profit rate has been higher than the industry average for several years running. The profit rate is usually measured as either the rate of return on assets (ROA) or the rate of return on equity (ROE). These farms are obviously performing better financially than the industry. A major part of internal analysis is striving to understand why some farms have sustained competitive advantage and how that can be continued and replicated on other farms.

Hill and Jones (1998) identify four building blocks of competitive advantage: efficiency, quality, innovation, and customer responsiveness. These building blocks need to be worked on, developed, and maintained in balance with each other. One, efficiency for example, cannot be emphasized over the other three without a detrimental effect on profitability and thus competitive advantage.

1. Superior Efficiency. Efficiency involves using inputs in the most productive way possible. It is commonly measured by the cost of inputs required to produce a given output. Increasing efficiency obviously decreases costs. Other ratios can be used to measure and monitor efficiency: asset turnover, output or sales per worker, yields per acre, productivity per animal, etc.
2. Superior Quality. Quality is defined as meeting and exceeding the customer's expectations. Quality products and services are reliable in the sense they do the job for which they were designed. Superior quality means providing higher quality in terms of product design, production, reliability, and service. However, note that the customer defines quality, *not the farmer*. A fuller description of quality measurement, management, and control is in chapter 6.

   Higher quality (that is, doing a better job of meeting customer specifications) will do several things for a farm. It has been shown in many companies that higher quality will increase efficiency and lower costs, thus helping another building block of competitive advantage. Higher quality will also create a good reputation for the farm and, thus, allow a farm to receive a higher price and (or) increase the market for its products. With higher quality, less time is spent making defective products, fixing mistakes, and hauling away mistakes and scrap. Thus, worker productivity improves, so costs decrease.

3. Superior Innovation. Innovation involves advances in products, production processes, marketing processes, management systems, organizational structures, and strategies. It is anything new or novel about the way a farm operates and the products it produces. Innovation that creates a unique product may allow the farm to differentiate itself and receive a higher price. Innovation can also allow a farm to lower its costs below competitors and thus improve profitability.

4. Superior Customer Responsiveness. To be responsive, a farmer must know what the customer needs and how to satisfy those needs. Customer responsiveness increases the value customers receive, the price they are willing to pay, and the likelihood of repeat and new business. If a processor has been pleased with a farmer's performance, that processor will likely be interested in bringing new business to that farmer compared to one they are not satisfied with. The ability to customize products and services (specific varieties, special harvest windows, specific animal characteristics, for example) to the unique needs of individual customers increases customer responsiveness and, thus, the ability to receive a higher price and (or) increase market potential. Faster response to a customer query (whether to rent more land or grow a certain product, for example) or for job completion (custom harvest, for example) is superior responsiveness. Superior responsiveness can improve both prices received and market share. Poor response is a major source of dissatisfaction and potential market loss.

What Are the Farm's Strengths, Weaknesses, Opportunities, and Threats?

The traditional SWOT analysis involves identifying and analyzing a farm's strengths, weaknesses, opportunities, and threats (fig. 2.4). A strength is something a business is good at doing or a characteristic that gives it an important capability. A core competency is something a company does especially well in comparison to its competitors. A weakness is something a company lacks or does poorly or a condition that puts it at a disadvantage. Strengths and weaknesses are internal conditions for the farm.

	Good things	Bad things
Internal to the farm	STRENGTHS	WEAKNESSES
External to the farm	OPPORTUNITIES	THREATS

Figure 2.4. SWOT analysis.

Some strengths and weaknesses may be identified when performing an analysis of the farm's financial condition and performance. Other strengths and weaknesses may be identified by reviewing the functional areas of the farm: production, marketing, finance, and human resources. In each of the functional areas, both the tangible resources (land, buildings, livestock, equipment) and intangible resources (reputation, technological knowledge, marketing knowledge) need to be evaluated as to whether any resources are obviously better or worse than the competitor's resources. Capabilities, such as the skills at organizing and directing resources or the ability of the farm's organizational structure to make use of those skills, should be compared to the competitor's capabilities. As described in the next section, benchmarking can show differences in prices, costs, and operational efficiencies and thus helps explain differences in performance that are not explained by differences in resources between the farm and its competitors.

Those strengths that cannot be easily duplicated by another farm are candidates for core competencies. For example, a hog producer may have these strengths: lower feed costs per animal, better performance, and better worker productivity. The first two strengths might be easily duplicated by competitors. The third may be a core competency due to the manager's ability to attract, hold, train, and improve good workers and the difficulty for other managers to achieve the same level of people skills. While not ignoring other strengths (and weaknesses), core competencies should be maintained and used to build and sustain the success of the farm.

Opportunities and threats are good and bad things that could happen to a business. Opportunities and threats are external conditions in the marketplace and, thus, should be part of the external analysis for the farm. However, they are analyzed again here in an internal analysis in terms of what an individual farm can take advantage of or needs to be protected from because (1) not all industry opportunities and threats are available to or threatening an individual farm, and (2) an individual farm may face opportunities and threats that are unique to its situation.

A SWOT analysis is not complete with the identification and listing of a farm's strengths, weaknesses, opportunities, and threats. A complete SWOT analysis needs to discuss five additional questions.

1. How should the strengths be used? Which should be moved on?
2. Which weaknesses are critical to success? That is, which need to be improved and which can be ignored for now? From which weaknesses does the farm need to be protected?
3. Which opportunities can be taken advantage of?
4. Which threats are potentially destructive?
5. Can the use of or response to any of these be done together to gather synergism, efficiencies, and other benefits?

Are the Farm's Costs Competitive with Rivals?

A major source of competitive advantage for farms is having costs per unit lower than competing farms. These competing farms may be next-door neighbors or across the oceans. The question is whether the specific farm has lower costs of production and delivery than other farms producing the same product for the same market. Costs per unit need to be estimated and then compared to other farms. This comparison can be done with USDA costs of production survey data and with the farm record associations present in many states.

As described more completely in chapter 7, cost comparisons can be done in several ways. The first comparison is a horizontal comparison across farms of the total costs per product unit (bushel, hundredweight, head, for example). A historical comparison shows how the costs have changed over time for both the specific farm and for all farms. A vertical analysis shows the importance or size of the various cost categories and identifies which areas show the largest potential for cost reduction. Vertical cost analysis leads into value chain analysis.

A value chain identifies the activities, functions, and business processes that have to be done in designing, producing, marketing, delivering, and supporting a product or service. Value chain analysis breaks down the whole process in detail so a manager can understand how costs are generated and how the process can be changed to improve efficiencies, increase quality, and decrease costs. A major goal of value chain analysis is to identify the farm's sources of competitive advantage (or disadvantage).

To perform a value chain analysis, costs need to be organized by activity rather than by broad category. That is, the list of costs in a whole farm statement or on a Schedule F for tax purposes is not sufficient to perform a complete value chain analysis. For farm management, a good place to start value chain analysis is enterprise budgets developed from the farm's own records. As described in chapter 4, enterprise budgets show the separate activities of, for example, tillage, planting, and crop protection. Allocating whole farm costs to separate activities can be difficult, but experience shows the information obtained is worth the effort. These activity-based costs can be compared to the potential value created by each activity to ascertain which activities need to be improved or changed to bring costs more in line with the value created.

The activity-based costs can also be used to compare the farm's cost structure and cost level to competing farms. This comparison can be done by using USDA survey data, but it should be noted that with survey data usually just an average is reported. For improvement to take place, comparisons need to be made with the best farms. These data are harder to come by but are available in the annual reports of farm record associations. Comparisons can also be done by benchmarking with other farms.

Benchmarking involves comparing costs and physical efficiencies of performing activities as well as the physical process of performing those activities. Through tours and private discussions, a farm can identify the best practices of other farms and then, after assuring that they are estimated using the same procedures, compare the costs of those best practices with his or her own practices.

Who would not only allow a competitor to come in and study the physical process but also open his or her books to that competitor? Farmers who realize they need to continuously improve their own farm and can learn from others, that's who. The cooperating farms agree (1) to share data, (2) to produce data on a time schedule, and (3) not to share the data with anyone else outside of the group. Benchmarking sessions tend to be organized by management consultants, farm record associations, and accounting firms. These groups can organize the connections between small groups of farms and either create or already have comparable data available. Some private accounting firms are industry specific and supply benchmarking studies and information on best management practices to their clients, but not to the general public.

How Strong Is the Farm's Competitive Position?

The ability of a farm to improve and(or) maintain competitiveness depends not just on the farm's past record but also on the strength of the position the firm is in. That is, strengths and core competencies found in the SWOT analysis need to be strong in relation to the trends present in the industry. Also, the firm needs to be making the correct moves to position itself to take advantage of the trends in the industry. Upon reviewing both the internal analysis done to this point and the external analysis done on the industry, the farm can be evaluated for the signs of competitive strengths and weaknesses using the following factors:

Signs of Competitive Strengths
- important core competencies
- distinctive strategies
- cost advantages
- good match of the farm's strategic product groups with the industry's growth areas
- above average profit margins
- taking advantage of cost economies
- above average technological and innovational capabilities
- creative, entrepreneurially alert management
- capable of capitalizing on opportunities
- possessing skills in key areas

Signs of Competitive Weaknesses
- competitive disadvantages
- losing ground compared to other farms
- below average growth
- short on financial resources
- poor strategic product groups compared to industry growth
- weak where best growth potential is
- high cost producer
- not able to take advantage of cost economies
- poor quality of and/or missing skills in key areas

What Strategic Issues Need to be Addressed?

At this point, the external and internal analyses are reviewed and put together to assess how well the farm is placed in the industry situation and what strategic issues or points need to be studied, improved, and changed. This identification of issues is the beginning of crafting strategy which is described in the next section.

Crafting Strategy

In this, the fifth and last element in strategic planning, a strategy needs to be crafted for the overall business and for each part of the business. Some people call this determining the **business model**, that is, how the farm will make a profit and meet other goals too. The term "crafting strategy" is used to describe that this is not a simple, linear process. Several steps and choices need to be made together to form a cohesive, solid strategy for a farm. The initial description of the strategy is written, evaluated, and rewritten perhaps many times.

First, we look at product development and process design as part of crafting a strategy. Then we show how to use generic strategies and begin to craft a set of strategies that may fit a farm's situation. Several tests that can be used for evaluating potential strategies are described. The section ends with some ideas for improving strategic planning.

Product Development and Process Design

Product development is identifying the specific physical and economic characteristics of the items that could be produced: soybeans, milk, hogs, and so on. These characteristics include the price that customers are able and willing to pay as well as the desired nutrient levels, taste, and baking characteristics, for example. **Process design** is specifying the inputs, methods, actions, jobs, machines, and steps to be used in the production process. By incorporating product development and process design into one process, we can improve the chances that the resulting product or service will have high production quality, the least expensive production, the most efficient use of resources, and the qualities the customer wants.

Farmers may think they don't need new ideas for products, that is, they may think that "corn is corn." But producing #2 yellow corn at 14.5% moisture is not the same as producing high-oil corn. Producing #2 yellow corn for a market that sets a price above $2.25 per bushel is not the same as producing for a market below $2. If the market has changed its definition of a product by lowering the offered price, a farmer has to realize that he or she needs to produce this new product defined not just by physical characteristics but also by cost characteristics.

A similar point could be made for any product. For example, some customers want skim milk, others want whole milk. Some consumers do not like the taste of milk produced from different feedstuffs—a very obvious process design problem. And the question of what price the market is willing to provide can change the complete definition of the milk product that a farmer needs to produce.

By making product choice and process design simultaneous, a better balance can be achieved between four objectives: cost—the cost of producing the product; quality—meeting the needs of the consumer; delivery—speed or cycle time of production; and flexibility—the ability to change products and processes as conditions change.

To incorporate all the appropriate facets of product choice and process design, Schroeder (2000) describes six steps for new-product development:

1. Idea generation. Three strategies for product development and idea generation are:
   a. Market pull: "make what you can sell"
      • What do consumers want? Organic? High oil corn? $2 corn or $2.25?
      • This is the primary basis for new product introduction.
      • Because this is what the consumer wants, little consideration is made to existing technology and processes (as opposed to the next strategy).
   b. Technology push: "sell what you can make"
      • New products are developed from current technology and existing equipment.
      • Even though they may not be new, superior products have a natural advantage in the marketplace. Even though the equipment is old, the net cost may be lower (in the short run) than buying new equipment.
   c. Interfunctional

- Many times new product introduction requires cooperation between the different functions of a business: marketing, operations, engineering, finance.
- This method can produce the best results and also be the hardest to carry out.

2. Product selection. At this point in the process, the purpose is to identify the best ideas, not to make the final decision. Selection may be based on subjective and limited information. Three tests need to be done as part of product selection at this step.
   a. market potential—what is the market size and pricing opportunities?
   b. financial feasibility—are the potential prices sufficient to cover costs?
   c. operations compatibility—can the farm produce the product consumers want?

3. Preliminary product design *and* preliminary process design. In this step, the purpose is to develop the best design of both the product and the process. A manager will consider many trade-offs between the objectives of cost, quality, delivery, and flexibility. This step is where the interfunctional idea of all parts of the farm working together is first used. It will be used again in the final design phase and in production. This joint design process is needed because the market price and the production cost for initial design rarely line up nicely for a producer. This preliminary design step should result in a product that is competitive in the market and producible by the farm. If it does not, this step either loops back for a new design or the process stops because of problems with feasibility and compatibility.

4. Prototype construction. In agriculture, this step is most likely done by the suppliers, processors, and universities. Demonstration plots, testing stations, experiment stations, and research centers are places where prototype construction is done for agriculture. Some farms may be involved in prototypes depending on their size or closeness to the final consumer.

5. Testing. Farmers are involved in testing when they try growing a new crop or livestock genetics or use a new technique on a small area or a few animals. Does the new product work or serve as planned? Can the design of the facilities and the process produce the product as defined by the customer at a profitable cost?

6. Final product design *and* final process design. The product and process are finally chosen for use in this last step. However, this "final" selection is final only until new information is obtained about new processes and new products.

In the early stages of **crafting strategy**, the goal is to develop a few strategies that may work for a farm. The final, chosen strategy is found at the end through a set of tests and comparisons. The final strategy may be an amalgamation of the early proposals due to the strengths and weaknesses found during subsequent evaluation and testing. The process of crafting strategy started in product development and process design, now we look specifically at how a farm can take the product and process designs and formulate a business strategy for the market. However, this is not a linear process and a farmer will probably go back to product and process design before all plans are finalized.

The actual strategy that a farm chooses is most likely a combination of elements found in what are called "generic strategies." These are a good place to start formulating potential strategies that a specific farm could follow.

Generic Strategies

Generic strategies are listed and explained below:

Low cost leadership: In this strategy, a farmer aims to develop a low-cost production position within the industry based on experience, size, and/or efficient operations. For most of agriculture, the only strategy available to producers is this low-cost strategy.

Growth: For many farms the strategy is to increase size as measured by sales, net worth, profit, acres, and/or animals.

Prospector: With this strategy, a farmer emphasizes marketing effectiveness and market development. A farmer who chooses this strategy is also likely involved in a differentiation, niche, or focused strategy.

Protector: Once a market position has been developed with a differentiation, prospector, or niche strategy, a farmer may adopt a strategy to maintain, protect, and fortify a secure market position.

Reactor: This is the strategy of not choosing a strategy because the manager is unable to develop a competitive advantage and cannot effectively compete in the marketplace. This strategy, if followed for long periods, will lead to reduced success at achieving visions, goals, and objectives—if not the death of the firm.

Differentiation: A farm following this strategy strives to create unique perceptions about its product(s) among its consumers. A farm selling directly to the public will most likely be including elements of this strategy in its chosen strategy. The recent interest in identity-preserved (IP) grains and meat and in contracts is one piece of evidence of farms striving to differentiate their products and to protect their markets.

Focus or niche: A farm with a focused strategy strives to serve a small but well-defined market niche. Examples of this include a supplier of organic vegetables to local markets and/or restaurants, a producer of rabbit or buffalo meat, and a grower of organic blue corn for a food processor.

Best-cost provider: With this strategy, a farmer may be producing a commodity (milk, for example) but supplying it to a certain market at a reasonable cost and, at the same time, meeting other characteristics (such as delivery, quantity) that the buyer wants.

Retrenchment: If the current strategy is not working or being accomplished, a farm could take a retrenchment strategy in either a strategic or operational adjustment. A strategic retrenchment involves redefining and reducing the goals upon which the original strategy was built. An operational retrenchment involves revising objectives and timetables toward more realistic levels. These retrenchments may be due to overly optimistic projections of either the external or internal environments or unanticipated changes in those environments.

The Process of Crafting Strategy

After reviewing current and potential products and processes and understanding generic strategies, one can begin the process of crafting strategy.

An obvious place to start is by identifying the current strategy being followed by a farm. The current strategy can be seen in the pattern of actions, choices, and approaches a farm uses in dealing with its internal and external situation (fig. 2.5). After identifying the current strategy, a new set of strategies can be crafted that better fit the situation described in the external and internal analyses and improve progress toward the farm's objectives and stakeholders' vision. This set of strategies can be developed using the ideas described in the rest of this section. The preferred strategy can be found using the tests described in the next section.

1. Actions to out-compete rivals
2. Responses to changing external circumstances
3. Efforts to broaden or narrow the product line, change product quality, and modify customer service
4. Actions to alter geographic coverage
5. Actions to merge with or acquire rivals
6. Efforts to integrate backward or forward or to outsource current internal activities
7. Actions to form strategic alliances and collaborative partnerships
8. Efforts to pursue new opportunities or defend against threats
9. Actions and approaches that define how functional activities are managed
10. Actions to strengthen resources and capabilities
11. Actions to diversify

Figure 2.5. Patterns of actions that define strategy.

Strategy crafting is the managerial process of deciding how to achieve the targeted results within the farm's physical and economic environment and its prospects for the future. Objectives are the ends; strategy is the means. A well-crafted strategy provides direction for day-to-day activities by defining "our way of doing business."

An effective strategy reflects organizational resources and capabilities and the competitive environment. Good strategies are based on the competitive advantages of a business, that is, on those points at which a business has an edge over its competitors.

In reality, strategy involves both proactive and reactive actions. Proactive strategies are deliberate and intended actions resulting from the careful, purposeful evaluation of the business environment. A farm's strategy also includes adaptive, perhaps even knee-jerk reactions to unanticipated changes, new developments, unexpected events, and so on. While proactive actions certainly sound the best, the rate of change in today's world means that staying with the planned strategy may be the wrong choice when new events occur and new information is obtained. Thus, well-reasoned reactive strategies can be just as valuable to the success of a farm as carefully planned proactive strategies.

External factors that shape a farm's strategy include:

- Societal, political, regulatory, and community citizenship considerations. These are usually limiting factors.
- Industry attractiveness; changing industry and competitive conditions. A farm's strategy ought to be closely matched to industry and competitive conditions.
- Specific market opportunities and threats. A well-conceived strategy aims at capturing a farm's best growth opportunities and defending against external threats to its well-being and future performance.

Internal factors that shape a farm's strategy include:

- Organizational strengths, weaknesses, and competitive capabilities. A farm's strategy ought to be grounded in what it is good at doing (its organizational strengths and competitive capabilities); it is perilous for success to depend on what it is not so good at doing (its organizational and competitive weaknesses).
- The personal ambitions, business philosophies, and ethical principles of managers. Managers stamp these on the strategies they craft.
- Shared values and culture. A farm's values and culture can dominate the kinds of strategic moves it considers or rejects.

Strategy should change and evolve as the business environment changes and evolves. However, if a strategy is rewritten or reworked too often, managers are probably guilty of erratic decision making and weak "strategizing." Large changes in plans and strategy can occur, especially in crisis situations, but, if changes are made too often, confusion will reign within management and employees and performance will suffer.

Farmers need to be entrepreneurial (creative, risk-taking, innovative) when crafting strategy as well as able to do outside-in strategic thinking. Outside-in thinking can be understood best when contrasted against its mirror: inside-out thinking. A manager who thinks inside-out is risk-averse, accepts currently acceptable performance, concentrates on current plans and operations to do them better, dismisses new trends because "they won't affect us," or studies new trends to death. Inside-out strategies tend to be traditional approaches, acceptable to internal coalitions, philosophically comfortable to the current management, and safe in terms of not disturbing current conditions. They are not very forward thinking.

Strategy needs to be crafted for the whole farm, but it also needs to be crafted for each enterprise on the farm. A strategy also needs to be developed for the separate functions of the farm business: production, finance, and marketing. These strategies for each enterprise and function need to be consistent with and supportive of the strategy crafted for the whole farm.

Tests for Evaluating Strategies

Alternative proposed strategies need to be compared using a set of tests, but before those tests are described and the strategies evaluated, each proposed strategy should be reviewed to determine whether it is written in a clear, concise fashion and is internally consistent. If it can't be understood or has conflicts between its parts, it needs to be either rejected or rewritten before it can be considered further.

To help choose which potential strategy may be the best for a farm, several tests exist to evaluate how each strategy would help a farm attain its objectives. Since each test evaluates alternative strategies from a different perspective, several tests should be used to provide a better overall evaluation and, hence, the choice of the most robust strategy.

Vision Consistency Test. How well does the proposed strategy fit with the business and personal vision of the farmer and other stakeholders? If a strategy does not fit with the visions and ambitions of the people involved, the chances of success are low because enthusiasm and attention will be low, so the strategy should be rejected.

The Goodness of Fit Test. How well does the proposed strategy fit with the external analysis of the industry? How well does the proposed strategy fit the internal analysis of the farm? Even though an idea or strategy sounds great, if it does not fit both the external and internal conditions of the firm, it needs to be rejected.

Building for the Future Test. How well does the proposed strategy help maintain and develop the building blocks of competitive advantage: superior efficiency, superior quality, superior innovation, and superior customer responsiveness? How well does the strategy contribute to value creation, low-cost processes, and product differentiation? Does the proposed strategy contribute to building resources and capabilities for the future? If a strategy uses but does not build resources, it should receive a low score on this test. A low score on this test may not require rejection, however. Consider a farmer with no heirs or partners who want to continue the farm business. This farmer may be very justified in following a strategy to use up depreciable assets (buildings and machinery, especially) at a rate correlated with his or her retirement plan.

The Performance Test. How well does the proposed strategy contribute to achieving the strategic and financial objectives of the farm? What are the predictions for income, rates of return, net worth growth, expansion in physical size, transition to the next generation, and so on? The score for performance should not be based solely on high income. The performance score should be based on a balanced view of the proposed strategy's contribution to the strategic objectives of the farm.

Importance Test. Are important issues identified in the external and internal analyses addressed by the proposed strategy, or is it focusing on the trivial? Does the proposed strategy feel comfortable (or enjoyable) but fail to explain how opportunities will be taken advantage of and threats will be defended against? Does it explain how it will build on strengths as well as improve weaknesses, or does it talk about new buildings? For example, does the proposed strategy aim to have the best Holstein herd in the region or the most advanced use of precision farming technology but ignore the need to achieve, say, cost objectives required by the industry to remain profitable? If a proposed strategy does not address the important issues, it should be rejected. Perhaps the basic idea is sound, but it may need to be rewritten to explain, in terms of the external and internal analyses, why and how the strategy meets the important issues.

Feasibility Test. Is the strategy amenable to programs that can effectively be implemented? Can it be broken down into programs and projects with measurable objectives? Can these smaller parts be accomplished? A well-written, grand strategy, which no one can figure out how to implement, is infeasible and should be rejected.

Resource Test. Are resources available to implement the strategy? Can people be hired to do the work needed? Can financing be obtained?

Confidence Test. How high is the confidence that the anticipated outcomes of the proposed strategy will occur? What is the risk that events will occur that will change the expected results—especially in a negative direction?

These tests can be used to identify how different strategies perform or satisfy different objectives and, thus, help improve the final selection process. To use these tests, a farmer and his or her stakeholders give each proposed strategy a subjective score for each test chosen. Their scores come from their opinion of how each strategy would help the farm perform according to different tests. For example, how well does each strategy fit the vision of the stakeholders or fit the external analysis of the industry?

In most instances, the scores range from 1 to 5 with 5 being the highest score. That is, 5 indicates the best a given strategy can do in terms of a specific test. Once a score is estimated for each test, the scores are summed across all tests for each strategy. The strategy with the highest score is the apparent best strategy for the farm. However, since the scores are subjective, a farmer could also drop from consideration those strategies with the lowest scores and spend more time evaluating the remaining strategies before making a final decision.

As an example of scoring and evaluating strategies, suppose a dairy farm has developed four potential strategies. The proposed strategies (and the generic strategies involved) are low-cost milk production (low-cost leadership), finding new markets for milk (prospector), merging with a neighbor (growth and cost leadership), and producing organic milk for an ice cream manufacturer (prospector and best-cost provider). After describing the strategies, estimating the financial performance of each strategy, and reviewing the farm's vision and objectives, the farmer and others involved gives each strategy a score of 1–5 for all the tests (table 2.1).

In this example, two strategies are tied for the top: merging with a neighbor and producing for the organic market. To choose between these two, the farmer and other stakeholders can evaluate the differences between the scores on individual tests, such as, vision consistency, performance, or confidence. The two strategies could be evaluated in more detail to help the decision.

Also, remember that crafting strategy is a dynamic and cyclical process. In this example, the farmer should not feel constrained to the current list of strategies. Since two were tied with the top score, a new, combined strategy could be developed and considered. In this example, the farmer and the neighbor could develop and consider the combination of merging and producing for the organic market.

Table 2.1. Scoring Proposed Strategies

Proposed strategy	Strategy Tests (high = 5, low = 1)								
	Vision consistency	Goodness of fit	Building for the future	Performance	Importance	Feasibility	Resources	Confidence	TOTAL SCORE
Low cost	2	3	2	3	4	4	3	2	23
New markets	4	3	3	4	2	2	2	3	23
Merger	3	4	5	3	4	4	3	4	27
Organic	4	3	4	4	3	3	3	3	27

Since these are subjective scores, there is obvious concern that the persons doing the evaluation could manipulate the scores in order to obtain a preconceived notion of what they think is the best. These concerns can be met by each evaluator striving to score without following preconceived notions of the desired result. If a group is evaluating the proposed strategies, each member of the group could estimate scores independently and then discuss them as a group. These and other potential problems and solutions are described in the next section.

Improving Strategic Planning

Hill and Jones (1998) have identified four problems or weaknesses that farmers may encounter with formal strategic planning: planning under uncertainty, ivory tower planning, planning for the present, and managers' biases. Let us discuss each of these and how they can be overcome plus three other ideas for improving strategic planning: devil's advocacy, dialectic inquiry, and using an advisory board.

1. Planning under uncertainty

   Problems due to uncertainty of the future show up in two ways. First, we think we can forecast the future accurately, so we do not incorporate risk and change into our plans. We choose a strategy, and as conditions change we may find that another strategy or an adaptation of the strategy would have been better, but it is too late to recover. Second, we know we can't predict future events accurately, so many people do not think planning is worthwhile. The resultant strategy is, at best, reactive and profitable by chance. At worst, the farm is put at risk as conditions change.

Both the problems of blind faith in forecasting and of rejection of planning due to uncertainty can be solved by using scenarios to develop pictures of possible futures. Proposed strategies are analyzed under each scenario, then robust strategies are designed to cope with different scenarios. Scenario planning tends to expand people's thinking and result in better strategies. This process is more fully described in the next section.

2. Ivory tower planning

In a large farm, managers may not stay in touch with the rest of the farm and the marketplace. Thus, their planning could easily be unrealistic in terms of what the farm can do and in what the market wants or how it will respond. This could also happen in a smaller farm, even in a one-person farm if the manager does not look realistically at what he or she and the farm can do and what is happening in the market.

The solution may be easier in a single-person farm, but the solutions are the same in all sizes of farms. Managers and people involved need to talk to people at every part of the farm and should consider having a person outside the farm, a nonstakeholder, interview them and their workers about what is happening. Benchmarking could provide an objective comparison of the farm's productivity and financial measures with other farms in the area and industry. Obtaining and listening to others' evaluation of the market and the economic environment will help open one's own thinking to alternative views of the marketplace and what needs to be done in all aspects of farming.

3. Planning for the present

A common problem in crafting strategy is using the fit model alone. We craft a strategy to fit the world and our farm as we see it, but don't incorporate any adaptations needed for the future. We study, analyze, present, discuss, list, and so on; it's all very structured and neat. We fit our existing resources into the current environment. But we don't spend enough time on developing new resources and capabilities. We don't spend enough time on creation and/or exploitation of future opportunities.

The solution is to develop a strategic intent, that is, a bold ambition for the future. But strategic intent is more than unfettered ambition. Developing a strategic intent helps focus managers' and workers' attention on the vision of what we want the farm to be, motivates people by communicating the value of the target, leaves room for individuals and teams to adapt to changing circumstances, sustains enthusiasm over time by providing new operational definitions as circumstances change over time, and consistently guides resource allocations.

In practice, both fit and intent are used. That is, set an ambitious goal, go through the strategic planning process, continually monitor the external environment and internal performance, and be ready to change if necessary to keep moving toward the bold ambition.

4. Managers' biases

Cognitive biases affect how we think and what we assume to be truth. As individuals, we all have cognitive biases; thinking we don't is a cognitive bias because it's not true. Cognitive biases and potential solutions to deal with them are listed in table 2.2.

The cognitive biases listed in the table are often evident in individuals. Groups can exhibit a problem called groupthink. When a group starts analyzing and making decisions without questioning their underlying assumptions of the situation, they are said to suffer from groupthink. This may be due to an overly influential member of the group or acceptance of one idea and then rationalization for accepting that one idea or plan.

5. Devil's advocacy

Devil's advocacy involves bringing up all the reasons that might make the proposed strategy unacceptable. This can be a very effective method of developing an improved, robust strategy to deal with different events in the future. This should be done in an objective way, however. If one group member is doing this, the group should not view this person in a negative light.

6. Dialectic inquiry

Dialectic inquiry involves developing a plan (thesis) and a counterplan (antithesis) that represent possible but conflicting approaches to running the farm. The plan and counterplan are discussed and poked at to find weaknesses in assumptions, actions, and so on. Out of this discussion, a better plan should emerge—either a third strategy that is a combination of the first two or a strengthened plan or counterplan is developed because of the discussion. An individual farmer can also benefit from dialectic inquiry

Table 2.2.   Cognitive Biases and Potential Solutions

Cognitive bias	Potential solution
Prior hypotheses or beliefs	Pay attention to evidence that refutes prior beliefs
Escalating commitment	Don't commit more if things aren't working
Reasoning by analogy (especially overly simple analogies of a complex situation)	Ask if the analogy really does make sense in the current situation
Representativeness of our knowledge (generalizing from a small sample or even one vivid anecdote)	Base decisions on large sample theory
Illusion of control	Accept that we don't control everything

by developing a plan and counterplan and then playing opposite roles poking and probing each strategy, trying to find weaknesses.
7. Using an advisory board

Having an advisory board of persons interested in and knowledgeable about the farm but not directly involved is one way a farmer can benefit from others evaluating his or her plans and ideas. The advisory board is discussed more fully in chapter 13. This advisory board can help an individual overcome many of the problems just discussed, although groupthink can still be a problem if it is not addressed directly.

Using Scenarios for Choosing a Strategy

Since the future is not known with certainty, we cannot evaluate the potential success of following one proposed strategy over another with certainty. As mentioned earlier, one solution to solving the problem of planning under uncertainty is to develop scenarios or pictures of the future, as described in chapter 11. Then we can predict the results of following a proposed strategy under the conditions described in each scenario. After we have estimated the potential results, we can use several methods to help decide which strategy best fits a farm and its situation. These methods are described below.

1. *Bet on most-probable scenario.* Decide which scenario is the most likely to happen or is happening and choose the strategy that positions the business in the best way for that picture of the future. If only one scenario is very likely to happen, this is the wisest and easiest choice.
2. *Bet on the scenario and strategy that is the "best" for the business.* This involves choosing a likely scenario and the strategy that is estimated to provide the best outcomes in terms of financial and strategic objectives in that likely scenario. However, choosing a strategy that is crafted for only one scenario may be expensive in the sense of what opportunities are given up by not following other strategies or by business failure due to following the wrong strategy when the future doesn't look like the scenario originally believed.
3. *Hedge on what will happen (i.e., keep a door open).* Choose a strategy and make your plans on the likelihood that a specific scenario will occur but keep evaluating what is happening and be ready to change if needed.
4. *Preserve flexibility (don't shut a door).* If two (or more) scenarios are considered equally possible, craft a strategy that allows flexibility and ability to make adjustments as the future unfolds.
5. *Influence the outcome (affect what happens).* If policy changes are needed, work with legislators and Congress to encourage those changes to be made. Use advertising and public relations to influence and change consumers' opinions and choices. Talk to the people and businesses involved; strive to

convince them to make decisions that would be beneficial to your business. Negotiate deals.

Please note that the idea of influencing outcomes is based on legal behavior. Some companies and persons have gotten into trouble by trying to influence outcomes in illegal ways such as price fixing, collusion, and other unlawful activities. Illegal behavior is not being encouraged here; it is illegal.

6. *Combine methods*. Bet on most probable, try to influence the outcome, but keep an eye on the future as it unfolds and be ready to move and change if necessary.

A robust strategy is viable despite which scenario happens. However, a robust strategy may be expensive due to the costs of maintaining options so financial goals may not be met. Also, developing a strategy of being ready to move as needed in all or several scenarios may seem robust, but the business may lose by ending up stuck in the middle!

Other factors can be considered in crafting strategies. These have obvious advantages but some disadvantages also.

1. *First mover advantage*. If you are the first to ask about the purchase of land, for instance, you may be the one who gets to buy the land. Being the first to say you are able to produce according to the proposed contract may provide you a chance to participate in a very beneficial venture. The obvious disadvantages are of moving too fast into uncharted waters. Success is not guaranteed to those who move first, nor to those who wait. Success comes when the choices are evaluated and decisions are made appropriate to the opportunity.

2. *Initial competitive position*. By being the first or among the first, a business may have considerable advantage over other, later entries. For example, processors do not share new ideas and profitable ventures with all farmers. The first farmers who receive a chance to consider new, potentially beneficial contracts are those whom the processor already knows.

3. *Costs or resources required*. A large initial investment and high operating costs create greater risks for the business. Smaller investments may allow a strategy to be followed with smaller rewards except that the financial cost is lower if conditions change enough to warrant a change in strategy.

4. *Risk*. Several aspects of risk can affect the decision on which strategy to follow: (a) The timing of the resource commitment. Having to invest a large amount up front increases the risk of losing that investment. (b) The degree of inconsistency of strategies for alternate scenarios. If alternate scenarios require different investments and operations, the cost of going with one can

be great if another scenario is the one that develops in the future. (c) If the probabilities of different scenarios are very different, then the business can see which is the more likely. However, if the relative probabilities of occurrence are very close, the decision is more difficult. (d) If the cost of changing strategies to adapt to future scenarios is low, management has an easier time in crafting robust strategies. However, if the cost of changing is high, management needs to spend considerable time and careful deliberation evaluating potential scenarios and crafting strategies to deal with different futures.

5. *Competitor's expected choices.* As a last note, management should also consider and monitor what competitors are doing. For farms this involves not only what close neighbors are doing, but also what producers in other regions and countries are doing. The producers of products with close substitutes (such as pork, beef, poultry, and other meat producers) need to consider their competitors' choices or potential choices as they make their own choices.

STRATEGY IMPLEMENTATION

Implementation is, in many ways, the most difficult part of strategic management. By comparing the current strategy identified earlier with what the strategic analysis points to as the preferred strategy, changes can be made to improve progress toward the farm's objectives and stakeholder's vision. To do this, implementation involves designing the structure of the organization, aligning functional strategies (such as production and marketing) with the chosen strategy for the whole farm, obtaining and directing the needed resources, and adapting the plan and implementation to the change that is inevitable. The difficulty of these tasks is compounded by the need to keep the overall vision in mind without getting immersed in and overwhelmed by the details of day-to-day operations.

Implementation is a hands-on job that includes the following activities (adapted from Thompson and Strickland 2003).

- Developing an organization capable of carrying out the strategy successfully. Training yourself and your employees and acquiring the necessary resources, for example. Hiring new staff. Reassigning duties.
- Allocating resources to those activities critical to strategic success. Budgets of money, labor, and time resources are needed.
- Establishing policies and procedures that support the chosen strategy such as better marketing practices and employee incentive programs.

- Motivating people and providing incentives to pursue the target objectives and perhaps modifying their duties and job behavior to better aim at financial and strategic objectives.
- Tying the reward structure to the achievement of the targeted results.
- Creating a company culture and work climate conducive to successful strategy implementation. Allowing employees to keep the vision in mind and to suggest changes for successful implementation.
- Installing internal support systems that enable company personnel to carry out their strategic roles effectively day in and day out.
- Instituting best practices and programs for continuous improvement.
- Exerting the internal leadership needed to drive implementation forward and to keep improving the way the strategy is being executed. Keeping the vision in mind.

Back to the first item on that list: the first step in strategy implementation is developing the correct organization. That is, a farmer needs to alter the organizational structure, staffing, leadership style, and employee development approaches to achieve strategic goals. If this step isn't done successfully, the probability of successful change is greatly lowered. On a farm with multiple operators, partners, or staff, assigning specific individuals to manage or supervise different parts of the farm can be a step toward achieving strategic objectives. Examples of changing organizational structure include assigning a specific person to be in charge of the milking herd and its nutrition, assigning a specific person to be in charge of scouting and pest management for crops, assigning a specific person to be in charge of marketing, and so on. These new assignments may necessitate new training. These changes may also require changes in how upper management (perhaps the older generation) manages the farm. Not relinquishing direct control may be one of the largest obstacles to expanding and achieving objectives because of the redundant management time spent on decisions and on the slowness of decision making if several stakeholders need to be consulted before decisions are made and implementation continues. These obstacles can be overcome by training of both new and old managers. To plan for these changes in the organization, specific steps (with target dates) need to be identified. Some of these steps may be identical to steps identified in the functional areas (marketing, production, finance, and personnel) and in programs and projects, as discussed next.

To implement strategy well, strategies for each of the functional areas should be designed to support the overall strategy. One way to do this is to identify important activities for each of these functional areas that will facilitate that area's contribution to the overall objectives and vision contained in the strategy for the whole farm. These functional strategies should include the points shown in table 2.3 for financing the expansion of a crop farm.

Table 2.3. A Functional Strategy for Financing a Crop Farm Expansion

Functional area:	Finance
Specific strategic steps needed:	1. Negotiate land purchase of 1,250 acres by, and a land rent contract with, local investor. 2. Negotiate machinery loan. 3. Consolidate and increase operating loan.
Reason/motivation for steps:	Investor with local connections is willing to buy land and rent to us if we can obtain financing for larger machinery line and operating needs.
Plan of action:	1. Negotiate and sign letters of agreements between current landowner, investor, and ourselves. 2. Use letters of agreement to negotiate and obtain financing for machinery and operating capital needs specified in business plan. 3. Talk to F&M Bank here, John Deere Leasing, AgriLease, Wells Fargo in Hutchinson, AgriStar, and perhaps other creditors. 4. If financing terms meet goals stated in projections, finalize and sign financing agreements, and then sign contracts for land purchase and rental.
Responsible person(s):	Mary to negotiate with current landowner and investor. Mary and Bill to negotiate with potential creditors.
Measurable objectives:	Letters of agreement, needed level of financing in agreements, and contracts to purchase and rent.
Target dates:	Letters of agreement by August 1. Finalize financing and contracts by August 25.
Budget and resources needed:	$1,000 for attorney fees; 120 hours estimated by Mary and Bill; use of car or truck and computer.

Another part of implementation is to identify programs and projects that are needed to achieve the vision and objectives identified as part of the planning process. These programs or projects are not the whole strategy but pieces or steps needed to accomplish or move forward with the strategy. Examples of these programs or projects are building a new milking parlor, seeking new markets, and leasing more cropland. These projects could be explicitly needed to achieve the stated goals of, respectively, a larger herd, diversified markets, or increased size. Under this method of implementation, the farmer (or management team) assigns a program leader (or specifies a certain amount of time by an individual farmer), sets specific measurable objectives with target dates, and allocates the needed human, physical, and financial resources to the program or project. This part of implementation can be planned by specifying, for each program and project, the points listed in the example project of building a new swine finishing barn (table 2.4).

As an example of a strategic program, consider a farmer's decision to become a certified organic producer and capture those premiums (table 2.5). This farmer knows he has to educate himself on both production and marketing of organic products before he actually starts the three-year certification process.

Table 2.4.   A Strategic Project Plan for Building a Swine Finishing Barn

Title of program or project:	Building a swine finishing barn.
Reason/motivation:	Achieve new operating efficiencies and increase capacity.
Plan of action:	Finalize contractor bid, obtain building permits, initiate and monitor building progress.
Responsible person(s):	Tim
Measurable objectives:	Bid finalized, permits obtained, building started, pit poured, roof and walls, building finished, & pigs in.
Target dates:	Bid by April 1; permits by June 1; building started by July 1; pit poured by July 20; roof and walls by August 15; building finished & pigs in by September 15.
Budget and resources needed:	$440,000 and 5 acres.

Table 2.5.   A Strategic Program to Become Certified Organic

Title of program or project:	Organic certification and marketing
Reason/motivation:	Estimated budgets show better profits
Plan of action:	Educate, start production, negotiate markets
Responsible person(s):	Steve and Mike
Measurable objectives:	Attend classes, join association of organic producers, finalize budgets and production process, contracts
Target dates:	Classes & joining, winter year 1
	Finalize plans & budget, December year 1
	Solidify market contacts, January year 2
	Produce & control, years 2, 3, & 4
	Attend association meetings, all years
	Become certified in year 4
	Market organic products in year 4
Budget and resources needed:	$5,000 for 4 years of training and related expenses; 600 hours for extra planning

Controlling implementation is very similar to controlling the overall strategy. The major difference is the shortness of the time frame for controlling implementation. To control implementation, actual progress toward the objectives identified in the planning of the organization, functional areas, programs, and projects is measured and evaluated. Corrective actions are taken as needed. This process is discussed in more detail in the next section.

STRATEGIC CONTROL

Since the world changes, strategic planning and implementation are not one-time exercises. Strategic control is needed to evaluate a farm's performance and results during and after implementation of the chosen strategy. New circumstances may call for corrective adjustments. The underlying situation may change or implementation may not go as planned, so each task of strategic management requires constant evaluation regarding whether to change direction.

Strategic control systems provide the ability to monitor, evaluate, and take corrective actions to assure that strategic objectives will be met. Control involves (1) choosing the key indicators that measure progress toward objectives, (2) establishing standards against which performance is to be evaluated, (3) creating measurement systems for the key indicators, (4) comparing actual performance to the established standards, (5) evaluating the results, and (6) taking corrective actions.

Molz (1988) describes strategic control as having two parts: *strategic product* and *strategic process*. For strategic product, actual outcomes are compared with targeted outcomes projected in the strategy implementation program. Causes of deviations are evaluated and corrective actions are taken as needed. For strategic process, a farmer needs to routinely reevaluate all assumptions and inputs used to develop the strategic plan and decide whether unanticipated deviations from expectations will alter the viability of the strategic plan. Thus, Molz is saying that a farmer needs to evaluate in two ways: whether the chosen strategy is being implemented as planned and whether the chosen strategy needs to be redesigned.

Hill and Jones (1998) suggest that a control system can be separated into five areas: (1) financial controls for rates of return, income levels, and so on; (2) output controls, such as productivity measures, efficiency measures, and cost measures; (3) behavior controls, such as budgets and standardization of inputs and processes, which can be used when it is difficult to monitor outputs; (4) values and norms embodied in organizational culture to assure that people will behave in the ways that the farmer wants them to behave; (5) reward systems designed to recognize behavior and actions that contribute to achieving strategic objectives.

CONCLUDING COMMENTS

Strategic management is a continuous process. Large companies have strategic planning departments that have an annual schedule of reviewing, monitoring, and evaluating the different parts of the business and its environment. When the year is over, the process does not stop but restarts at the top of the list. There is some flexibility in this schedule due to the changing economy and environment, but the basic idea of strategy as a continual process can always be seen.

For a farmer (most of whom do not have a strategic planning department, I have noticed), the continual part of the strategic management process comes after the planning is done formally the first time. A farmer then needs to scan the external environment, evaluate the impact of new conditions and events, and choose any needed corrective actions. Scanning the environment means paying attention to agricultural and general news media, monitoring what the government and other institutions are doing, watching what competitors are doing, and networking with others who are also watching and monitoring the farming environment. Evaluation of potential impacts is done on paper in a formal estimation of the impacts of events or using mental models of how the business and its environment interact. The mental models may be used to evaluate whether the potential problem or opportunity needs a formal analysis.

SUMMARY POINTS

- Strategy is the pattern of actions used by a farmer to accomplish goals and objectives.
- Having a strategy helps a farmer make reasoned, cohesive, and consistent choices among alternative courses of action in an uncertain world.
- Actual strategy is a mixture of planned action and relative quick reactions to changing circumstances.
- Strategic management consists of planning a strategy, implementing the strategy, controlling the outcomes of the strategy implementation, and adjusting the strategy over time as conditions change.
- Strategic planning involves the identification of the stakeholders, selection of the vision, mission, values, and objectives of the farm; analysis of the external environment; analysis of the internal environment; and crafting the best strategy for the farm in its current environment.
- Strategy implementation involves designing the organizational structures and procedures needed for the chosen strategy and obtaining and directing the resources needed to put the strategy into action.
- Strategic control involves designing control systems, comparing actual results to goals and objectives, monitoring the business environment, and modifying the organizational structure, implementation plan, or even the chosen strategy as needed to meet goals and objectives.
- The need for a disciplined strategic management process is as true for a one-person or one-family farm as it is for a multipartner, multiemployee farm.
- Both strategic and financial objectives need to be specified.
- Answering eight key questions provides a comprehensive external analysis.
- Porter's five forces describing competition are risk of entry by potential competitors, rivalry among established farms, bargaining power of buyers, bargaining power of suppliers, and substitute products.
- Key success factors (KSFs) are measures and performance standards that must be met to be competitive. They are "key" because other results will follow if the KSFs are met.

- Answering five key questions provides a comprehensive internal analysis.
- A farm has a sustained competitive advantage when its profit rate has been higher than the industry average for several years running.
- Four building blocks of competitive advantage are efficiency, quality, innovation, and customer responsiveness.
- SWOT analysis will identify a farm's strengths, weaknesses, opportunities, and threats.
- Crafting strategy can start with adapting one or more generic strategies to a farm's specific conditions.
- Crafting strategy is not a linear process but one with many iterations.
- The current strategy of a farm can be seen in the pattern of actions, choices, and approaches to dealing with its internal and external situation.
- Several tests are available for evaluating potential strategies.
- Scenarios can be used to evaluate potential outcomes under alternative scenarios.
- The first step in strategy implementation is building an organization capable of carrying out the chosen strategy successfully. Several other steps are also involved.
- To implement the chosen strategy well, strategies for each business functional area as well as identification and planning the projects and programs are needed for success.
- Strategic control is needed to monitor actual results in achieving goals and objectives and making corrective actions as needed.
- Strategic management is a continuous process because the world changes.

REVIEW QUESTIONS

1. What is strategy?
2. Why is strategy sometimes called "management's game plan for the business"?
3. Why will strategy be both proactive and reactive?
4. What is strategic management?
5. Why bother with strategic management when you want to just start working?
6. What is a company's vision? Its mission?
7. Why should a farm set objectives? Why both financial and strategic objectives?
8. What are the eight questions for external analysis?
9. If I have already done an external analysis, why should I do an internal analysis?
10. Describe Porter's five competitive forces. How can a farmer use this information for crafting strategy?

11. What are key success factors (KSFs)?
12. What are some key success factors for a dairy farm? A hog farm? A crop farm?
13. What are the five questions for internal analysis?
14. When we say a farm has a sustained competitive advantage, what do we mean?
15. What are the four building blocks of competitive advantage? Give an example of each.
16. Describe SWOT analysis. Discuss each part and give examples of each category.
17. What are the four objectives that need to be considered as product and process choices are made?
18. Why is the following statement false? *The same design for a livestock building can minimize production costs and maximize flexibility for adapting to future changes.*
19. Describe each of the nine generic competitive strategies. Which strategy must most farmers follow?
20. How can a farm's strategy be identified?
21. How can a farmer develop potential strategies for his or her farm?
22. How can a farmer choose the best strategy from among several alternatives?
23. Describe the process of strategy implementation.
24. What is strategic control?

3

Marketing Plan

A primary concern of a business is: *Will someone buy my product or service?* The answer to this question can be found by developing a marketing plan. While the rest of the whole-farm plan needs to be developed simultaneously with the development of the marketing plan, the questions of whether there are buyers and what price they are willing to pay are crucial to the success of the business and should be addressed quite early in the whole business planning process.

The first section of this chapter reviews the components of a marketing plan. The rest of the chapter is a brief review of historical price movements, price forecasting methods, and the rudiments of setting up a marketing control system. This chapter will introduce marketing terms, methods, and fundamentals. The goal is not to explain how to watch markets and make marketing decisions. Other texts, classes, and materials are more appropriate for those discussions.

MARKETING PLAN COMPONENTS

The components of a marketing plan can be described as a set of seven questions (or decisions). Answering these questions will help a farm manager decide if there are indeed people who will buy the products. In addition, what, how, when, and where do they want to buy?

What and How Much Will Be Produced?

Listing this production question first points out the need to answer many questions simultaneously or in an iterative process. Before a farm manager can decide how to market products, he or she needs to know what products will be produced and how much of each. And, before a farm manager can decide what to produce and how much of each product to produce, he or she needs to know what price each product can be sold for and what the costs of production are estimated to be. These issues are addressed in detail at several points in this text and will not be addressed here. Understanding the industry and the potential for products is discussed in chapter 2, "Strategic Management." Calculating the costs of production is explained in chapter 4, "Budgeting." Choosing enterprises and the level of production is discussed in chapter 5, "Production and Operations Management."

When to Price?

Most agricultural products do not need to be physically delivered to be priced. With the increased understanding and increased opportunities to forward contract and use the futures market, pricing does not have to take place solely upon delivery or after storage. The promise to deliver is often sufficient to be able to obtain a price. So the timing question has expanded to whether to price the product before, upon, or after physical delivery of the product. A forward contract for milk or corn, for example, is a contract that sets a price for those products and specifies delivery dates in the future. Delivering grain to the local elevator may involve satisfying a forward contract, selling some at that day's cash price, and putting some in storage to be sold on a future date. The timing question also involves whether to price the product before or after input purchases. After checking the costs of feeders and estimating her own feeding costs, a cattle feeder may decide to use the futures market to price the finished steers even before an agreement is made to buy feeders. Seed and fertilizer purchases are often priced and paid before delivery to take advantage of sales and price incentives.

Note that obtaining a price is not the same as receiving a check for the product. The timing question for obtaining a price is a question of how to obtain or protect a profitable price. A forward contract sets the price to be received upon delivery at a future date. The physical delivery to the buyer involves one price that determines that payment to the farmer. If the farmer was using the futures market, another payment may come from that use. The net price received is a combination of the cash price and the net price received from the futures market.

Where to Price?

Farmers have both decreased and increased opportunities for pricing. The number of local product buyers and input sellers has decreased in many areas. However, the ability to sell and buy regionally has increased. Some groups of farmers have even made international sales directly. Electronic markets are also more prevalent and will continue to increase in number. Where to price is a geographical question. Should the farmer sell locally, regionally, nationally, of internationally? What type of business should be used, a cooperative or a private business?

What Marketing-Related Services to Use?

This question concerns inputs mainly. For example, who should deliver the input to the farm? Will the dealer apply the fertilizer or the farmer? Does the fertilizer dealer do the soil testing or does an independent lab do the testing? When machinery is purchased, should a repair contract be purchased as part of the package? What is the trade-off between paying points and the interest rate when obtaining a loan?

What Form, Grade, or Quality to Deliver/Purchase?

Buyers may want or be willing to buy different grades of the same product. For instance, the price of corn is based on certain standards of moisture, foreign matter, and so on, but corn does not have to meet those standards to be sold. Farmers have a choice of selling corn at higher moisture levels and paying the elevator to dry the corn or taking a cash discount and selling it upon delivery. Farmers could also dry the corn themselves before selling it. So the farmers' decision revolves around the current versus future price, the cost of drying at the elevator, and the cost of drying on the farm. Livestock feeders also have a choice of the weight at which to sell the animals. A cow-calf operator can decide whether to sell calves or keep them longer and sell yearlings or even fed cattle. Crops need nitrogen but farmers can choose whether to deliver that nitrogen in the form of anhydrous ammonia, urea, manure, or another form of nitrogen fertilizer.

How to Price: Cash, Contracting, Hedging, Options, Other Methods?

Selling a product at the cash price available in the market may be the easiest way to price a product but the cash price is not always the best price. On the other hand, the cash price is also not always the worst price. While all products will eventually be sold at a cash price, many other pricing methods exist to help a farmer improve on that cash price. These methods include forward contracts, hedging, futures options, and other contractual arrangements. These options can be used for both selling products and buying inputs.

Forward Contracts

Forward contracts are contractual obligations for the farmer to deliver a specified volume of a product (grain, milk, livestock, for example) at a certain time in the future and for the buyer to pay a specified price for that volume of product. Forward contracts specify the desired quality attributes of the product and any discounts and premiums that may occur if the actual product deviates from the specified levels. The price specified in a forward contract is fixed and will not change with changes in the market.

Hedging

Hedging involves purchasing a futures contract to reduce the risk of adverse price movements for either products or inputs. A futures contract is a legal and binding contract to deliver or accept delivery of a specified commodity on or before a specified date in the future. Contracts are available at several locations, the two most recognized are the Chicago Board of Trade and the Chicago Mercantile Exchange.

Hedgers already hold (or will hold shortly) what are called cash positions, that is, cattle on feed, growing grain, and so on, and enter the futures market to hold the opposite position. For example, when a hog feeder buys hogs, he sells a futures contract to deliver hogs at a future date. When the hogs have reached market weight, the hog feeder buys back the futures contract and sells the hogs in the cash market. The feeder receives the cash price plus the difference between the selling and buying prices on the futures contract. The net price will in all likelihood not equal the price for which the original contract was sold since physical delivery most likely takes place at a location different from that specified in the contract. The difference between the local cash price and the futures price is called the basis. This hog feeder has hedged the price for his live hogs by trading the risk of fluctuations of the cash price for the smaller risk of fluctuations in his basis.

This same hog feeder may hedge his price for feed by using the futures market. In this case his cash position is that he needs feed and thus buys a futures contract. When the feed is actually bought locally, the feeder sells the futures contract. Again, the feeder is hedging the cost of his feed by trading the risk of fluctuations in the cash market for the smaller risk of fluctuations in his basis.

Speculators do not have any cash position. They do not own livestock, grain, or any other commodity. They are gambling that they know which direction the futures market will move and position themselves to take advantage of that market movement. Speculators provide liquidity to the futures market by being willing to speculate and take chances thus providing more buyers and sellers to the marketplace. Unhedged producers are essentially speculators in the cash market.

Futures Options

A futures option is the right but not the obligation to buy or sell a futures contract at a specific price on or before a certain expiration date. Options can be used to protect a price for a product or input for a specific cost that is known at the time of the transaction. As with a futures contract, the opposite position is taken in the cash and options market. To protect his price, a hog feeder could buy a put option (that is, the right to sell hogs in the future). As the cash market moves, the value of the option adjusts. When the hogs are sold, the hog feeder would sell back the put option bought earlier. One advantage of options is that the buyers and sellers of options are not subject to the margin calls that buyers of futures contracts may receive if prices move against their position.

Other Pricing Methods

Other pricing methods are available such as basis contracts and delayed sales contracts. Some producers of fruit and vegetable crops market directly to consumers through roadside stands and farmers' markets. One option used by a few growers, especially organic growers, is community-supported agriculture (CSA). A person buys a share in a CSA farm and then receives a share of the production that year. The CSA shareholder is not guaranteed specific quantities, only that share of the actual production obtained that year.

When and How to Deliver?

This question concerns the physical delivery of the product. To answer this question, a farmer needs to know what potential transportation and storage costs are. With these cost estimates, the farmer can then evaluate the price available locally versus regionally and analyze market information and trends to decide whether it would be best to market now or in the future. Estimating the costs of direct retail and marketing is needed to evaluate the benefits of these options.

HISTORICAL PRICE PATTERNS

Many factors affect prices. Supply and demand forces affect how the marketplace behaves, and where the market price is discovered. Since these forces change and move over time, we also see prices change and move over time. It is with these movements over time that we start to understand what determines changes in prices and how we can forecast prices. We start with systematic or regular price movements and move to irregular price movements or shocks.

Systematic Price Movements

A *trend* is a gradual and sustained increase or decrease in prices over many years. In the last 20–30 years, price trends have been associated with technical change, population growth, and increases in international trade. Changes in consumer preferences (such as the increase in poultry consumption) can also create a price trend.

Recognition of long-term trends in prices and demand can help a manager make long-run plans and decisions. Profitability projections are different if the price trends are up, stable, or down. If price trends are up or stable, older, less efficient units may remain profitable. Downward price trends will force the older, less-efficient units out of production. However, expansion may still take place in face of decreasing demand and lower prices, because new technologies may allow new units to be profitable at lower prices.

A long upward trend from 1970 to 2001 can be seen in the U.S. average wheat prices even though the price also varies around that trend (fig. 3.1). The U.S. average corn price does not exhibit any consistent price trend from 1980 to 2001 (fig. 3.2). The average prices for slaughter steers show an upward trend from 1970 to 2002 (fig. 3.3).

Cyclical movements are repeated patterns of increasing and decreasing prices over several years. These patterns are longer than one year and may be several years long. They are related to lags in production response to price changes and overreactions of producers to price changes.

Twenty years ago, cycles were obvious in hog and beef prices as producers would slowly build up their breeding herds in response to good prices and then slowly decrease them in response to poor prices. Together, producers would build the total breeding herd too large in response to good prices and cut their herds too much in response to poor prices. This slowness to respond and then

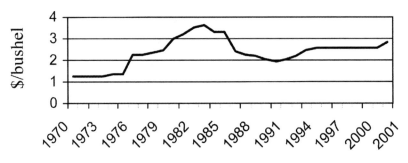

Figure 3.1. U.S. wheat price, 1970–2001 marketing years. (Source: USDA)

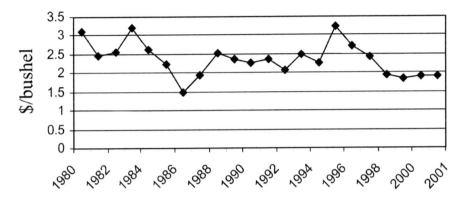

Figure 3.2. U.S. average corn price, 1980–2001. (Source: USDA)

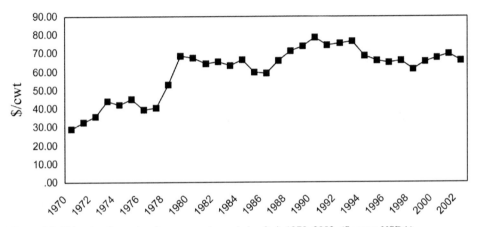

Figure 3.3. Nebraska direct slaughter steer prices, choice 2–4, 1970–2002. (Source: USDA)

the overreaction to price-level changes caused prices to cycle up and down over several years. The beef cycle was longer than the hog cycle due to the breeding and replacement cycle being longer for beef than for hogs. In recent years, these cycles have become less obvious for several reasons, including the increasing size of production units that have more stable production levels.

Although cycles are not as obvious today as they once were, a contrarian investor or manager can benefit from knowledge of cycles and behaving in the opposite manner from the behavior causing the cycle. That is, as others are divesting and decreasing production due to a cyclical downturn in prices, a contrarian manager would be increasing to be ready with production when the price turns up. However, this strategy requires more knowledge than the hope that prices may rise. The underlying demand for the product (and any changes occurring) needs to be understood as well as the length of the production cycle and other forces causing the cyclical behavior in the market.

Boom and bust cycles are examples of severe price cycles. Often, the prices never recover from the bust; there is only one cycle. Rapid increases in profitability (or promises of profitability) need to be studied carefully to see whether they are sustainable or without long-term support.

Seasonal movements are patterns of price changes within the year associated with seasonality in supply and/or demand. Weather and production patterns cause prices to move in similar patterns each year. In a normal weather year, crop prices are at their lowest point in the year at harvest. This can be seen in the average monthly prices for soybeans from 1974 to 2000 at Worthington, Minnesota (fig. 3.4). Holiday demands for turkeys and hams create higher prices during parts of the year. For some products, like Easter lilies or Christmas trees, the demand is so seasonal that, immediately after the holiday, demand disappears and price approaches zero.

With storable products such as grains, farmers may be able to benefit from seasonal movements in prices. As we just noted, harvest prices are typically the lowest of the year, but a few exceptions to this pattern do exist. In years following poor crops, prices tend to be high in the fall and decrease through the year. This pattern is clearly seen in the average monthly prices in Worthington for the six marketing years after poor soybean crops (fig. 3.4). In years following poor crops, a manager may decide to change the normal marketing plan by selling earlier those crops that normally would have been sold later in the year. Tax considerations may affect this decision (especially if two crops would be sold in one tax year), but the decline in prices after a poor harvest is dramatic and should cause a reevaluation of marketing plans.

Seasonality can be seen in hogs and cattle also. Price data from 1970 to 1999 show prices of slaughter steers in Nebraska to be highest in the spring

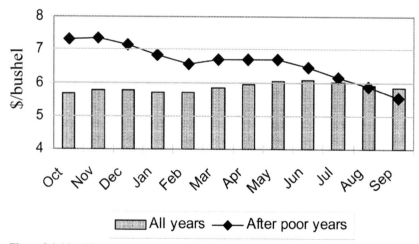

Figure 3.4. Monthly cash soybean prices, Worthington, Minnesota, 1974–April 2000. (Source: adapted from Weness 2002)

and lowest in September with another low in January (fig. 3.5). The (now discontinued) price data for barrows and gilts in Iowa and southern Minnesota also show hog prices to be the highest in the summer and lowest in November (fig. 3.6).

Irregular Price Movements (Shocks)

Irregular price movements result from "sudden and unforeseen" shocks to market supply or demand. Outlook information is especially important to monitor events that may be unusual and thus shocks to the market. Shocks can be local, regional, national, or global and have natural or human causes. Floods, winds, hailstorms, and other destructive storms are examples of weather changes that can cause sudden market shock. Since it develops

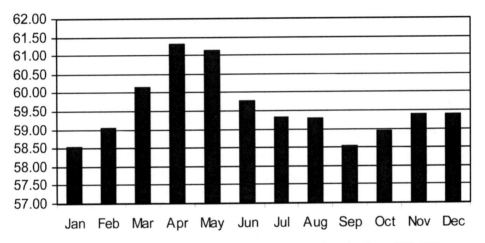

Figure 3.5. Average monthly prices for slaughter steers, choice 2–4, Nebraska direct, 1970–1999. (Source: USDA)

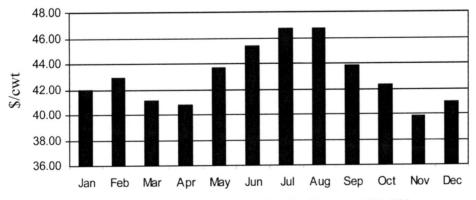

Figure 3.6. Average monthly prices for barrows and gilts, Iowa/So. Minnesota, 1970–1998. (Source: USDA)

slowly, a drought is not a sudden shock. However, since it is not expected to last from one year to the next, a drought in the United States is a shock to the market in that year. Political changes, both at home and abroad, can also cause market shocks and sudden price movements. Disruptions in the normal transportation flow of a product, such as a train wreck, labor strike, or damage to locks, dams, or bridges can cause local and regional fluctuations in prices.

GENERAL PRICE FORECASTING METHODS

Forecasting prices can be done by five general methods: balance sheet, econometric modeling, time series analysis, charting, and historical price movements. Understanding these methods can help a manager understand basic market forces and mechanisms and be better able to formulate his or her own forecast.

Before starting forecasting, a farm manager needs to decide whether a specific price forecast or a forecast price direction is needed. The answer to this question will determine how much work and data are needed and which method should be used.

Balance Sheets

Balance sheets are developed by the USDA to show the "balance" of supply versus disappearance for a commodity. In a balance sheet, supply of a commodity comes from (1) estimates of the commodity carried over from the previous year, (2) current year production, and (3) imports during the current year. For example, in December 2002, the USDA forecast the total supply of soybeans in the United States would be 2.9 million bushels for the marketing year starting September 1, 2002 (table 3.1). This total mostly comes from production (2.69 billion bushels) but also comes from carryover from the previous year and imports. Utilization comes from (1) soybean crush for oil and meal, (2) exports, and (3) seed and feed. For the marketing year starting September 1, 2002, USDA projected total use or disappearance to be 2.725 billion bushels. Ending stocks or the estimated carryover to the next year was estimated to be 175 million bushels.

Commodity balance sheets can be viewed as the national version of an individual farm's estimates of beginning inventory plus production plus purchases compared with estimates of on-farm uses (fed to livestock, for example), sales, and ending inventory. However, the estimates for the national balance sheets involve more uncertainty than the farm version.

Table 3.1. Balance Sheet for Soybeans: U.S. Supply and Disappearance (million bushels)

Year beg. Sept. 1	Supply				Disappearance				
	Beginning stocks	Production	Imports	Total	Crush	Exports	Seed feed residual	Total	Ending stocks
2000/01	290	2758	4	3052	1640	996	168	2804	248
2001/02[a]	248	2891	2	3141	1700	1063	170	2933	208
2002/03[b]	208	2690	2	2900	1660	900	165	2725	175

Source: Oil Crops Outlook, OCS-1202 text and tables, Economic Research Service, U.S. Department of Agriculture, December 11, 2002, http://jan.mannlib.cornell.edu/reports/erssor/field/ocs-bb/2002/ocs1202t.pdf, accessed December 13, 2002.
[a]Estimated
[b] Forecast

For crops, national production estimates are made from surveys of, first, planting plans and then crop condition as the growing season progresses. While large companies do their own forecasting, USDA's reports of production estimates are highly anticipated and markets can make sudden shifts if the estimates are different from expected levels. Domestic use is estimated from trends in consumption and population and the estimated impact of any price changes. Exports and imports are projected from past trends and from the conditions in international markets. Carryover from the past year is estimated from actual levels of production and use in that previous year. Carryover from the current year to the next year is estimated as a residual amount.

Analyzing balance sheets involves looking at the fundamentals of supply versus disappearance, of supply versus demand. When disappearance is closer to supply, ending stocks decrease and prices for the crop tend to increase. Expressing the ending stocks as a percentage of production can be used as an indicator of the pressure on prices. A higher percentage indicates pressure for prices to move down. A lower percentage indicates pressure for prices to move up.

Changes in the balance sheet can also be used to estimate changes in future prices. Based on past data and adjusted for any seasonal and cyclical effects and assuming other conditions (such as other production levels and consumer preferences) are unchanged, a 1% change in total corn supplies is estimated to cause a 2%–3% change in the seasonal average corn price in the *opposite* direction (Futrell 1982). Conversely, a 1% change in use will change prices by 2%–3% in the *same* direction.

A similar effect can be seen in other crop and livestock commodities. For example, a 1% change in total soybean supplies will cause a 2%–2.5% decrease in the seasonal average price. A 1% change in wheat supplies is estimated to cause a 1%–1.5% decrease in wheat prices.

Econometric Models

Econometric models of supply and demand are estimated statistically and used to estimate changes in both physical levels and prices. Econometric models are often used to estimate some components of the balance sheets. Econometric models can also be used to estimate the complicated effects and cross-effects of changes in some factors underlying demand and supply for a specific commodity and its substitutes.

Time Series Analysis

Time series analysis is a statistical approach to monitoring changes over time. It is interested mostly in the changes and any patterns in those changes. Time series analysis does not have as strong an economic foundation as the balance sheet or econometric modeling. Thus, time series analysis, while useful for predicting price changes based on past data, cannot explain the underlying behavior of producers and consumers. It also cannot predict the impact of new market shocks because these new shocks have not been seen in earlier data.

Charting

Charting is the plotting of prices over time. It allows us to see patterns and trends. Especially in the futures markets, these graphical patterns are used as signals of future price changes. Since it does not have a strong foundation of economic reasoning behind its predictions, charting cannot predict the impact of new changes. However, since so many people understand and follow charting, the market does follow some "rules of charting." Thus, it can be useful to follow and understand charts especially for immediate marketing decisions.

Historical Price Movements

Simple statistics and graphs of historical price movements can provide an understanding of past market behavior and the impact of certain factors. For example, historical data show that the average cattle price in February is higher than the average January price (fig. 3.5). In further analysis of the cattle prices for Worthington, Minnesota, Weness (2000) found an increase occurred in 87% of the years. In 13% of the years, the price remained steady or moved down between January and February. The average June cattle price also was lower than the average May price in 87% of the years. Similarly, Weness found that in 80% of the years analyzed, the average May hog price was higher than the average April hog price. This information on how often product prices follow the monthly averages information can be useful in both marketing decisions and in production scheduling decisions.

Simple statistics can also be useful to understand where a current price is in the historical range. For example, in Worthington, Minnesota, the cash corn price was greater than $2.75 per bushel 10% of the time in the ten years from 1991 to 2002 and not at all in the five years from 1997 to 2002 (Weness 2002). The cash corn price was greater than $2.25 per bushel 36% of the time from 1991 to 2002 and 20% of the time from 1997 to 2002. This frequency information tells an astute manager that if a cash price greater than $2.25 were to become available, a manager might need to seriously consider selling corn. The fundamentals of the corn market may suggest that not all corn should be sold, but the uncertainties of the market and the certainty of obtaining a historically high price may cause at least some corn to be sold. Similar information can be developed for other commodities and markets.

MARKETING CONTROL

Control is one of the four main management functions (planning, organizing, directing, and controlling). Control is the process of determining and implementing the necessary actions to make certain that plans are transferred into desired results. Through effective control, we are better able to achieve the goals and objectives established in the plan.

In marketing, control takes place in three steps: establishing standards, measuring performance, and, if needed, taking corrective actions. Standards are established during planning (especially during enterprise and whole-farm planning). Examples of marketing standards include expected production levels and prices, needed product quality levels, timing of production, and needed input quantities. Measuring marketing performance takes place in two ways: monitoring market conditions and recording actual performance by the farm.

Three kinds of corrective actions are used in market control:

1. Rules to change the marketing plan if market conditions change. Forward contract more if prices move up, for example.
2. Rules to change the implementation of the marketing plan. Change marketing location if prices improve at locations other than the original, selected location, for example.
3. Change the goals and standards if market conditions change drastically. Raise the expected price for milk when market conditions improve, for example.

Marketing control uses three types of control: preliminary, concurrent, and feedback. When used before crop or livestock production begins, hedging, forward contracting, and options are examples of preliminary control for obtaining expected prices. Concurrent control consists of monitoring of current and expected market conditions and taking corrective actions, such as entering the options market to protect prices. Feedback control would involve, first, analyzing how marketing was done in the past production cycle and how well expectations were met and, second, deciding whether changes in the marketing plan are needed to better meet goals in the next production cycle.

SUMMARY POINTS

- A marketing plan can be viewed as answering seven questions: (1) What and how much will be produced? (2) When to price? (3) Where to price? (4) What marketing-related services to use? (5) What form, grade, or quality to deliver/purchase? (6) How to price: cash, contracting, hedging, options, other methods? (7) When and how to deliver?
- Systematic price movements include trends, cycles, and seasonal patterns.
- A trend is a gradual and sustained increase or decrease in prices over many years.
- Cyclical movements are repeated patterns of increasing and decreasing prices over several years.
- Seasonal movements are patterns of price changes within the year associated with seasonality in supply and/or demand.
- Price shocks or irregular price movements result from sudden and unforeseen shocks to market supply and demand.
- Outlook information and other news need to be monitored continually in order to see these unforeseen events as early as possible.
- Forecasting prices can be done by five general methods: balance sheet, econometric modeling, time series analysis, charting, and historical price movements.
- Marketing control consists of three phases: establishing standards, measuring performance, and, if needed, taking corrective actions.

REVIEW QUESTIONS

1. Why is it wrong to say a farmer only needs to worry about producing a good product?
2. What are the seven decisions that make up the marketing plan?
3. Compare and contrast the three systematic price movements: trend, cyclical, and seasonal.

4. How can a commodity balance sheet be used in price forecasting?
5. When we need to make decisions based on what we think prices will be in the future, why would we study historical prices?
6. Once the marketing plan has been developed, why should a farmer develop a marketing control plan?
7. How are marketing control standards set?
8. How are corrective actions developed for marketing control?

4

Budgeting

Management is a dynamic process that needs information to be effective. Budgeting can help a manager by providing economic information for decisions concerning a production period, an annual plan, or a long-run plan. Budgeting can provide details about individual enterprises and information about the whole farm or ranch. Budgeting is used, in a variety of ways, in all phases of management: planning, organizing, directing, and controlling.

Budgeting provides information that can be used to support a variety of management tasks. Budgeting:

1. Helps the manager select the best crop and livestock enterprise combinations.
2. Can be used to refine organizational and operating structures. It also forces a manager to develop a production and marketing plan.
3. Forces the manager to uncover cost items that might otherwise be overlooked.
4. Allows the possible outcomes of a change to be studied before resources are actually committed to the change.
5. Can be used to test the economic and financial feasibility of alternative production technologies and management practices.
6. Can be used to develop and organize information that will be useful to lending agencies when the business needs operating, intermediate, or long-term loans.
7. Can help the manager select among investments, when credit is limited, by estimating both the profits and the impacts on cash flow of each investment.
8. Provides information that the manager can use to compare the projected and actual results of implementing a plan.

GENERAL BUDGET TERMINOLOGY

To use budgets effectively, the manager must understand the terminology used to describe and explain them. If these terms are not understood, information may be used incorrectly, and the result may be wrong decisions.

A **budget** is a projection of income and expenses used for planning the future. A budget can be either for a single enterprise or a whole farm or ranch. In farm management, this definition of a budget is different from the customary "accounting" budget, which tells how much is available to spend.

In farm management, an **enterprise** is a common name for any alternative such as corn or dairy. On some farms, enterprises may be defined in more detail such as corn production, corn marketing, the milking herd, or even a support activity such as a machinery shop. A combination of enterprises describes the **whole-farm** business.

Throughout this text, the term **data** is used to denote raw, discrete, unprocessed data gathered from original sources. **Information** is processed, interpreted data. For example, land-market data show current land prices. These data become information for a purchase decision when these prices are compared with historical land-price trends and the land's potential income. In the budget development process, data are provided by the manager and systematically used to develop budget information for the manager's use.

The two basic components of budgeting are **income** and **expense**. A proper understanding and use of income and expense concepts is the basis of effective budget development. Income is a value that is or should be received in return for goods or services. An expense or cost is a charge that is or should be made for an item used in the production of goods or services. Note that "expense" and "cost" can be and are used interchangeably here.

Several groups of terms or distinctions may be used to clarify the significance of income and expense items in specific situations. These distinctions include:

1. Variable versus fixed costs
2. Direct versus overhead costs (and listed costs)
3. Cash versus noncash income and expense
4. Opportunity costs
5. Long-run versus short-run
6. Economic versus accounting values
7. Total, average, or marginal income and costs

The distinctions may not be absolutely clear in all situations, but the ideas embodied by each term help us understand the use (and misuse) of budgets and the information within budgets. The terms and distinctions are discussed in more detail in the following sections.

Variable Versus Fixed Costs

The distinction between variable and fixed values is more applicable to costs than to income. Most farm income will vary with production, but very little income is fixed in the sense that a cost can be fixed. Thus, the discussion in this section concerns only costs.

Which costs are variable depends in large part on the planning horizon of the farm. The length of the planning horizon depends on the decision being made. In the long run, all inputs are variable, so all costs are variable. For example, if the decision is to expand the dairy herd and build a new farmstead, very few costs are fixed. In the short run, some costs will be fixed and some costs will be variable. For example, if the land and equipment are already owned or rented and the decision is which corn varieties to plant, many costs are fixed.

Fixed costs are those that occur no matter what or how much is produced on a farm. Examples of fixed costs include taxes on land and buildings; interest on the investment in land, buildings, and machinery; and depreciation on buildings and machinery. Many costs, such as office expenses or supervisor's wages, are fixed costs in the sense that they will be incurred no matter what enterprise is involved. The operator's labor and family labor also may be considered fixed costs under certain circumstances: the cost of these items could be the same no matter how much is produced.

With a given set of fixed expenses, production increases will lower the fixed costs per unit of production. From this comes the expression, "spreading your fixed costs." For instance, harvesting more acres with the same combine spreads the fixed costs of combine ownership over more acres. However, not all costs decline as acreage increases. Although the fixed costs per acre decline, the labor and fuel costs per acre, for example, remain the same no matter how many acres are harvested. That is why they are called variable.

Variable costs are those that are more directly associated with the volume of business. The costs of seed, fuel, machine repairs, fertilizer, herbicides, and so forth, are examples of variable costs in crop production. Hired hourly labor, feed costs, veterinary expenses, and marketing expenses are examples of variable expenses of livestock production. Interest on operating loans is also a variable cost. These costs vary in total with the size of the farm business.

The level of some inputs or operations, such as weeding or cultivating, may not be changed as easily as the fertilizer level. However, these costs are still considered variable costs because they are specific to a certain enterprise and would not be incurred if that enterprise were not produced.

Labor costs may be fixed or variable depending on how long the employment or work relationship is expected to be. The cost of a full-time, permanent employee is usually considered a fixed cost. In contrast, the cost of a temporary or seasonal worker is usually considered a variable cost. Unpaid family labor that works at the farm full-time is considered a fixed cost. Unpaid family labor that works only part of the year (say a college student home for the summer) is usually considered a variable cost, but that may vary with the person's commitment to the farm over time. If labor is hired for a specific job, especially if it is directly related to a product (e.g., crop weeding, plowing, hoof trimming), it should be considered a variable cost.

The distinction between fixed and variable costs is important because of how each affects current decisions. By definition, fixed costs are fixed—they have to be paid no matter what happens or what is decided today or in the future. Although fixed costs are not affected by current decisions, a manager's decision to act (and thus incur a variable cost) or not to act, may be influenced by fixed costs. Past investments can affect the response to and the value of the current decision to spend money. For example, the decision to apply anhydrous ammonia to corn in the spring depends on the cost of the anhydrous and the application compared with the expected value of the crop response to the nitrogen. Expenses for fall fertilizer and starter fertilizer should not be explicitly included in the anhydrous decision because those expenses have already been spent and become fixed costs. However, the presence of that previously applied fertilizer can affect the response to and the value of the anhydrous and thus the decision to apply anhydrous. In another example, the past investment in feeding facilities greatly affects the decision to buy feeders this year, but the loan on the facilities has to be paid whether feeders are bought or not.

Many farmers try to cut their per-unit variable costs of production by replacing labor with machinery. These farmers may be substituting higher fixed costs (depreciation, interest, and property taxes on machinery) for lower variable labor costs. Decreasing variable costs while increasing fixed costs does not guarantee profits and may decrease a farmer's flexibility to change with changes in the economy.

All costs become fixed once they have been used or committed to use. These are sometimes called **sunk costs**. Once seed is planted, fertilizer is applied, or gasoline is used in the tractor, they become fixed, that is, sunk, although they were variable costs before they were used. The cost of the seed becoming a sunk cost as the seed is planted is a good visual picture of a sunk cost. Since sunk costs are fixed costs, they do not affect production decisions, as discussed earlier.

In some years, when it becomes obvious after planting that the future income from an enterprise will not cover total crop costs, an understanding of sunk costs will help the manager reduce losses. For example, a dry period during pollination can severely reduce yields to a point where the farmer will lose money on the crop. The farmer still has some variable costs, such as the harvesting expenses, which may or may not be incurred. The harvest decision will depend upon whether the potential value of the crop will exceed the variable harvest costs. If the potential value is less than the variable harvesting costs, the farmer should leave the crop in the field and not harvest it because doing so will increase the financial loss. If the gross income exceeds the variable harvesting costs, the farmer will reduce total losses by harvesting. In this second case, the final accounting may show a loss for the enterprise, but the actual loss would have been higher if the crop had not been harvested. In decisions such as this one, any remaining variable costs become the key to the decision; variable costs already committed are now fixed or sunk costs.

Direct Versus Overhead Costs (and Listed Costs)

In most published budgets from extension offices, costs are listed as either direct or overhead. The distinction between direct and overhead costs can be described in terms of the ease of assigning a cost to a certain crop or livestock enterprise.

Direct costs are most likely also variable costs used directly in the production of a specific crop or livestock. Corn seed is used to grow corn not soybeans; so, obviously, it is a direct cost for corn. Similarly, hog feed is for hogs, veterinary expenses for milk fever are for dairy, and so on. The fuel used to dry corn is obviously a cost of corn production. Although the fuel used for tractors and other machinery may come from the same fuel tank, we can use engineering data to estimate and then allocate how much fuel is needed to combine an acre of corn, for example, and how much is needed per hour to haul hog manure.

Overhead costs are those costs that are hard to assign directly to a particular enterprise. Most fixed costs are overhead costs. The depreciation and interest costs for machinery used in several enterprises are harder to allocate to each of those enterprises—unless a detailed log is kept that shows when each machine is used and on what enterprise. We may still make an arbitrary allocation to the enterprises, but since we do not know for sure if that allocation is correct, we classify the cost as an overhead cost. A similar case can be described for the cost of a full-time employee who works on both crops and livestock. We cannot directly allocate the cost of this worker to the specific crops and livestock without detailed work records, so we make an arbitrary allocation and list the cost as an overhead cost in each budget.

Published extension budgets often call the total of all costs "**listed costs**" or "sum of listed costs." While they are very accurate, published budgets may not contain or list all the costs that may occur on a specific, individual farm since they are developed for a representative farm. The term "listed costs" is used to alert the user to the fact that, while the budget includes most costs, the total may not include every cost for a specific farm.

Cash Versus Noncash Income and Expenses

The cash versus noncash distinction is based on the nature and timing of a transaction. Income was just defined as "a value that is or should be received." Similarly, an expense or cost was just defined as "a charge that is or should be made." These definitions do not require that an actual cash transaction be involved for the income or expense to be listed for a specific time period. Examples of noncash items include unpaid family labor, the value of farm products consumed at home and by employees, crops fed to livestock, manure applied on crops, depreciation, and interest on equity. Although they do not affect the cash in your wallet or bank account today, these **noncash income and expenses** may be important tomorrow or next year. Thus, both cash and noncash items need to be estimated to make better decisions for the future and to prepare more accurate assessments of a farm's financial condition and performance.

For example, the feed raised by a farm and fed to the livestock on that farm is a noncash expense for the livestock enterprise and as a noncash income for the crop enterprise. No cash is exchanged. The livestock enterprise does not buy the feed, and the crop enterprise does not sell the feed. However, to obtain an accurate assessment of the profitability of each enterprise, the noncash value of the feed needs to be assigned as income to the crop enterprise and an expense to the livestock enterprise. If this is not done, the profit will be overstated for the livestock enterprise and understated for the crop enterprise.

The estimate of whole-farm profit will still be accurate if noncash items are left out of both enterprises, but a manager will not have complete information to make decisions. Suppose that a farmer has a sufficient income that danger signs are not readily apparent. However, the farmer may not realize that more profit could be made buying the feed in the market and using the land to produce another crop. A farmer may also realize that his or her skills are better used managing livestock, renting the land to another farmer, and buying feed from that farmer or the market. Without the estimates of noncash transactions, the resulting management information and, thus, opportunities are lost.

In a real-life example, a California rancher estimated his cow-calf and hay budgets plus some alternative crop budgets. After considering this information, he realized, "I could raise more hay per acre by planting olive trees!" That statement may sound crazy because cows cannot eat olives instead of hay. However, his budgets showed that, when the cows paid market prices for the hay he produced, hay was less profitable for the ranch than raising olives for cash and buying hay in the market. If he had not included the noncash transactions of "selling" hay to the cows, he would not have seen the opportunity for increased profits and decreased risk by diversifying his income sources.

Buying land is another example of needing both cash and noncash items for complete information. Today's land purchase involves the cash expenses of a down payment and loan fees today and the cash payments for interest and loan principal in future years. In those future years, the buyer plans and hopes to make a profit in two ways: first, by receiving sufficient cash income to pay all cash expenses (including principal and interest) plus some cash for family living, and second, by enjoying the appreciation in land values. The appreciation in value is a noncash item until the land is sold. However, if this noncash item is not included in budgets, a potentially profitable purchase may not be made. This noncash appreciation also needs to be included in the buyer's financial statements (along with potential noncash selling costs) to have a complete picture of the buyer's financial condition when the land is held and not sold.

Opportunity Costs

Opportunity cost is a concept used to specify a cost for using resources. It is very important in farm management but easily misunderstood. The opportunity cost of a resource is the income that could be received from the best alternative use of that resource. While this may seem like a vague and useless concept, it has real applications.

Labor, land, and capital resources can be used in several ways. Each alternative may generate a different income from those resources. For example, an acre of land could be used to produce several crops. Each of these crops would produce a certain income, but only one crop can be grown at a time. Therefore, the income from other crops is forgone.

When any resource is used in one way, the income from using it in any other way is lost. The income from the best alternative use is called the opportunity cost of that resource because the farmer has given up the opportunity to earn that income. So the "cost" of using the resource in the chosen use is its opportunity cost (that is, the income that it could have earned in the next-best alternative). For resources that are not purchased for each production period, such as equity capital or owner's labor, the opportunity cost of that resource can be used to evaluate whether the chosen use is the best use of that resource.

Opportunity costs are a factor in nearly all farm management decisions. The farm manager must not only ask, "will this use of capital, labor, or land be profitable?" but she must also ask, "Would this capital, labor, or land produce a higher income if it were used in another way?"

Long-Run Versus Short-Run

A manager's planning horizon affects how prices and quantities are chosen. The planning horizon relates to the length of time affected by the current decision. A manager who is deciding whether to sell steers today, tomorrow, or next week has a very short planning horizon for that decision. The same manager would have a longer-run horizon for a decision concerning the purchase of a neighbor's feedlot. Because of the different planning horizons, this manager might use two different beef prices in developing the budgets involved with these analyses.

The length of the planning horizon affects not only values but also the process of choosing which prices and quantities to use in budgeting. For situations with longer horizons, the manager needs information that reflects a longer period, a broader geographical base, and a larger, more political economy.

A very short horizon dictates the use of current prices and quantities. That is, today's or next week's values may be used even if they are much higher or lower than what is normally expected. The very short-run decision is concerned with profitability in the short-run horizon.

The planning horizon for a farmer deciding which crops to grow next year is too long for him to rely on projections based only on today's prices. This manager needs to evaluate economic and environmental factors that may affect profitability next year. For such an intermediate term, projections that emphasize local information from the recent past may be sufficient.

Investment decisions require the longest planning horizons. Obviously, land and perennial crop decisions involve a longer commitment of capital than annual crop decisions. Machinery decisions require a longer horizon than fertilizer decisions. Thus, the former decisions require a longer horizon. The longer the horizon is, the more information is needed to project prices and quantities into the future. For sizable, long-term investments, a manager may need to evaluate historical data, long-run population trends, world climatic conditions, political conditions on both a regional and an international view, and other information.

In summary, the planning horizon affects the size and complexity of the information base that should be used by managers. As the planning horizon lengthens, the sources of information need to come from farther away in time, space, and markets.

Economic Versus Accounting Values

The distinction between economic and accounting values is made by considering three basic questions. First, how is the value determined? Second, how is the value allocated between enterprises? Third, is the value cash or non-cash?

The method for determining value can vary with the purpose of the budget. For example, the financial cost of owning land should be estimated one way when we want to estimate the impact of existing loans on cash flow and another way when we want to compare the return from farming with the potential return from alternative investments. To evaluate the impact of existing loans, we would use the accounting cost of owning land that uses the actual interest rate on land loans. For comparing the returns from farming with potential returns in other investments, we would use the economic cost of owning land as the opportunity cost of the money tied up in the land. The money tied up in the land is both the actual loan and the equity in the land including any market price appreciation. When land prices and interest rates change, the difference between economic and accounting values can be large and make a large difference in how we interpret the budget we are estimating and, thus, the answer we develop.

The method of allocation can also cause a difference between economic and accounting values. In an economic sense, the crop may be allocated only part of the interest costs on land. The rest is allocated to a land investment enterprise. This method recognizes that the value of land comes from two sources: farming and speculation. The accounting process may place the full cost of holding land on the crop because until the land is sold, the crop is the only source of revenue. However, some accounting systems may be able to allocate land costs between farming and speculation.

The third distinction relates to the difference between cash and noncash values discussed earlier. If the owner works on the farm, but does not explicitly pay herself, her "pay" is a noncash cost. An economic enterprise budget would include the owner's implicit pay because it is a cost of production. An accounting budget may not include the owner's pay as a cost. It might instead label the "bottom line" as the return to owner's labor, management, and risk. Including the owner's pay can make the analysis of the financial status of the farm or ranch more accurate, since the owner's labors would be a cash cost if they were done by someone else.

Total, Average, and Marginal Income and Costs

In analyzing income and expenses, comparing like units is important. Five different quantity relationship measures are (1) total income or expense per farm (or enterprise), (2) average income or expense per acre, (3) average cost or return per unit of production, (4) marginal return or cost, and (5) marginal input cost.

Total Income and Expense Per Farm

Total income and expense per farm are the total for the whole farm of the particular item. Total crop sales per farm, total custom hire income or expense per farm, total fuel cost per farm, total fertilizer cost per farm, or total feed cost per farm are examples. These figures are important to the calculation of whole-farm cash flow and income statements.

The totals are interesting as a measure of the size of the business, but by themselves, such summary figures are not very useful in a detailed business analysis. For example, a farm may have spent ten thousand dollars on fertilizer in the current year. But the farmer cannot determine whether enough, too little, or too much money was spent on fertilizer unless the farmer specifies the crops and the acreage of each, how the fertilizer was allocated between these crops, the types of soil involved, and several other items. Even where all farm costs are incurred by a single enterprise, the use of these figures is limited unless other measures are used that consider the size of the business.

Average Income or Expense Per Acre

Average income or expense per acre—income and expenses shown on a per-acre basis—are more useful in business analysis than the farm totals. Per-acre cost figures for individual items are particularly useful in analyzing the efficiency of the use of these items. Labor and machinery costs per acre, seedbed preparation costs per acre, weed control costs per acre, and harvesting costs per acre are examples of useful measures in business analysis. Average net income or net return per acre is useful for comparing alternative crops.

Average Cost Per Unit of Production

Cost quoted on a per-unit of production basis (per bushel of corn produced, per ton of grapes, etc.) is very useful in business analysis. These average costs per unit of production are directly comparable to product prices. More effective marketing decisions can be made based on this information.

Marginal Return or Cost

The concept of margins is also important in farm decision making. Marginal returns and costs are used to choose the optimal level of production. Simply stated, the marginal return or cost is the additional return or cost, respectively, resulting from the production of an additional unit of output. The key to using this concept is the word "additional."

A farmer may wish to decide, for example, whether to increase corn production. The question is whether the added revenues will be greater than the added costs. The marginal revenue is the value of the additional production. Marginal cost is the additional cost of obtaining the additional production. It is the cost of the combination of additional fertilizer, pesticides, labor, and any other inputs needed to increase production. If the marginal cost (added costs) of the additional inputs exceeds the marginal return (added returns) from the additional production, obviously the farmer should not increase production.

Marginal Input Cost

Marginal input cost, also called the marginal factor cost, is different from the marginal cost concept. The marginal input cost is the cost of using an additional unit of a specific input. Knowing, for example, the cost of applying an additional 25 lb. of fertilizer per acre or the cost of cultivating an additional time, may be helpful for some decisions.

The greatest application of the marginal input cost concept is in determining the amount of an input to use when there are varying amounts that can be used. That is, as more input (say, fertilizer) is added, at what level does the marginal input cost become greater than the value of marginal production from that same unit of input.

The terms we just discussed are used throughout economics and budgeting. They are important for the correct development, interpretation, and understanding of budgets and the information within those budgets.

TYPES OF BUDGETS

The result of the budgeting process is information. This information is assembled into budget reports so that it can be understood and properly used. Four basic types of budgets are used in management: whole-farm, enterprise, cash flow, and partial.

Whole-Farm Budget

The whole-farm budget provides a summary of the major physical and financial features of the entire farm business, actual and planned. Whole-farm budget development is the process of identifying the component parts of the total farm business, determining the relationships among the different parts, and assessing the influence of each component upon the other components and the business as a whole. The whole-farm budget is a compilation of the enterprise budgets multiplied by the chosen number of acres or animals plus any whole-farm receipts and expenses not accounted for in the enterprise budgets. The procedures for choosing what enterprises will be included and specifying the activity levels for the chosen enterprises are described in chapter 5, "Production and Operations Management." The whole-farm information is essentially the same as that in projected financial statements, and this discussion is covered in chapter 8, "Financial Management."

Enterprise Budget

An enterprise budget is a statement of expected returns or receipts, expected costs, and the resulting net returns. It is not a historical accounting; it is a projection used for planning the future. An enterprise budget is never for the whole farm, it is always developed for one crop acre, one animal, or similar units of production. An enterprise budget shows what is expected if specific methods and inputs are used to produce a specified amount of a product or products. It is based on the economic and technological relationships between inputs and outputs. An enterprise budget lists the expected products, prices, and gross revenue; the inputs and machinery operations needed, their prices, and total costs; and net returns. Operating or variable costs are usually listed separately from ownership or fixed costs. Some enterprise budgets also contain information such as break-even yields and prices, the timing of income and expenses, and machinery and equipment requirements.

For an example of a budget for a crop enterprise, consider a budget for growing corn following soybeans in northern Iowa/southern Minnesota (table 4.1). The budget shows a potential yield of 160 bushels per acre valued at a cash price of $2.25 per bushel with a government payment of $26 per acre for estimated total receipts of $386 per acre. The direct expenses of growing corn total $354.65 per acre; the overhead costs total $34.47 per acre. So the listed costs total $389.12 resulting in a net return of $–3.12 per acre. A charge for unpaid labor and management ($15) is listed as an expense in this adapted budget. If charge for unpaid labor and management is included as a cost, the resulting net return is $–18.12 per acre. The negative net return after including the charge for labor and management means that the receipts are not large enough to pay all expenses including the desired pay for labor and management.

In this example, total listed costs per bushel are estimated to be $2.43 per bushel. By adjusting the total costs for the estimated government transition payment, the resulting cost of $2.27 per bushel is an estimate of the cash price the farmer needs in the market to pay all costs including the charge for unpaid labor and management.

For an example of an enterprise budget for livestock, consider a cow-calf budget from the Coastal Bend area of Texas (table 4.2). Notice that the receipts are not one-calf only. The authors have made this a realistic budget by showing the receipts for an average cow in a cow herd. For this budget, the average cow in this herd is expected to produce 0.43 steer calves, 0.3 heifer calves, and 0.13 heifer replacement calves plus the herd sells 0.12 cull cows per average cow. Conception and calf death rates are incorporated by having a lower average number of cull cows sold per cow and by producing only 0.86 calves per cow. Total receipts (or gross income) are estimated to be $460.80 per average cow.

Variable costs for this cow-calf budget are based on the estimated costs for feed, machinery, labor, and interest on carrying these expenses until income is received. The total variable costs are estimated to be $204.37 per average cow. The return over variable costs is estimated to be $256.43. This is calculated by subtracting the variable costs ($204.37) from the estimated total receipts ($460.80).

Fixed costs included the depreciation, taxes, insurance, and interest on

Table 4.1.   Enterprise Budget for Corn Following Soybeans, Iowa/Minnesota, 2002

	Unit	Price	Quantity	Total
GROSS RECEIPTS				($/acre)
Corn	bu.	2.25	160	360.00
Government payment				26.00
TOTAL RECEIPTS				386.00
DIRECT EXPENSES				
Seed	kernels	$1/1,000	30,000	30.00
Nitrogen	lb.	0.21	140	29.40
Phosphate	lb.	0.25	60	15.00
Potash	lb.	0.13	50	6.50
Lime (yearly cost)				6.00
Herbicides				31.00
Crop insurance				6.00
Drying (LP gas)	gal.	0.85	25	21.25
Machinery (fuel, oil, repairs)				24.36
Hauling and handling				2.45
Labor				20.80
Land rent				145.00
Interest on direct expenses	annual rate	7.50%	8 mos.	16.89
TOTAL DIRECT COSTS				354.65
RETURN OVER DIRECT COSTS				31.35
OVERHEAD COSTS				
Machinery and equipment (depreciation and interest)				27.90
Dryer (depreciation and interest)				1.42
Hauling and handling (depreciation and interest)				5.15
TOTAL OVERHEAD COSTS				34.47
TOTAL LISTED COSTS				389.12
NET RETURN				−3.12
Charge for unpaid labor and management				15.00
NET RETURN after labor and management charge				−18.12
TOTAL LISTED COSTS per bushel				2.43
Cost per bushel (net of government payment)				2.27

Source: Based on Duffy and Smith 2002; and Weness and Anderson 2001.

machinery, equipment, and livestock and the lease and ownership of the pastures. The lease expense for the coastal pasture is included as a fixed cost in this budget because it is a long-term lease and separate from the owned pasture. Fixed costs are estimated to total $208.76 and the total listed costs are $413.13 per cow. With total receipts of $460.80, the net return is estimated to be $47.67 per average cow. Since the owner's labor and management is not listed explicitly, the net return is the return to the owner's unpaid labor and management.

Table 4.2. Enterprise Budget for One Cow-calf Unit, Coastal Bend Area, Texas, 2003

	Unit	Price	Quantity	Total
GROSS RECEIPTS				
Beef cull cows	9.5 cwt./head	$45/cwt.	0.12 head	$51.30
Heifer calves	5.25 cwt./head	84/cwt.	0.30 head	132.30
Heifer replacement calves	5.5 cwt./head	90/cwt.	0.13 head	64.35
Steer calves	5.5 cwt./head	90/cwt.	0.43 head	212.85
TOTAL RECEIPTS				460.80
VARIABLE COSTS				
20% protein supplement	cwt.	$9.75	4.84	6.80
Hay	roll	30.00	2.5	75.00
Salt and minerals	lb.	0.30	100	30.00
Vet. medicine	head	6.00	2	12.00
Fuel				7.35
Lube				1.10
Repair				3.06
Machinery and equipment labor	hrs.	5.77	3.79	21.87
Other labor	hrs	5.25	0.36	1.89
Interest on Operating Expenses	annual rate	3%		5.30
TOTAL VARIABLE COSTS				204.37
RETURN OVER VAR. COSTS				256.43
FIXED COSTS				
Machinery and equipment (depreciation, taxes, and insurance)				31.35
Machinery and equipment (interest)				19.85
Livestock (depreciation, taxes, and insurance)				13.94
Livestock (interest)				22.37
Pasture, coastal, annual lease	acre	25	1	25.00
Pasture, owned, annual taxes	acre	1.75	7	12.25
Pasture, owned, interest	annual rate	3%		84.00
TOTAL FIXED COSTS				208.76
TOTAL LISTED COSTS				413.13
NET RETURN				47.67

Source: Adapted from "Cow-calf Production," 2002.

A Cash Flow Budget

A cash flow budget shows the ability of a business to generate sufficient cash inflows to cover cash outflows during a specified period. Cash inflows include product sales, borrowed money, withdrawals from savings, sale of capital items, and similar items. Cash outflows include cash expenses, principal and interest payments on debt, capital purchases, salaries, family living expenses, and similar items. A cash flow budget shows the timing and size of operating loans required and computes the resulting interest costs; it shows when (or if) operating loans can be paid back. A cash-flow budget is usually prepared for a year by months. A cash flow budget shows the potential liquidity and solvency of the business. The cash flow budget is discussed in more detail in chapter 8.

A Partial Budget

The partial budget is used for estimating the effects of changes in the business. This change aspect is what sets the partial budget apart from the whole-farm budget and the enterprise budget. As just discussed, the whole-farm budget is set up to help plan the organization of the entire farm business and the enterprise budget is set up to help plan the organization of an individual enterprise.

The need for changes in the original plan can occur daily in farming, and the partial budget can be useful in determining whether considered or observed changes will contribute to profits. Only those items that are subject to change are considered in partial budget analysis. For example, a change in the market price of corn might lead a farmer to do a partial budget analysis of changes in the planned irrigation of corn. A price increase may signal the need for more water and a price decrease may be large enough to suggest a decrease in the water applied. This type of budget is a budgeting shortcut and, thus, a time-saver.

As an example, consider a farmer who plans to sell a group of hogs this week but wonders if feeding them another week before selling them would be more profitable. The partial budget for this example is shown in figure 4.1. During this week, the hogs are expected to gain 7 lb. from 220 lb. to 227 lb. The sale of the 227 lb. hog next week is an additional receipt for this partial budget. The change in plans means that a 220 lb. hog will not be sold this week; this is a reduced receipt. Another reduced receipt is the opportunity cost of not selling this week. This is estimated as the lost interest for one week that would (or could) have been received if the hog had been sold this week. This opportunity cost is estimated using the value of the 220 lb. hog and an annual interest rate of 12% for one week. Additional expenses due to the change are additional feed, labor, and other costs, and the chance of some hogs dying during the next week. The farmer does not estimate any reduced expenses for this change. The net effect of this change is estimated to be $1.02 per hog.

So far in this example, the price was assumed not to change between this week and next, but the effects of a price change could be estimated easily. From the information in the partial budget, we can see that if the total value of a hog sold next week were $96.59, the net effect of feeding for another week

Date of estimation: _October 2002__   By: __Joe Farmer_____

CHANGE: *Feed pigs for one more week rather than sell now.*
The pigs will gain 7 pounds from 220 pounds to 227 pounds.

UNIT: *One hog*

POSITIVE EFFECTS:

1. ADDITIONAL RECEIPTS
    Sell 227 pound hog next week: 2.27 CWT. @ $43/cwt. = 97.61

2. REDUCED EXPENSES
    None

> Total Positive Effects = *$97.61*

NEGATIVE EFFECTS:

3. REDUCED RECEIPTS
    Won't sell 220 pound hog this week: 2.2 CWT. @ $43/cwt. = $94.60
    Loss of interest on 220 lb. hog sold this week:

         12%/year for 1 week = $0.22

4. ADDITIONAL EXPENSES
    Additional feed expense for 7 pounds: $1.33
    (based on a feed cost of $19/CWT. gain)
    Labor and other costs for one week: .35
    Potential death loss: (0.1%)   .09

> Total Negative Effects =   *$96.59*

NET EFFECT = Positive Effects minus Negative Effects =

> *$ 1.02/hog*

NOTES: *This is based on no change in the price between this week and next week. If everything else remained the same, the price next week would have to drop only 33 cents from this week's price to make the additional weight not profitable. Marketing and trucking costs are also assumed to be the same.*

Figure 4.1. Partial budget for a change in market date for hogs.

would be zero. By dividing this value ($96.59) by the expected weight (227 lb.), we see that next week's price would have to drop from $43 to $42.55 per cwt. or $0.45 per cwt. for this farmer to be indifferent between selling now and next week. However, other factors such as the uncertainty of the price change and the potential need for cash now may affect the decision also.

MICROECONOMIC PRINCIPLES FOR MANAGEMENT

One major task for a manager is to allocate available resources between competing uses to meet the goals of the business. All problems of resource use involve one or more of three fundamental economic principles.

1. Increase the use of an input if the value of the added output (that is, income) is greater than the added cost.
2. Substitute one input for another input if the cost of the substituted input is less than the cost of the replaced input and production level is maintained. This substitution can be a complete replacement or simply a change in the mix of inputs.
3. Substitute one product for another product if the value of the new product is greater than the value of the replaced product and the total cost is constant. This substitution can also be a complete replacement or a change in the mix of products.

These principles would be sufficient for all planning if the manager had unlimited resources, no concern about time, and perfect knowledge. Since this is not true, three additional economic principles must be introduced as aids to the decision-making process:

4. If resources are limited, use each unit of resource where it will give the greatest returns.
5. When choices involve different time periods, compare the alternatives based on the present values of the resulting cash flows.
6. When risk and uncertainty cloud predictions, different levels of prices, costs, and yields should be used to evaluate the potential variation in expected income and cash flow.

In the next subsections, the first two principles are discussed in more detail. The third and fourth principles are discussed in chapter 5, "Production and Operations Management." The fifth principle is discussed as part of the investment analysis in chapter 9, "Investment Analysis." The sixth principle is discussed in chapter 11, "Risk Management."

How Much of Each Input Should Be Used?

If the goal is only profit maximization, each usage level of an input (such as fertilizer or herbicide) should be evaluated to figure out if its cost is more than the income expected because of its use. For example, *each* increment of fertilizer use (say, 25 lb.) is evaluated for profitability. The optimal level is the highest fertilizer rate at which the cost of applying the last unit is less than the value of the resulting yield increase. An astute manager would not add the next increment past the optimal level because it would cost more than the value of the additional yield.

This approach to choosing the optimal input level can be seen in an example of wheat response to nitrogen fertilizer (table 4.3). This experimental data from California shows the response of irrigated durum wheat to preplant anhydrous ammonia. The experiment measured yields at nitrogen levels of 0, 80, 160, 240, and 320 lb. per acre. The physical yield reaches a maximum yield of 95.4 bushels per acre with 240 lb. of nitrogen.

The optimal level of nitrogen should be chosen based on the value of the marginal product compared with the marginal input cost (that is, the value of the increased yield compared with the cost of the fertilizer that produces the increase). The process of calculating these values and costs is explained in the following paragraphs.

The marginal product is the difference in yield due to one fertilizer increment. In this example, the marginal product of the first 80 lb. of nitrogen is 24.3 bu. of wheat. This is the difference between 79.5 bu. (the yield with 80 lb. of nitrogen) and 55.2 bu. (the yield without any nitrogen applied). The marginal products for the next three 80 lb. increments are calculated in the same way. The highest level of nitrogen (320 lb.) is too high from even a physical standpoint because the yield decreased due to the overapplication of nitrogen. Thus, the marginal product of increasing from 240 lb. to 320 lb. of nitrogen is negative.

The value of the marginal product will vary with the price chosen for the

Table 4.3.   Wheat Response to Nitrogen Fertilizer

Nitrogen applied	Wheat yield	Marginal (or additional) product	Value of marginal product	Marginal input costs	Marginal net return
lb/acre	bu/acre	bu/acre	$/acre	$/acre	$/acre
0	55.2				
		24.3	102	27	75
80	79.5				
		13.4	56	20	36
160	92.9				
		2.5	11	20	−9
240	95.4				
		−8.3	−35	20	−55
320	87.1				

analysis. In this example, the price of $4.20 per bushel is used. The value of the marginal product is the price times the marginal product. With a price of $4.20 per bushel, the value of the marginal product from the first 80 lb. of nitrogen is $102 per acre (24.3 bushels × $4.20 per bushel). Values of the other marginal products are calculated in the same way.

The marginal input cost is the price of the input times the input quantity for the increment (plus any application cost in this example). Using a price of $0.25 per pound, the marginal input cost of 80 lb. of nitrogen is $20. For the first increment there is an application cost of $7 per acre, but no additional application fee is assessed as the fertilizer level increases. So the marginal input cost for each increment is $20 per acre except the first increment which is $27.

With these experimental data and the prices and costs used, the optimal level of nitrogen fertilizer is 160 lb. per acre. The value of the marginal product from the second increment (from 80 to 160 lb. of nitrogen) is $56 compared with a marginal input cost of $20. The value of the marginal product from the third increment (increasing to 240 lb.) is $11 compared with a marginal input cost of $20. This optimal level also can be calculated by increasing the nitrogen rate if the marginal net return (gross income minus nitrogen costs) is positive.

The 80 lb. increment used in this experiment is large and nature does not jump that suddenly from one yield level to another. When using these data for actual field application, the manager may want to use his or her judgment and evaluate nitrogen rates between 160 lb. and 240 lb. In this example, a plot of yields and nitrogen levels would show that over half the yield increase between 160 and 240 lb. probably occurs in the first 40 lb., that is, between 160 and 200 lb. of nitrogen. The value of one-half of the marginal product between 160 and 240 lb. is just over $5, which is less than the marginal input cost of 40 lb. of nitrogen (that is, $10). So although the 80 lb. increment is large, further analysis shows that the initial selection of 160 lb. is still the optimal nitrogen rate. This optimal rate is the one to use in budget development.

What if the separate increments had not been evaluated? If the maximum yield was chosen as a target, the value of the increase in the wheat yield from 0 to 240 lb. ($170 per acre) is greater than the increase in the input cost ($67). By that analysis, the nitrogen rate should be 240 lb. While that level (240 lb.) is profitable because total return is greater than total costs, it is not the optimal rate. It is not the optimal or most profitable rate because, as we can see in the table, the step from 160 to 240 lb. of additional nitrogen increases the cost higher than the value of the yield increase. That is, the value of the marginal product ($11) is less than the marginal input cost ($20).

For another example of choosing the economically optimal level of one input, consider the case of Gene Burns and his decision of how much nitrogen to apply to his continuous corn. Finances were tight, and the price-cost squeeze was very tight. The soil test lab's recommended fertilizer level was 150 lb. of nitrogen (N) per acre. He wondered if this was correct. He did not want to spend more than required.

Gene remembered that he had received some data on corn response to nitrogen at the fall field days at the University of Minnesota's Southern Research and Outreach Center near Waseca. Since his farm was on soils similar to the center's soils, he had kept that table of responses. After a little search, he found the paper and started looking at the information. It was a report on the effect of nitrogen on corn based on 12 years of data. Using a corn price forecast of $2.25 per bushel, a nitrogen price of $0.30 per lb., and an application cost of $7 per acre, Gene estimated the marginal net return that he might receive with each level of nitrogen using the university's data (table 4.4).

Based on the university's information, the optimal rate is 160 lb. of nitrogen per acre. Below that level each additional step of 40 lb. of nitrogen has a positive marginal net return. But the step from 160 to 200 lb. has a negative marginal net return even though the yield is estimated to increase. So, Gene decides to follow the lab's recommendation of 150 lb. of nitrogen, which is more specific to his farm's conditions that year.

When the decision involves two inputs, the decision process is similar but has a few more steps. As a first step, the best level of the first input is found when the second input is "fixed" at a low level. Then the first input is "fixed" temporarily at the "best" level just chosen and the "best" level of the second

Table 4.4. Continuous Corn Response to Nitrogen Fertilizer

Nitrogen applied	Corn yield	Marginal (or additional) product	Value of marginal product	Marginal input costs	Marginal net return
lb/acre	bu/acre	bu/acre	$/acre	$/acre	$/acre
0	75				
		25	56.25	19	37.25
40	100				
		15	33.75	12	21.75
80	115				
		10	22.50	12	10.50
120	125				
		8	18.00	12	6.00
160	133				
		3	6.75	12	−5.25
200	136				

level is found. This process is repeated until neither input is increased again. This process can be visualized as working from the upper left corner to the lower right corner of a table showing yield response, for example, to both phosphate and nitrogen fertilization.

As an example of choosing the optimal levels of two inputs, let us consider another of Gene Burns' fields of continuous corn. The soil test lab has recommended 150 lb. of nitrogen and 225 lb. of phosphate. To test whether these recommendations were right, Gene pulled out another table from the university's research and outreach center showing the effects on corn yields of different rates of nitrogen and phosphorus (table 4.5).

As described above, choosing the optimal combination of nitrogen and phosphorus starts by finding the best level of phosphate when nitrogen is "fixed" at the lowest level, 125 lb./ac., in this table. We use the same procedures used earlier when only one input was being analyzed (see tables 4.3 and 4.4). Using a corn price of $2.25, a phosphate price of $0.24 per lb., and an application cost of $6 per acre, the best level of phosphate (with nitrogen "fixed" at 125 lb./acre) is 200 lb. per acre (table 4.6).

Now, with phosphate "fixed" temporarily at 200 lb. per acre, the "best" level of nitrogen is estimated using a corn price of $2.25 per bu., a nitrogen price of $0.30 per lb., and a nitrogen application rate of $7 per acre. The row of yields with the phosphate rate at 200 is transposed into a column to fit the same format used in the earlier examples (table 4.7). This shows that with a phosphate rate of 200 lb. per acre, the best nitrogen rate is 175 lb. per acre.

At this point, we need to revisit the rate of phosphate to decide if it should be increased due to a better response at higher nitrogen levels. As shown in the original table, the estimated yield with 175 lb. of nitrogen and 200 lb. of phosphate is 135 bushels per acre (table 4.5). So, the marginal yield response from

Table 4.5. Effects of Rates of Nitrogen and Phosphate on Continuous Corn

Phosphate applied	Nitrogen (lb/acre)					
	125	150	175	200	225	250
(lb/acre)	Corn yield (bu/acre)					
75	88	93	96	97	96	93
100	99	105	108	110	110	107
125	108	114	118	121	121	119
150	115	121	126	129	129	128
175	119	126	132	135	136	135
200	122	129	135	138	140	140
225	122	130	136	140	142	142

Table 4.6. Continuous Corn Response to Phosphate Fertilizer with Nitrogen "Fixed" at 125 lb/ac

Phosphate applied	Corn yield	Marginal (or additional) product	Value of marginal product	Marginal input costs	Marginal net return
lb/acre	bu/acre	bu/acre	$/acre	$/acre	$/acre
75	88				
		11	24.75	12.00	12.75
100	99				
		9	20.25	6.00	14.25
125	108				
		7	15.75	6.00	9.75
150	115				
		4	9.00	6.00	3.00
175	119				
		3	6.75	6.00	.75
200	122				
		0	0	6.00	-6.00
225	122				

Table 4.7. Continuous Corn Response to Nitrogen Fertilizer with Phosphate "Fixed" at 200 lb/acre

Nitrogen applied	Corn yield	Marginal (or additional) product	Value of marginal product	Marginal input costs	Marginal net return
lb/acre	bu/acre	bu/acre	$/acre	$/acre	$/acre
125	122				
		7	15.75	14.50	1.25
150	129				
		6	13.5	7.50	6.00
175	135				
		3	6.75	7.50	−0.75
200	138				
		2	4.50	7.50	−3.00
225	140				
		0	0	7.50	−7.50
250	140				

200 to 225 lb. of phosphate (with 175 lb. of nitrogen) is 1 bushel (136 minus 135). The value of the marginal product is $2.25 per acre, the marginal input cost is $6 for 25 lb. of phosphate, and the marginal net return is $-3.75 per acre. Thus, increasing phosphorus to 225 lb. is not worthwhile.

The resulting recommendation is 200 lb. of phosphate and 175 lb. of nitrogen per acre using the yield response data in table 4.5 and these prices and costs. These rates are a little different from the soil test lab's recommendations of 225 lb. of phosphate and 150 lb. of nitrogen. So, for this field, Gene Burns may need to probe deeper into the differences between his field and the research plots to see if those differences support the different fertilization recommendations.

The analysis of marginal input costs and marginal product values has wide applications. This method can be used for analyzing incremental levels of an input (such as fertilizer and feed) and it can be used to evaluate the use or nonuse of such inputs as herbicides and cultivations. The basic approach is to estimate the biological or physical responses to inputs and then choose the appropriate prices and costs to apply. The concept of marginal analysis is critical at several points in farm management: developing enterprise budgets, evaluating the current plans, or analyzing a potential new enterprise.

What Mix of Inputs Should Be Used?

More than one combination of inputs will often produce the same quantity of output. A common example is the mix of forage and concentrate feedstuffs for cattle. Another example is the different combinations of mechanical cultivation, herbicides, hand hoeing, and other controls that will achieve the same level of weed control. A third example is the different methods of pruning: hand, mechanical, or both.

The optimal combination of inputs that produce the same output is the least-cost combination of those inputs. If the number of alternatives is not large, the simplest way to choose is to calculate the cost of each mix of inputs and select the lowest cost mix. The lowest cost mix will be the maximum profit combination also if the output level does not change. If output levels do change, a partial budget can be used to include the changes in the value of the output and the changes in inputs.

For a simple example of input substitution, suppose a farmer can estimate from his records that for the same level of weed control, one herbicide spray will substitute for four hours of hand hoeing. If the total wage cost is $11 per hour, the hand weeding would cost $44 per acre (4 hours per acre × $11 per hour). If the herbicide spray costs $27 per acre, that is the least-cost weed control method and without considering other factors would be the farmer's choice. Other factors that may affect this decision include potential changes in the weed population dynamics and environmental concerns.

As an example of choosing the optimal feed mix, let us consider four possible combinations of corn and supplement to feed a pig from 40 lb. to 80 lb. (table 4.8). If we temporarily ignore potential time differences due to different protein levels, these different combinations of corn and supplement are estimated to produce the same 40 lb. gain in a pig. Using feed costs of $2.30 per bushel for corn and $160 per ton for the supplement, the cost of each ration is calculated. The corn cost per pound ($0.041) is calculated by dividing the price per bushel by the standard weight of 56 lb. of corn per bushel. The supplement cost per pound ($0.08) is calculated by dividing the supplement price per ton by 2,000 lb. The cost of each ration is the sum of the cost of corn plus the cost of the supplement. For the first ration, the cost of the corn is $4.26 and the supplement costs $1.04 resulting in a total cost of $5.30 to feed the pig from 40 lb. to 80 lb. This simple process is done for the other three rations. Although the differences are small, the third ration is the lowest cost ration at $5.17 and would be the choice if only the feed cost were considered.

Suppose the corn price increased to $2.52 per bushel and the supplement price fell to $130 per ton, the ration costs would change, but would the choice change? The total cost would be $5.53 for ration 1, $5.32 for 2, $5.24 for 3, and $5.30 for 4. Thus, in this case, ration 3 would still be the least cost ration even though the new prices make corn relatively more expensive than the supplement.

The choice of a ration may also be affected by the type of operation. In the previous example, the ration changed but we assumed output did not because all rations were estimated to produce a 40 lb. gain. However, animals gain faster with a ration higher in protein. If we consider the time of gain, these rations do not produce the same output. So managers concerned with the speed at which animals gain these 40 lb. would need more information on the rate of

Table 4.8.   Combinations of Corn and Supplement Estimated to Produce a 40 lb Gain in a 40 lb Pig

Ration	Corn (lb/hd)	Supplement (lb/hd)	Cost of corn	Cost of supplement	Total ration cost
1	104	13	4.26	1.04	5.30
2	95	16	3.90	1.28	5.18
3	89	19	3.65	1.52	5.17
4	86	22	3.53	1.76	5.29

gain and the potential for alternatives of the space. These managers could use a partial budget to analyze their choices.

In the next section we learn how these principles for choosing input use can be used to help develop budgets and thus support the goals and objectives of the farmer.

BUDGET DEVELOPMENT

So far we have discussed the purpose of budgeting and budget terminology. This section builds upon that information and explains how enterprise budgets are developed using farm records, economic logic, and your knowledge of agriculture; how resources, products, and costs are described; and how budget information for enterprises and whole farms or ranches are developed.

Usually, the process of developing budgets involves:

1. Describing the resources, inputs, and products for the farm or ranch. Some general sources of information that can be used for resource descriptions are described in the section "Sources of Data," and in the other references cited in that section.
2. Describing how these resources fit together into an enterprise budget. This step is described after a discussion of the proper budget units (that is, sizing up the budget).
3. Calculating receipts and costs and preparing the enterprise budget reports.
4. Describing how the enterprises are combined in a whole-farm or ranch plan and preparing whole-farm or ranch reports.

Sources of Data

Good data are needed to develop good budgets. Without good data, the information in budgets will be incomplete or inaccurate. If the information is not good, any decisions based on that information may be wrong. Wrong decisions can result in problems for the business, including the loss of profits.

The GIGO principle is valid for budgeting and all analysis: if you put Garbage In, you will get Garbage Out. Budgeting involves the input of large amounts of data to produce detailed and accurate budgets. Much of these data must be specific to an individual farm (such as machinery and equipment use, irrigation use and efficiency). Other data (such as yields, input rates, prices) may be more effective if they are farm specific, but more general sources can be used. In this section, several sources of farm and ranch data are discussed.

Many sources of information and data exist for budgeting. Income tax records and historical records for individual farms and ranches provide very localized information on yields, production, weather, fertilizer use, and so on.

National, state, and local government units collect and distribute price, production, and acreage data for the major crops. Recommendations for fertilizer rates, seeding rates, chemical rates, and so on, are available from the manufacturers, dealers, government and university research and extension units, and other groups. Earlier in this chapter, we discussed the economic principles for making input and product decisions. Some farmers hire independent consultants to recommend the right rates for inputs; this is done especially for pesticides and other chemicals that are often more strictly regulated than other inputs.

Labor requirements, wage rates, and benefits can be determined from several sources. The "going rate" (that is, the most common wage) in the local area is often the most-used source. Labor contractors often provide workers for a job for one total fee. Union contracts may set wage rates, benefits, hours, and other conditions for some workers. The best estimates of labor productivity are from individual farm and ranch records. Productivity estimates can be obtained from government, university, and private sources also.

Budget Units

While an enterprise budget provides bottom-line information about a large component of a farm business, such as "growing corn," or "dairy operations," or "whatever we are doing with the south forty," budget calculations are based on smaller standard units of measurement, not the present size of the enterprise. Most enterprise budgets should be described by the smallest rational unit of that enterprise.

For example, most crops are described on a per-acre basis, and livestock budgets are usually estimated on a per-head basis. However, other units may be rational for a particular enterprise, or for specific conditions being studied. Sometimes a suggested unit may be too small to be practical (e.g., one laying hen), so a budget unit should be a larger multiple (e.g., one hundred or five hundred laying hens).

The "smallest rational unit" depends upon the enterprise being analyzed, the conditions surrounding that enterprise, the units commonly used to describe similar enterprises (for example, data in USDA or Extension reports), and the manager's preferences.

Rather than using the present size of the enterprise (e.g., two hundred acres or two hundred sows), the smallest rational unit (e.g., one acre or one litter) is used to describe enterprise budgets for three reasons.

1. The amounts and values are expressed in constant and understandable terms (e.g., pounds per acre, pounds per head).
2. The smaller units make it very easy to evaluate different combinations of potential enterprises because of the ease of adjusting activity levels.

3. The smallest rational units are usually the units used by USDA, Extension, credit sources, and other agencies that develop representative budgets. Thus, by using these standard budget units, the manager can easily compare his or her estimates with other sources of information.

Developing Enterprise Budgets

Enterprise budgets can be developed from either whole-farm records or economic engineering. In the first method, whole-farm receipts and expenses are divided and allocated to each enterprise. This budget shows historically how each enterprise contributed to the whole-farm performance. It also can be used to evaluate a farm's performance in comparison to the historical enterprise budgets of other farms available through farm record associations and other public reports. With some updating, these budgets can be used to forecast performance under future conditions. The second method, economic engineering, starts by listing all the inputs, labor, operations, equipment, buildings, and land needed to produce a product. Economic engineering is needed when a farmer has no experience with an enterprise, wants to expand, or wants to create a benchmark to compare actual performance. The steps involved in both methods are described in following sections.

Developing Enterprise Budgets from Whole-Farm Records

When enterprise budgets are developed for the first time, many managers will start with records for the whole farm. While whole-farm records may be very accurate, all of the costs and returns are not easily identified with specific enterprises. Whole-farm costs and returns can be allocated to individual enterprises by using the following eight steps:

1. *Determine the costs of separate items for the whole farm.* If your whole-farm records are in good order, they should show expense figures by individual item (i.e., seed corn, fertilizer for soybeans, feed for the dairy calves). If these records are not up to date or individual items are not specified, these data must be gathered and organized before costs can be allocated to specific enterprises.

2. *Identify the enterprises on the farm.* Most farms grow more than one crop and/or raise more than one category of livestock. If only one crop is grown, allocation is very easy. If the farm business involves more than one crop or livestock enterprise, these enterprises have to be identified and listed. At this time, the farmer should decide whether to separate some enterprises into their parts. For example, the dairy herd could be divided into two subenterprises: the milking herd and all other dairy cattle, such as calves and replacement heifers. The farrow-to-finish hog enterprise could be split

into feeder pig production and hog finishing. Corn production could be split into different land types or different tenure arrangements. These divisions are done to have a better idea of how each part of the business is doing. The divisions can be made even though all of the activity takes place on the same farm.

3. *Classify the costs as direct or indirect.* Direct costs are those attributed to a specific enterprise. Examples are fertilizer applied to wheat and feed fed to dairy cows. Indirect costs are those that cannot be associated with a specific enterprise. These would include costs for a truck used for several crops (or general farm duties), and fencing used for several types of livestock.

4. *Allocate the direct costs and returns.* Direct costs and returns are used or produced by an enterprise directly. Corn harvesting costs are allocated to corn. Weed control for wheat is allocated to wheat. Veterinary expenses for dairy cows are allocated to dairy.

5. *Determine the best way to allocate indirect costs.* There are three main ways to allocate indirect costs:
   a. On the basis of use: Machinery-use hours can be used to allocate fuel, repairs, and other machine costs to the appropriate enterprise. Factors such as total crop acreage, total herd value, and number of head may be used as the basis for allocating costs such as insurance or erosion control to the various enterprises which benefit from them. General farm liability insurance may be allocated to crops based on an average cost per acre.
   b. On the share of gross income: Some office and professional expenses (e.g., telephone, an accountant) may be related to the farm's total expected income. Thus, an enterprise's contribution to total gross income for the farm may be the best basis for allocating those costs. For example, office expenses can be allocated between enterprises based on each enterprise's share of total expected gross income.
   c. On the share of variable costs: Some costs may be allocated based on that enterprise's share of total variable costs for the whole farm. For example, an employee may spend more time on those enterprises that have the largest costs. Thus, the employee's salary should be allocated based on each enterprise's share in total variable costs—rather than gross income. The share in costs more accurately reflects each enterprise's share of the employee's salary.

d. With a combination: Choosing among the three main methods is an arbitrary decision. The goal is to obtain accurate enterprise budgets for analysis and planning. Sometimes, a combination of the three methods will provide the most accurate allocation between enterprises.

6. *Calculate the percentage of total use, gross income, and variable costs for each enterprise.*
   a. Shares of total use (e.g., machinery hours) are calculated by: (1) determining the actual (or expected) use for each enterprise; (2) calculating the total use for the whole farm; and (3) calculating each enterprise's share in the total use.
   b. Shares of gross income are calculated by: (1) determining the actual (or expected) acreage, production, and price for each enterprise; (2) calculating the gross income for each enterprise; and (3) calculating each enterprise's share in the gross income for the whole farm.
   c. Shares of variable costs are calculated by: (1) determining the actual (or expected) variable costs for each enterprise; (2) calculating the total variable costs for the whole farm; and (3) calculating each enterprise's share in the total variable costs.

7. *Allocate the indirect costs.* The whole-farm indirect costs are allocated to the enterprises by multiplying the total whole-farm cost by each enterprise's shares of use, gross income, or variable costs.

8. *Calculate the costs per unit.* The per unit cost in each enterprise is calculated by dividing the enterprise's share of the total costs by the number of units of that enterprise (e.g., the number of corn acres or cows in the milking herd).

As an example of allocating indirect, whole-farm costs, consider how a dairy-corn-alfalfa farm could allocate general office expenses ($15,436) and an employee's salary and benefits ($24,878).

Since most of the office expenses are used for marketing or for those enterprises that have the higher potential income, the farmer decides these expenses should be allocated based on gross income shares. As shown in table 4.9, the farmer calculates the potential gross income for each enterprise, and the share of total, and allocates the office expenses based on that share. The allocated office expenses are $154.36 per dairy cow, $20.58 per acre for corn, and $18.52 per acre for alfalfa. These costs can be used in the respective enterprise budgets.

Table 4.9.   General Office Expense Allocation

Enterprise	Size	Total gross income	Enterprise's share of gross income	Enterprise's share of general office expenses	General office expenses per unit
Dairy	70 cows	$126,300	70%	$10,805	$154.36/cow
Corn	180 acres	43,560	24%	3,705	20.58/acre
Alfalfa	50 acres	11,000	6%	926	18.52/acre
Whole Farm:		180,860	100%	15,436	

While the employee's time is also related to the potential income, the farmer has decided that the share in variable production costs is a more accurate method to allocate the employee's salary and benefits between the enterprises. As shown in table 4.10, the farmer calculates the costs are $252.33 per dairy cow and $34.55 and $19.90 per acre for corn and alfalfa. These costs can be used to itemize the employee's salary and benefits in the respective enterprise budgets.

Sample Budget Using Whole-Farm Records
As an example of developing an enterprise budget from whole-farm records, let us follow Ron Purcell as he estimates a soybean enterprise budget from his whole-farm information. He grows corn and soybeans only and has no livestock. Mr. Purcell has 870 tillable acres. Last year he had 420 acres of corn and 450 acres of soybeans. His corn yield was 151 bushels that he sold for an average of $1.68 per bushel. His soybean yield was 46 bushels that he sold for $4.57. He also received an average of $41.90 per corn acre and $42.80 per soybean acre from the government. Knowing he would need them later, he calculated each crop's percentage share of the farm's gross income. This is shown in table 4.11.

Table 4.10.   Employees' Salary and Benefits

Enterprise	Size	Total variable costs	Enterprise's share of variable costs	Enterprise's share of employee's salary & benefits	Employee's salary and benefits per unit
Dairy	70 cows	$73,640	71%	$17,663	$252.33/cow
Corn	180 acres	25,560	25%	6,220	34.55/acre
Alfalfa	50 acres	4,000	4%	995	19.90/acre
Whole Farm:		$103,200	100%	$24,878	

Table 4.11.   Ron Purcell's Gross Income Shares by Crop

	Corn	Soybean
Acres	420	450
Yield	151 bu	46 bu
Price per bushel	$1.68	$4.57
Other income/acre	$41.90	$42.80
Gross income/acre	$295.58	$253.02
Total gross income for the farm	$124,144	$113,859
Crop's share of total gross income	52.2%	47.8%

From his records, Mr. Purcell easily divided his seed, fertilizer, chemical, crop insurance, and drying fuel expenses between corn and soybean (table 4.12). For repairs and fuel and oil, he considered the total hours of machinery operations required for each crop and decided to allocate 54% to corn and 46% to soybeans. So he wrote $3,749 of the fuel and oil and $7,775 of the repairs in the soybean column. After looking at the list of miscellaneous expenses, he estimated 35% (or $904) of those expenses were due to soybeans. He decided to split his operating interest between corn and soybeans on the basis of each crop's share of all other direct expenses. Since soybeans' share of all the other direct expenses was 37.5%, he wrote that percentage or $2,646 in the soybean column. When he was done allocating the totals, he divided each column of total crop expenses by each crop's acreage (420 for corn and 450 for soybeans) to calculate his direct costs per acre.

Mr. Purcell realized that the costs in his records that he could not easily allocate directly to either corn or soybeans were his overhead costs. He decid-

Table 4.12. Ron Purcell's Allocation of Direct Expenses

Direct expense item	Whole farm total	Corn expenses Farm total	Per acre	Soybean expenses Farm total	Per acre
Seed	$23,490	$15,750	$37.50	$7,740	$17.20
Fertilizer	20,346	16,296	38.80	4,050	9.00
Chemicals	19,647	10,332	24.60	9,315	20.70
Crop insurance	7,011	3,276	7.80	3,735	8.30
Drying fuel	1,260	1,260	3.00	0	0
Fuel and oil	8,150	4,401	10.48	3,749	8.33
Repairs	16,902	9,127	21.73	7,775	17.28
Miscellaneous	2,583	1,679	4.00	904	2.01
Operating interest	7,056	4,410	10.50	2,646	5.88
Total direct expenses	106,445	66,531	158.41	39,914	88.70

ed to allocate 46% of the machinery depreciation to soybeans as he did for the direct machinery expenses. Purcell also realized the real estate taxes should be allocated on an average per acre basis. He decided to allocate the rest of his overhead costs to each crop by using each crop's share of gross income. Once he made the allocation of the totals to each crop, Mr. Purcell again calculated the per acre costs (table 4.13).

Mr. Purcell also realizes that his records only contain his cash expenses. He does not pay himself for his own labor and management or for his equity in the farm. To obtain the most accurate estimate of his soybean costs, he decides to estimate these too. In return for his labor and management, Mr. Purcell wants to earn about $30,000 from his farm for family living expenses. Since his crops are the only source of income, he divides the $30,000 by his tillable acres (870) to calculate his labor and management cost is $34.50 per acre. He bought his land many years ago and has steadily paid off his debt. He also has a mix of new and used machinery. His balance sheet shows he has $1,850 in equity capital per acre. This equity is the total for land, buildings, and machinery. He would like to receive a 5% return on this equity from farming and hopes to receive more when he retires and sells the land. He calculates his desired return to equity from farming as 5% of $1,850 or $92.50.

Now Mr. Purcell develops his soybean enterprise budget by putting his estimates of his receipts and costs per acre into one report (table 4.14). Since his desired returns to labor, management, and equity are noncash expenses, he places them at the end of the budget and calculates a second "net return." Since this budget comes from whole-farm records, he does not have any specific quantity information except his yield. Mr. Purcell estimates his total listed costs are $171.48 per acre and $3.73 per bushel. Based on his yield and price information, he has a net return of $81.54 per acre.

Table 4.13.   Ron Purcell's Allocation of Overhead Expenses

Overhead expense item	Whole farm total ($)	Corn expenses ($)		Soybean expenses ($)	
		Farm total	Per acre	Farm total	Per acre
Hired labor	5,118	2,672	6.36	2,446	5.44
Term interest	36,108	18,848	44.88	17,260	38.35
Machinery depreciation	15,312	8,268	19.69	7,044	15.65
Real estate taxes	11,484	5,544	13.20	5,940	13.20
Farm insurance	3,468	1,810	4.31	1,658	3.68
Utilities	2,298	1,200	2.86	1,098	2.44
Miscellaneous	3,771	1,968	4.69	1,803	4.01
Total overhead expenses	77,559	40,310	95.98	37,249	82.78

Table 4.14.   Mr. Purcell's Soybean Enterprise Budget from Whole-Farm Records

	Unit	Price	Quantity	Total
GROSS RECEIPTS				
Corn	bu.	$4.57	46	$210.22
Government payment				42.80
TOTAL RECEIPTS				253.02
DIRECT EXPENSES				
Seed				$17.20
Fertilizer				9.00
Chemicals				20.70
Crop insurance				8.30
Drying fuel				0
Fuel and oil				8.33
Repairs				17.28
Miscellaneous				2.01
Operating interest				5.88
TOTAL DIRECT COSTS				88.70
RETURN OVER DIRECT COSTS				164.32
OVERHEAD COSTS				
Hired labor				$5.44
Term interest				38.35
Machinery depreciation				15.65
Real estate taxes				13.20
Farm insurance				3.68
Utilities				2.44
Miscellaneous				4.01
TOTAL OVERHEAD COSTS				82.78
TOTAL LISTED COSTS				171.48
NET RETURN				81.54
Charge for unpaid labor and management				34.50
Charge for equity in land and machinery				92.50
NET RETURN after equity, labor, and management charge				−45.46
TOTAL LISTED COSTS per bushel				3.73
Cost per bushel (net of government payment)				2.80

However, after subtracting his desired charges for his equity, labor, and management, he now estimates a loss of $45.46 per acre. So while Mr. Purcell appears to be doing well with a positive net return after paying all listed costs, he does not have sufficient income to meet his noncash expenses for labor, management, and equity capital. This may be due to lower than normal prices and yields, higher than normal expenses, unrealistic expectations, a changed business environment, or a combination of these factors. Mr. Purcell will now have to spend some time considering these numbers and conditions and decide how to continue.

Developing Enterprise Budgets from Economic Engineering Studies

When a farmer has no experience in a particular enterprise or with a new set of technology, an enterprise budget can be developed from engineering data from manufacturers, university reports, and other sources. The steps in this process are:

1. Prepare a process map.

   List all the products, operations, and resources associated with that enterprise. (This list is the "process map" and is described in more detail in chapter 5.) Products include the obvious ones such as wheat, tomatoes, and beef, and not-so-obvious products such as stubble for grazing and manure for fertilizer. Operations and resources include everything done for or because of that enterprise. For crops, operations include tillage, planting, irrigating, spraying, harvesting, marketing, and management. For livestock, operations and resources include land, buildings, feeding, veterinary costs, trucking, horses, management, etc.

   The object of this step is to write down a reminder of every receipt and expense. The most logical way to ensure that this list is complete is to start at the beginning and think of everything that happens through the production period. Breaking this list into major stages may be helpful. Crop production might be divided into tillage, planting, growing, harvesting, and marketing. Stages in a livestock business might be feeding, calving, moving, and marketing.

2. Identify specific resources to be used.

   The next step is to identify the specific resources to be used for each activity listed in the first step. Which plow will be used to plow? Which tractor will be used to pull that plow, and who will drive it? What kind of fertilizer will be applied as a plow-down fertilizer? What kind of insecticide will be applied? What medicines and veterinary services will be needed? Which trucks will be used to haul the cows to the range, and who will drive the trucks? Farm operations need to be defined to link resources (e.g., tractor, range seeder, driver, seed, and fertilizer) in one operation. The object of this step is to identify the specific names of all of the resources used in the enterprises; finding the quantities, prices, and costs is done later.

3. Specify input use levels.

   Step 3 is detailing specific physical information about the resources. How much fertilizer will be applied in the fall? How many seeds will be planted per acre? How much hay and concentrate do the cattle eat? The object of this step is to quantify your estimates of resource use. (The economic principles used in these resource and input decisions are discussed earlier in this chapter.)

4. Select prices and costs.

The next step is to search for prices and costs. This is a critical step toward obtaining correct answers to the questions being asked. Some estimates may merely involve a review of your records, or calls to dealers and brokers. This seems easy, but even this information must be selected carefully. A budget being prepared to help decide what to plant this year may require very different prices from a budget being prepared to help decide whether to buy land or build buildings. The difference goes back to the difference between short run and long run (discussed earlier in this chapter). Planting decisions for the current year need price forecasts for the current year. Building and investment decisions need a longer outlook of where prices will be rather than where they are at the time of building. The sources of price and cost information should be noted so the final budget can be more clearly understood and interpreted.

Other costs, such as machinery and building, may be harder to estimate. For machinery and building costs, an average value based on farm records or published reports may be sufficient. In other situations, the annual costs of machinery and buildings can be estimated using standard engineering and economic formulas and the machine's or building's list price, purchase price, annual use, and useful life. A few computer programs and spreadsheet applications are available to estimate these costs for a specific farm and situation (e.g., Laughlin and Spurlock 2001, Lazarus 2002, Center for Dairy Profitability 2003, Ferreira 2001).

Annual machinery costs are divided between ownership costs and operating costs. Ownership costs are fixed once the machine is purchased. They include depreciation, interest, housing, insurance, and taxes. An alternative to calculating depreciation and interest separately is the annualized capital recovery cost that incorporates both the original cost and a charge for having money tied up in the machine, much like calculating the annual loan payment. If the machine is leased, the lease payment is used in place of the capital recovery or depreciation and interest costs. Operating costs vary with the use of the machine. They include fuel, lubrication, repairs, maintenance, labor, and custom services. Different methods are used to calculate each category of costs. An example of calculating machinery costs is in appendix A1 at the back of this book.

The annual costs of building services can be remembered by the acronym DIRTI (depreciation, interest, repairs, taxes, and insurance). An example for calculating building costs is in appendix A.

5. Assemble reports.

The last step is to calculate receipts and expenses and assemble reports. Luckily, computer programs and spreadsheets can eliminate much of the drudgery of the calculations and report preparations (see, for example, Laughlin and Spurlock 2001, Center for Dairy Profitability 2003, Ferreira 2001).

Sample Budget Using Economic Engineering

As an example of developing an enterprise budget using the economic engineering method, let us follow Richard and Louise Vansickle as they estimate the returns for buying and feeding medium no. 1 yearling steers. They have farmed and fed cattle for many years in northern Iowa so they have many records from past years. However, they also know that cattle and feed prices can vary greatly from year to year and the condition and weight of the cattle can vary too. So they always estimate what their costs and returns will be for each group of yearlings before they buy them and bring them to their farm.

As they start their calculations, the market for 850 lb. yearlings is near $80 per cwt. in Sioux Falls, South Dakota. They anticipate they could sell them in about six months weighing 1,350 lb. Based on the current future prices and expected basis, their expected cash price at market time is $68 per cwt. They expect a 0.5% death loss.

Based on past experience and their expectations of feeding this group for six months, they estimate the feed they will need for one yearling is 60 bushels of corn, 270 lb. of hay, 0.8 ton of corn silage, and 150 lb. of supplement. Prices for these feedstuffs can be locked in at $2.20 per bushel for corn, $90 per ton for hay, and $350 per ton for supplement. They grow their own corn silage and think they could sell it for $15 per ton.

They estimate their other costs by updating their previous expenses based on current conditions. Veterinary and medical costs are estimated to be $15 per head, and utilities and other operating costs are estimated to be $3 per head. Hired labor costs are estimated to be $9, and transportation and marketing costs to be $8.

Current interest rates are 6.5% for operating loans. They would borrow the cost of the feeder for six months and estimate that they would carry the rest of the operating costs an average of three months.

For fixed costs, they have their facilities and their own labor. They have estimated the depreciation and interest on their 15-year-old facilities to be about $0.15 per day. For the estimated six-month feeding period depreciation totals to $27 per head. They would like to earn a minimum of $10 per head for their labor, management, and risk.

After preparing these individual estimates of costs and revenues, the Vansickles compile the data in a standard enterprise budget format for feeding yearlings (table 4.15). Based on their calculations, they estimate they would lose $46.62 per head after paying for the feeder and all their variable and fixed costs. If they ignored their fixed costs for facilities and did not pay themselves, they would still lose $9.62 per head. If they do not buy the feeders, they would lose the $27 per head for facilities depreciation and interest. They could use their time (that is, the $10 per head) in other enterprises. Based on these estimates, they decide this group is not worth buying.

Budgets, Productivity, Competition, and Structural Change

To grow or only to remain viable, a manager has to be as efficient as the industry requires to remain competitive. Since change will always be with us, this knowledge is extremely critical for any business planning to expand or just stay in business. The critical benchmark that should be used in budgets is not today's productivity and efficiency requirements, but what will be required in, say, 10 to 15 years when the buildings and equipment (and investment) are still

Table 4.15   The Vansickles' projected enterprise budget for feeding yearlings from 850 lb. to 1,350 lb.

				Total
GROSS RECEIPTS				
Fed cattle	13.5 cwt/head	$68/cwt	0.995 head	$913.41
TOTAL RECEIPTS				913.41
DIRECT EXPENSES				
Yearling	850 lbs/head	$80/cwt	1 head	$680.00
Corn grain		$2.20/bu	60 bu	132.00
Hay		$90/ton	270 lbs	12.15
Corn silage		$15/ton	0.8 tons	12.00
Supplement		$350/ton	150 lbs	26.25
Vet & medical				15.00
Other operating costs				3.00
Hired labor				9.00
Transportation & marketing				8.00
Operating interest (6.5% on yearling & half of other var costs, 6 mos)				25.63
TOTAL DIRECT COSTS				923.03
RETURN OVER DIRECT COSTS				−9.62
OVERHEAD COSTS				
Facilities depreciation and interest				$27.00
Vansickles' labor, management, & risk				10.00
TOTAL OVERHEAD COSTS				37.00
TOTAL LISTED COSTS				960.03
NET RETURN				−46.62

(Adapted from F. Norton, Yearling Finishing Budget, Minnesota Extension Service, February 1, 2002.)

capable of production. Also, the critical benchmark is not the average but the competitive edge.

A simple estimate of future productivity can be made by extending past trends. This can be done by estimating the annual percentage change and subsequent progression of future percentages. It can also be done by graphical extrapolation of past trends.

Care must be taken, however, to project a benchmark productivity level required to be competitive. Projecting the average productivity will not provide a good benchmark. Future producers are not among the average producers today. In the strategic analysis of the industry, the best producers must be identified and benchmarks developed based on their abilities. For example, average milk production per cow for Minnesota has been increasing for many years. We could take those data—they are easily available—and we could quickly make a projection into the future. However, past data contain many farms in the average that are not expected to be competitive in the future: farms with smaller herds, older equipment, less productive management techniques. Thus, planning to compete against today's projected averages will be planning to compete against producers who will not be in business tomorrow. Data must be obtained that show the productivity of today's top producers because they will be tomorrow's competitors.

This analysis of productivity and efficiency trends needs to be done for different regions of the country (and the world) to be sure we understand which regions, farms, and technologies will have competitive advantages in the future. With this knowledge we can better understand the forces shaping structural change and their impact on our decisions. This is part of external analysis within strategic management described in chapter 2.

Summary Points

- Budgeting provides economic information for a variety of management tasks and decisions.
- A budget is a projection of income and expenses used for planning the future.
- An enterprise is a common name for a part of the farm business.
- Fixed costs are those that occur no matter what or how much is produced. Variable costs are those that vary with the volume of business.
- Direct costs are those costs that are associated directly with production of a specific product or activity. Overhead costs are those costs that are hard to assign directly to a particular enterprise.
- Cash income and expenses involve an actual transfer of cash. Noncash income and expenses do not involve a cash transfer.

- The opportunity cost of a resource is the income that could be received from the best alternative use of that resource.
- Long-run decisions need long-run information. Short-run decisions need short-run information.
- The distinction between economic and accounting values is important due to different questions being asked.
- Total, average, and marginal income and costs provide different pieces of information for different kinds of decisions.
- The whole-farm budget provides a summary of the major physical and financial features of the entire farm business.
- An enterprise budget is a statement of expected receipts, costs, and net returns for a specific enterprise.
- A cash flow budget shows the ability of a business to generate sufficient cash inflows to cover cash outflows during a specified period.
- The partial budget is used for estimating the effects of changes in the business and includes only those items that are affected by the change.
- The economically optimal level of input use is where the last increment of input use benefits the farm more than its cost but the next increment costs more than its benefit. That is, increase input use as long as the value of the marginal product is greater than the marginal input cost.
- Substitute one input for another if the cost of the substitute is less than the cost of the input being replaced (and output remains the same).
- Enterprise budgets can be developed in two ways: from whole-farm records and by economic engineering.

REVIEW QUESTIONS

1. How can enterprise budgets be used in farm management?
2. What is the difference between fixed and variable costs?
3. Why is the difference between fixed and variable costs so important?
4. Why is the distinction between cash and noncash values important? Between long-run and short-run?
5. What is an opportunity cost? How is it used?
6. Describe the four types of budgets and how they are used?
7. Why are published budgets useful? How might they differ from an individual's budget?
8. Since published enterprise budgets do not always have costs classified as variable and fixed, what type of costs in these budgets can be used as a proxy for variable costs?
9. What does the term "marginal" mean? Why is it important in farm management?

10. Given the fertilizer response data for corn following soybeans shown in table 4.16, a corn price forecast of $2.25 per bushel, a nitrogen price of $0.30 per lb., and an application cost of $7 per acre, what is the economically optimal nitrogen rate for corn after soybeans?

Table 4.16.   Corn Yield Response to Applied Nitrogen Following Soybeans

Nitrogen applied	Corn yield	Marginal (or additional) product	Value of marginal product	Marginal input costs	Marginal net return
lb/acre	bu/acre	bu/acre	$/acre	$/acre	$/acre
0	109				
40	134				
80	146				
120	153				
160	158				
200	158				

11. What are the two ways an enterprise budget can be developed?
12. With the following information, use economic engineering to develop an enterprise budget for corn following soybeans in Indiana in 2002. Based on Purdue extension information, the income and costs are estimated as follows. The expected yield is 138.8 bushels of corn per acre and the expected price is $2.10 per bushel. The expected government payment is $25 per acre. The corn is estimated to need 133 lb. of nitrogen at a price of $0.16 per lb.; 51 lb. of P_2O_5 at a net price of $.23 per lb.; 58 lb. of K_2O at $0.13 per lb.; and 398 lb. of lime at $14/ton. Seed costs are expected to be $30 per acre. Herbicides are expected to cost $18 per acre. Diesel fuel use is estimated to be 8.5 gallons with a price of $0.95 per gallon. Repairs are estimated to be $9 per acre; hauling, $8; and miscellaneous expenses, $11 per acre. Interest on operating expenses is 6.5% per year and the operating loan on expenses is expected to be an average of eight months long. Overhead costs are estimated to be $52 per acre for machinery depreciation and interest, $7.2 per acre for drying and handling equipment, $37 per acre for family and hired labor, and $122 per acre for cash rent.

13. Which of the following are ownership costs for machinery?
    a. capital recovery
    b. housing
    c. insurance
    d. taxes
    e. all of the above

14. Which of the following are *not* operating costs for machinery?
    a. fuel
    b. lubrication
    c. repairs and maintenance
    d. labor
    e. interest on the loan
    f. All are operating costs

5

Production and Operations Management

Planning for production and operations takes place at many levels and points in a farm business. In this chapter, the processes and tools for planning production at both an enterprise level and at the whole-farm level are described. Examples of these decisions are included. With production and operations management, many questions need to be answered, such as:

What methods and technologies are being used?
What products or services should we produce?
What production process should we be using?
How can current processes be improved?
What raw materials or inputs are needed?
How much of these inputs are needed?
What facilities and equipment are needed?
What types of labor are needed?
How much of each labor type is needed?
When will materials, facilities, equipment, labor be needed?
What is the best schedule for the next few weeks?
Which jobs have higher priorities than others?
What will be the impact on the environment?
Are these plans helping accomplish the overall strategic goals?
Do the plans fit the vision of the owners?

Each of these questions can be complex and require careful, deliberate thought to ensure a good decision. Furthermore, the complexity increases because these questions often need to be answered simultaneously. For example, we cannot make a final choice about which or how much raw materials are needed until we have chosen the products to be produced and the process by which we will produce them. However, we cannot make a final choice on the products and processes until we have studied the potential products and processes and their needs for raw materials and labor. To deal with this simultaneity problem, a manager needs to understand how each of these questions can be answered individually and how they relate to each other. This simultaneity problem also means that the decision process will be an iterative process with steps back and forth and around as the questions are answered in a manner consistent with each other.

The decision process is not linear (as noted in chapter 1), and decisions made on the whole-farm level may change the assumptions and conditions used to plan individual enterprises. Managers need to review and perhaps modify many decisions as subsequent decisions are made and new information is obtained. Planning is not a simple task, and mistakes can be devastating.

Thus, this chapter starts with process mapping. The analogy to a geographical map is appropriate. By mapping the processes on a farm, we can understand how each part works, how the parts are linked together and affect each other, and how the farm is linked to its economic, political, and physical environment. With this understanding, a farmer can make better decisions in today's risky, complex world.

PROCESS MAPPING

A process map is a description of a method or process of accomplishing a task. It can be a geographical mapping of physical movements, but it isn't necessarily a physical map. A process map may be a list of the movements made, materials required, and equipment used. It may show the flows of both materials and information. In general business, process mapping is seen in several ways: assembly drawings, assembly charts, parts diagrams, routing sheets, flow diagrams, physical layouts, to name a few. A flow diagram for computer programming is a process map. In farming, the process can be a production process, such as crop production or livestock; a service process, such as custom work; or an internal process such as paying bills or gathering market information.

Although these maps can seem simple, they are a good way for management to understand its own business. They enable current management to see the business with "fresh eyes." A process map can be used to understand the process for any level of a business. A farmer, in his or her role as the general manager, sees one level of the processes involved in a business and needs to consider the problems or opportunities at that level. As the supervisor or worker, a farmer sees the business from a different perspective and so needs a different map to see the problems and opportunities at that level.

A process map for a whole farm should show the inputs and outputs of a business and how the business is organized. As an example, consider a farm that grows corn and soybeans for both sale and feed, produces feeder pigs to finish or to sell, and does some custom field work (see fig. 5.1). The internal organization of the business is shown by the small boxes inside the large box. Each of these small boxes represents an activity center of the farm: corn production, finish feeder pigs, livestock marketing, and so on. Since the other activity centers require their services, labor, machinery, and building services are organized as a separate center. The internal flow of materials, services, and information are shown by the lines within the process map of this level. This level of detail may be sufficient for a farmer interested in improving his or her data collection and reporting within the business and the economic efficiency of production and marketing decisions.

The external boundary of the farm is the large box. The environment is everything outside of the large box. The inputs are listed on the left and the outputs are listed on the right. Both controllable inputs (e.g., land and capital) and uncontrollable inputs (e.g., prices and weather) are listed to provide a complete picture of what affects a farm. Similarly, both good outputs (e.g., crops and income) and bad outputs (e.g., runoff) are listed to provide a com-

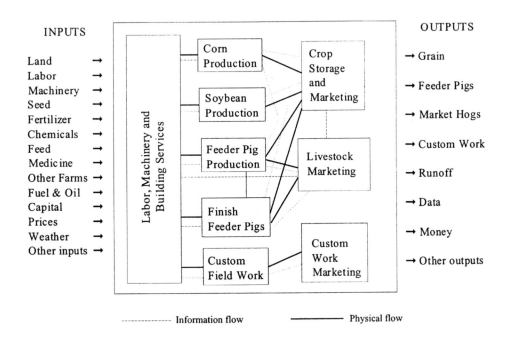

Figure 5.1. Process map of a whole crop-livestock farm.

plete picture of what the farm produces. Ignoring uncontrollable inputs or bad outputs can lead to bad decisions on the farm.

Farmers already use this simple view of their farm as a mental model when considering new products. To decide if a new product is even feasible to consider in detail, a farmer looks at the quantity and quality of the resources available and the potential products and by-products. The farmer evaluates whether the new product and process fits into his or her view of the farm, that is their process map of the whole farm. For example, when considering whether to sign a contract to finish hogs, a crop farmer does not need to consider the minute details of growing corn but does consider whether the capital and labor supplies are sufficient, the building location would be susceptible to runoff problems, and so on. In another example, a farmer may view the farm from this simple perspective when considering the impact of governmental and institutional requirements for reporting labor use, income and expenses, and chemical use. More complicated decisions and situations require more detailed maps of the farm and its environment.

Enterprise budgets (discussed in chapter 4) are process maps. They describe the inputs and operations needed to produce a crop or raise livestock. An enterprise budget for producing soybeans is an example (table 5.1). However, an enterprise budget does not include all the steps needed to produce a crop or raise livestock. A more descriptive list of steps, inputs, and equipment needed to produce soybeans is shown in the fuller process map (fig. 5.2).

As another example, a process map of a cattle feeding operation shows time and equipment requirements and feed ingredient requirements, storage locations, retrieval processes, and mixing and feeding instructions (table 5.2). Each step is identified by its type of activity, the time required, distance traveled, and equipment and materials needed.

In this cattle feeding example, a farmer can evaluate each step to see if that step is actually needed or can be modified or eliminated to increase efficiency. A farmer can see where the time requirements are the greatest or where the delays are longest and can evaluate whether the process can be changed to decrease time requirements. In this example, the farmer may ask (and find the answer to) why loading and grinding takes so long, why the operator must wait for the auger before loading, or why the whole feeding process takes over an hour.

Another map shows the steps involved in reloading a corn planter during the planting operation (table 5.3). This job of reloading the planter is a very specific job; thus, the map is very focused. It does not cover the whole planting process but only the process of reloading the planter once the day's planting has begun. As in the previous example, the farmer may not realize how long the reloading takes (i.e., 33 minutes) until the map is completed. Evaluating this information and alternative procedures before the planting season could save valuable time during the planting season because of the resulting equipment investments or modifications of the steps that decrease time requirements.

Table 5.1.   Enterprise Budget for Soybean Production in Southern Minnesota

GROSS RECEIPTS FROM PRODUCTION

Main Product:	Unit	Price	Quantity	Total
Soybean grain	bu	5.25	45	236.25
		TOTAL RECEIPTS		236.25

DIRECT EXPENSES	Unit	Price	Quantity	Total
Soybean seed	thous. seeds	0.1000	165	16.50
Treflan	lb. a.i.	1.000	7.84	7.84
Pursuit	lb. a.i.	0.063	225	14.18
Trucking	bu.	0.150	45	6.75
Operating costs				
Stock shredder	trip	1	1.39	1.39
Chisel plow	trip	1	1.09	1.09
Field cultivator	trip	2	0.62	1.24
Sprayer	trip	1	0.27	0.27
Planter	trip	1	1.19	1.19
Sprayer	trip	1	0.27	0.27
Cultivator	trip	1	0.72	0.72
Combine	trip	1	6.98	6.98
Labor costs				
Stock shredder	trip	1	1.35	1.35
Chisel plow	trip	1	0.96	0.96
Field cultivator	trip	2	0.48	0.96
Sprayer	trip	1	0.98	0.98
Planter	trip	1	1.49	1.49
Sprayer	trip	1	0.98	0.98
Cultivator	trip	1	0.96	0.96
Combine	trip	1	2.99	2.99
Land rent	acre	100	1	100.00
Miscellaneous direct costs	$	—	—	1.00
Interest on direct costs	annual rate:	10.00%	170.085	8.50
		TOTAL DIRECT COSTS		178.59
		RETURN OVER DIRECT COSTS		57.66

OVERHEAD COSTS

	Unit	Price	Quantity	Total
Stock shredder	trip	1	4.35	4.35
Chisel plow	trip	1	3.54	3.54
Field cultivator	trip	2	1.62	3.24
Sprayer	trip	1	0.76	0.76
Planter	trip	1	4.81	4.81
Sprayer	trip	1	0.76	0.76
Cultivator	trip	1	2.33	2.33
Combine	trip	1	12.50	12.50
Misc. overhead costs	$	—	—	4.30
		TOTAL OVERHEAD COSTS		36.59
		TOTAL LISTED COSTS		215.18
		NET RETURN		21.07

Total listed costs per unit of main product:	$4.78 per bu.
Net return per unit of main product:	$0.47 per bu.

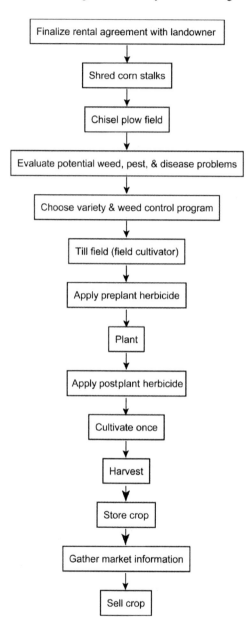

Figure 5.2. A process map for producing soybeans (following corn).

The type of decision that a farmer needs to make determines what level of process map is needed. When a farmer is making decisions about the future direction of the farm, he or she would spend considerable time considering the whole-farm map: how the farm is organized, what products are produced, what external forces are affecting the farm, and so on. For these decisions, a farmer would spend very little time worrying about process maps such as the

Table 5.2.   A Process Map for Feeding Cattle

Step	Description	Activity type	Time(min.)	Distance	Equipment/Materials
1	Pick up feed order	O	2		
2	Inspect order for correctness	I	1		
3	Walk to machine shed	T	3	100 ft.	
4	Attach mixer to tractor	O	5		95Hp tractor & mixer
5	Drive to feed shed	T	2	150 ft.	Tractor & mixer
6	Load vitamin mix	O	8		4 bags premix
7	Drive to corn bin	T	3	150 ft.	
8	Wait for auger	D	5		Auger
9	Inspect corn	I	1		
10	Load & grind corn	O	20		2.5 tons corn & auger
11	Inspect feed	I	1		
12	Drive to feedlot	T	2	300 ft.	Tractor & mixer
13	Unload in bunker	O	3	100 ft.	Tractor & mixer
14	Clean mixer	O	1		Tractor & mixer
15	Drive to machine shed	T	2	600 ft.	Tractor & mixer
16	Park mixer & tractor	S	4		Tractor & mixer
17	Walk to office	T	2	100 ft.	
18	Complete work order report	O	3		

Note: Activities are classified into five types: O = Operation, the actual work being done; T = Transport, moving the worker, the input, or the product; I = Inspection, checking the work in progress or at the end of the process; D = Delay,

feeding or planting examples. However, if feeding or planting efficiency needs to be improved due to those same external forces, for example, a prudent manager would spend considerable time looking for steps that can be eliminated or improved. A farmer who is planning to expand his hog operation is probably evaluating two levels of maps—at least two. The whole-farm map shows how the different units would work together (hog and crop production, for

Table 5.3.   A Process Map for Reloading the Corn Planter

Step	Description	Activity type	Time	Distance	Equipment/Materials
1	Stop at end of field	O	0		
2	Inspect supplies on planter	I	1		Planter
3	Walk to supply truck	T	3	100 ft.	
4	Drive to planter	T	1	100 ft.	Truck
5	Check seed variety	I	1		
6	Load seed	O	13		Seed corn
7	Load fertilizer	O	8		Starter fertilizer
8	Check equipment	I	4		Tractor & planter
9	Move supply truck	S	2	15 ft.	Truck
10	Start planting	O	0		Planter & tractor

Note: Activities are classified into five types: O = Operation, the actual work being done; T = Transport, moving the worker, the input, or the product; I = Inspection, checking the work in progress or at the end of the process; D = Delay, any down time for any reason; and S = Store, storing the product or work in progress.

example). The same hog farmer would study other maps of the expansion in more detail, such as how pigs, feed, and animal waste would flow through the buildings and farm in both space and time.

Farmers have many reasons for mapping a process. If a new process, such as no-till crop production, is being considered, the farmer needs to understand how it will affect other parts of the business. Process maps (written, verbal, or both) can be used to train a new employee. An expanding farm needs to evaluate new ideas for information gathering, machinery storage and use, and communication between people. Process maps can help a farmer find potential efficiencies to counter the continual cost-price squeeze. This use is discussed more in the next section, "Improving the Current Process."

As with any management technique, the value of the potential information needs to be compared with the cost of obtaining the information. When developing process maps, the cost (in both time and money) is greatest for new maps, but the benefits of mapping and thus understanding new processes can be great. The greatest benefits may come from mapping complex systems that cannot be grasped easily, although these systems also would have greater costs in preparing the maps.

Farm managers might be tempted to say, "I understand what needs to be done. Why bother to draw a map?" (This is especially tempting to say shortly after starting to map a process.) However, two points should be remembered. The first is how much information people say they learn when they develop a map. The second is how complicated farming is now and is expected to become. By learning to draw and understand the concepts of a simple map, this mental model of process mapping will be easier and more useful when complicated situations develop. At that point, the choice of putting the map on paper can be made on the basis of the number of people involved and the amount of detail to be understood.

IMPROVING THE CURRENT PROCESS

As conditions, technology, institutions, and markets change, the pressure to improve and cope with these changes is high. Keeping abreast of these changes and finding ideas that will improve a process are demanding tasks. In this section, we look at different ways to adjust the current process and how to evaluate whether the ideas will result in an improvement or not.

Ideas for improvement are not found in easy lists. They come from knowing the system and what it can or could be doing. Ideas also come from a general willingness to ask questions about the process: why is it done a certain way, how is this input used, who is responsible for doing certain activities, and so on. Ideas show up as process maps are being developed and discussed. Attending extension and dealer meetings, joining marketing and managing clubs, reading articles and reports, and just talking with neighbors are traditional ways of hearing about new ideas and their potential use. Although he wasn't talking about improving processes, Rudyard Kipling has, perhaps, the best thought about finding ideas and education, in general (Thompson and Strickland 1992, 428).

> I keep six honest serving men
> (They taught me all I knew);
> Their names are What and Why and When;
> And How and Where and Who.

Ideas for improvement can be found in five general areas: reducing input use, substituting inputs, increasing productivity, expanding, and reorganizing.

Reducing Input Use

No farmers will knowingly spend more money for inputs they do not need. But when new technologies come along, new knowledge is developed, and institutions change, the traditional production methods may become inefficient in their use of inputs. Examples of reducing input use include reducing pesticide application rates and numbers, reducing the number of tillage passes, changing to no-till equipment, decreasing the level of antibiotics in feeds, and so on. Banding herbicides instead of broadcast applications have been shown to decrease costs in many areas and crops. Herbicide costs can also be reduced by using weed maps and scouting information so herbicides are applied only where and when needed. Investing in feeders and systems that reduce feed wastage is another traditional method for reducing input use and cost.

An example of changing technology and knowledge is the development of new tillage equipment and how to use it. A study by West, Vyn, and Steinhardt at Purdue University found that the combination tool for one-pass produced corn and soybean yields that were statistically the same as the traditional tillage with a chisel plow in the fall and secondary tillage in the spring (West et al. 2002). Their study also showed that using the combination tillage tool in the fall, not using any spring tillage, and planting into a stale seedbed produced corn yields that were no different from systems that used spring tillage. These results can be used by farmers in similar geographical areas to reduce operating costs by decreasing the number of tillage operations and to reduce ownership costs by decreasing the amount of equipment needed.

Sharing machinery ownership is another way that can decrease input use not by reducing the number of operations but by decreasing the amount of money tied up in owning a machine. Owning a combine with a neighbor is a very quick way to cut investment costs. Of course, multiple ownership requires more input in terms of management time for communication about how costs will be allocated between owners and how the owners will determine priority of use.

Substituting One Input for Another

Examples of substituting one input for another include replacing labor and management for purchased inputs (or vice versa), using custom services instead of owned machinery, using manure for commercial fertilizer, changing the type of tillage equipment, and so on.

As an example of how to adjust a current plan by substituting one input for another and evaluating the impact, consider the process for soybean production shown in figure 5.2. Suppose this farmer is considering the substitution of a second mechanical cultivation for the postplant herbicide application. This is shown on the right-hand side of figure 5.3. This potential change is evaluated using the partial budget framework as shown in figure 5.4. By not applying the postplant herbicide, the estimated savings is $16.19. But the potential yield loss of four bushels per acre valued at $5.25 per bushel and the cost of the additional cultivation ($4.01) is estimated to create a total negative effect of 25.01. So the net effect of this potential change is a loss of $8.82 per acre. If this farmer considers only the income effect, the change is not an improvement but a potential detriment to farm income.

Another example of substitution is the choice of whether to own equipment or to hire a custom operator. The substitution is a change in who owns the machine, it is not necessarily a change in the physical operation. Owning a machine obviously obligates a farmer to both operating and ownership costs. However, when a farmer hires a custom operator, the cost of the custom operator is an operating cost for the farmer, but no ownership costs are incurred by the farmer. Also, owning a machine entails having an asset and possibly a loan on the farmer's balance sheet. Hiring a custom operator does not affect the farmer's balance sheet directly.

Using standard engineering cost relationships and equations, the costs of owning a 30 ft. grain drill and a 130 HP tractor were estimated in appendix A1 at the back of this book and are listed here in table 5.4.

The cost of owning and operating, $7.22 per acre, can be compared with custom rates in the area to determine which method provides the lowest cost for planting. If custom rates were at or below $7, hiring the grain drilling is probably better than owning the equipment. Other factors need to be considered before a final decision would be made. These other factors include timeliness and availability of a custom operator, other uses of the equipment, value of the owner's labor in other uses, impact on the balance sheet, and so on.

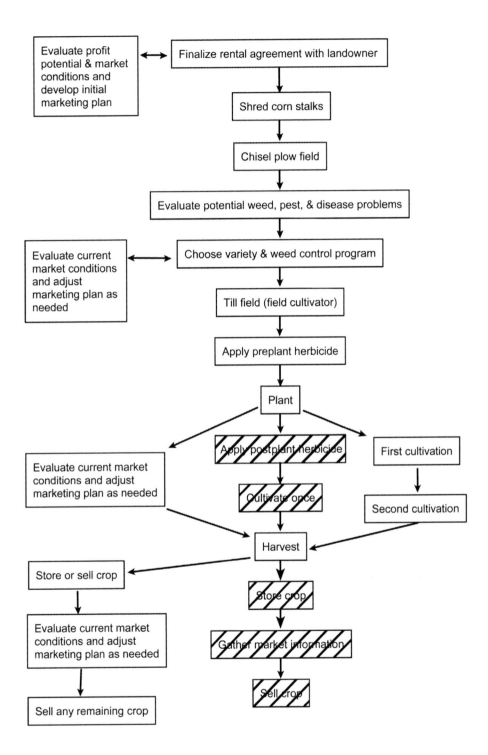

Figure 5.3. A revised process map for producing soybeans (following corn).

137

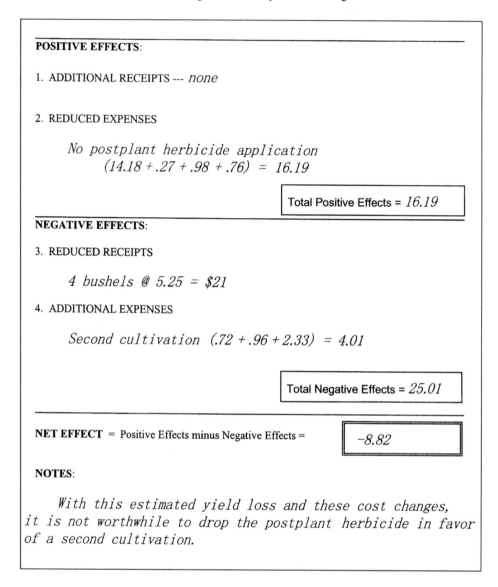

Figure 5.4. Partial budget of substituting a second mechanical cultivation for the postplant herbicide application for producing soybeans.

Improving Productivity

Examples of improving productivity are applying herbicides at the optimal plant and weed development stages (as well as optimal time of day), changing livestock rations for faster and/or better feed efficiency, monitoring animal heat stages for optimal breeding periods, and so on. Hiring seasonal equipment operators can improve machine productivity by increasing the hours that tillage equipment is used per day and thus over the entire season.

Table 5.4.   Estimated Costs of Owning and Operating a Grain Drill

Cost category	Cost per acre ($)
Ownership	5.12
Operating	1.34
Labor	0.76
Total	7.22

Expanding

Examples of expanding include buying or renting more land in order to spread the fixed costs of machinery, labor, and management, adding new building capacity, adding new crops to the cropping plan, and so on.

Reorganizing

Examples of reorganizing the business include rescheduling tasks to fully utilize labor and machinery, hire more labor to more fully utilize machinery, reschedule the marketing process, reassign labor and management resources to better utilize talents and abilities, and so on. An example of reorganizing the marketing function for soybeans is shown on the left side of figure 5.3. In this process improvement, market evaluation and market plan development is started very early and continues through the production process versus the method in the old process map (fig. 5.2) where the soybeans were produced first and then market information was gathered and pricing decisions were made.

  Combinations of these four methods may also be used. Any expansion will also entail improvements in productivity and reorganization of the business.

ENTERPRISE SELECTION

The optimal combination of products is influenced by the technical relationships between the products. There are three relationships that are important for product selection: competitive, supplementary, and complementary.

1. Products that compete with each other for the same resources are classified as *competitive products*. The selection between these products will be based on each product's potential income.
2. Products that do not affect each other are *supplementary products*. Row crops on Class I land and cattle range on poorer soils are supplementary products because a change in one doesn't affect the other. Supplementary products could become competitive products if a resource (such as capital or labor)

started to limit production. The decisions regarding a product are independent of its supplementary products unless and until they become competitive.

3. Products are *complementary* when one contributes to the production of the other. Honey is a complement to any product that requires pollination. Small grains are complements to livestock production when the stubble is grazed after grain harvest. Complementary products are chosen and expanded together until they utilize a resource to the point of becoming competitive.

The best combination of products is the one that best meets the farmer's objectives. Usually that is first measured as the combination that yields the highest return to the resources used in production. Supplementary and complementary products (as long as they are profitable) are chosen and expanded until they become competitive products. When evaluating the choices between competitive products, the decision becomes a process of resource allocation between potential products.

The optimal mix of products is found by analyzing the alternatives available in terms of their potential income and use of resources. Several methods are available for choosing the best mix of crops, livestock, and other products. Some methods are not very good: trial and error or the coffee shop, for example. Linear programming is a mathematical analysis tool that is very useful but requires high initial costs in terms of learning and data acquisition. The alternative of gross margin analysis for choosing enterprises is a simple, useful method with a pragmatic, structured procedure. Gross margin analysis can be used for decisions where the constraining factors are few in number and relatively simple. The gross margin is also called the contribution margin in some reports.

Gross Margins

The gross margin (GM) for an enterprise is defined as that enterprise's gross income (GI) minus its variable costs (VC):

$$GM = GI - VC$$

When using gross margins to choose the enterprises for a farm, the crop mix is usually chosen first by following the steps listed below.

1. Select the crop with largest GM/acre. Increase the acreage of this crop up to the acreage limit or another limit (e.g., disease or institutional limit).
2. Select the crop with the second largest GM/acre. Increase the acreage of this crop up to the acreage limit or another limit (e.g., disease or institutional limit).
3. Continue selecting crops in this manner until the acreage limit is reached.

4. Repeat the process in steps 1 through 3 using the GM/hour instead of GM/acre. If this results in a different crop mix, choose a compromise that balances the maximum of GM per acre and per hour.
5. Add livestock to the farm's enterprise mix if labor and other resources are available.

When a farmer first selects (or already has) a livestock operation, this farmer may, as a first step, determine the cropping mix needed to produce the feed requirements of the livestock. If any land is not used to meet the feed requirements, the crop selection for that acreage would follow the steps listed above. However, this livestock-first approach can ignore the opportunities provided by more profitable, nonfeed crops. If livestock is expected to pay the market price for feedstuffs (adjusted for transportation and storage costs) and there are no differences in quality, the simple steps listed above will provide the profit maximizing mix of crops and livestock for the farm. If the crop mix chosen by these steps does not produce the feed needed by the livestock, another crop must be more profitable, that is, the advantage of producing one's own feedstuffs is less than the advantage of purchasing those feedstuffs. As a California rancher once said, "I found I could grow more hay by raising olives on that land." That is, with the profit from the olives, he could buy more hay per acre than he could grow per acre.

Using Gross Margins: Gable Farm

As an example of using gross margins, let us consider the farm of Dave and Louise Gable. They have 700 tillable crop acres and 180 dairy cows. Besides corn, soybeans, and spring wheat, the Gables have also grown sweet peas and sweet corn for a local processor in recent years. Using their records and making some adjustments for next year, the Gables have estimated the gross margins for their farm (table 5.5).

Table 5.5. Estimated Gross Margins for the Gables' Farm

	Corn	Soybeans	Spring wheat	Sweet corn	Sweet peas
Yield	144 bu	46 bu	51 bu	5.8 tons	5,500 lb
Price ($/bu, ton, or lb)	2.15	5.25	3.00	46	.07
Other income ($/acre)*	25	8	40	25	18
Gross income ($/acre)	335	250	193	292	403
Variable costs ($/acre)	164	85	76	106	57
Fixed costs ($/acre)	97	84	74	55	84
Gross margin ($/acre)	171	165	117	186	346
Hours required	2.8	2.3	1.9	2.4	1.3

*Other income includes government payments and straw sales.

They maintain a rotation of 50-50 corn-beans or, in more general terms, 50% non-legumes and 50% legumes. If wheat or peas are planted, wheat would be counted on the corn side of the 50-50 mix, and peas would be counted with soybeans. Since they have 700 tillable acres, they plan to plant a total of 350 acres of corn, sweet corn, and wheat and a total of 350 acres of soybeans and sweet peas. For next year, they could have contracts for 155 acres of peas and 128 acres of sweet corn, but they have not signed them yet.

Using gross margins analysis, the Gables first choose to increase the sweet pea acreage up to the contractual limit of 155 acres since peas have the largest gross margin per acre ($346). Then, since sweet corn has the second highest gross margin ($186), they choose to plant sweet corn up to the contractual limit of 128 acres. For their 50-50 rotation, they tally 155 acres of their 350-acre limit of legumes and 128 acres toward their limit of corn and wheat.

Corn has the third highest gross margin ($171). The acreage of corn is also limited by the 50-50 rotation rule and the earlier decision to plant 128 acres of sweet corn under contract. So they set their corn acreage at 222 acres.

Soybeans have the fourth highest gross margin ($165). Since they do not want to plant more than 50% of their 700 tillable acres to soybeans and peas, they limit their soybean acreage to 195 acres since they already have decided to sign the pea contract and plant 155 acres of peas.

This plan uses all of their 700 tillable acres. So for next year, they will plant 155 acres of sweet peas under contract, 128 acres of sweet corn under contract, 222 acres of corn, and 195 acres of soybeans. They will not plant any spring wheat since it had the lowest gross margin and they did not need it to keep their 50-50 rotation in balance.

INPUT SUPPLY MANAGEMENT

Most often, farmers do not worry about whether input supplies will be available. The input supply industry has developed an excellent ability to have available or very quickly deliver most inputs the day they are ordered or the next day. These inputs include feed, fuel, implement parts, and minor repairs. Often even major items, such as tractors or combines, are available if the dealer has not had unanticipated sales. Other inputs are not set up for such a rapid delivery system, such as feeder livestock purchased directly from the grower.

There can be, however, benefits of anticipating and planning for input needs. There can also be costs of not planning for input needs. An obvious example of a benefit to planning is the discount price usually given to early seed orders. Other examples, which don't involve such direct monetary benefits, are the assured quality level in feeder stock and the reduction in stress in knowing that the credit line has already been secured. An example of the cost of not planning is the higher cost incurred for feed if prices rise and a feeder did not purchase or secure the price of future needs. Not having custom crop operations done in a timely manner is another example of the cost of not planning, and thus, not getting the contacts made early enough. In one sense, the efficient, dependable supply system has allowed farmers to rely on that system and not on planning.

The simple method for planning future input requirements is to take the input requirements listed in the enterprise budgets and multiply those amounts by the selected size of the enterprise. By using both the quantities needed and the timing of those needs, the needed size and timing of purchases or orders can be forecast. If the farm is small enough or the production plan has been stable, this can be done in the manager's head; but there are computer programs that can simplify the process for more complex operations.

As an example of planning for a sufficient labor supply, let us return to the Gables, who used gross margins to choose their cropping plan for next year. Now let us help them compare the amount of labor needed with their own supply of labor and estimate any additional labor that would be needed.

Mr. Gable is willing to work 60 hours per week in a normal week and 80 hours per week in the busy planting and harvesting seasons. If we give him 2 weeks of vacation, he is willing to work 320 hours during the planting season (80 hrs/wk × 4 weeks); 320 hours during the harvest season (80 hrs/wk × 4 weeks); and 2,520 hours during the rest of the year (60 hrs/wk × 42 weeks). Adding these up, we see that he is willing to work a total of 3,160 hours during the entire year. (Mrs. Gable has a full-time, off-farm job. So she is not available on a regular basis.)

This supply of labor needs to be compared to the estimated labor needed to grow the crops chosen and to take care of the dairy cows. Using the hours shown in table 5.5, the Gables need a total of 1,580 hours for their 700 acres of crops: 202 hours (155 acres × 1.3 hours/acre) for sweet peas, 307 hours (128 × 2.4) for sweet corn, 449 hours (195 × 2.3) for soybeans, and 622 hours (222 × 2.8) for corn. They estimate they will need 40 hours per cow, so with 180 cows, this farmer needs an estimated 7,200 hours for the dairy. In total they estimate a need for 8,780 hours during the year.

Compared to his willingness to supply 3,160 hours, they obviously need to hire labor. They need to hire an additional 5,620 hours. Two workers who each work 55 hours per week for 50 weeks per year would supply 5,500 hours. These two workers and some part-time, seasonal labor may meet the needs of the Gables. Staffing and organization is discussed more in chapter 13.

SCHEDULING OPERATIONS

Scheduling is an allocation decision. Scheduling allocates the resources already available due to earlier investment and hiring decisions. The schedule shows *what* is to be done, *when* it will done, *who* will do it, and *what* equipment will be needed. Scheduling is usually concerned with the short run: the next few months or weeks and even today and the next few hours.

Scheduling has three distinct objectives: *cost* (minimizing costs or staying within the budget), *schedule* (completing the jobs by the due date or earlier), and *performance* (the performance of the product or service being produced: meeting or exceeding customer expectations).

These are competing objectives. Usually two of the objectives can be met, but the third will not be met. Planning ahead can avoid some of the competition and provide a better balance of the three objectives. The overall vision, strategy, and goals of the farm should provide a guide in balancing these three competing scheduling objectives. Any farmer will also say that, when it comes down to the crunch, weather may also play a big part in scheduling decisions. However, the impact of weather may also be balanced with, say, the cost objective if that is restrictive for a specific farm or situation.

Two scheduling techniques can be useful for farming: sequencing and dispatching. Sequencing is a procedure for scheduling people, jobs, equipment, and other resources over several weeks or months. Dispatching rules can help a manager decide which jobs have the highest priority and should be done first on a certain day or next.

Sequencing

Sequencing is concerned with the exact order of operations for processing several jobs. The value of scheduling is the ability to plan and forecast on paper the needs for resources and the location and timing of how those resources are used. Ensuring the proper scheduling of equipment will allow operations to be done on time, inputs to be applied at the proper time, and resources to be used in the most efficient manner. Various assumptions on the number of days available for field work can be tested for the amount of machinery needed to produce a crop optimally. Thus, within some boundaries for uncertainty, a farmer can avoid having both too much and too little machinery. Planning workloads and working hours can improve the use of hired labor, lower the labor cost per unit of product, and, perhaps, lower labor turnover by more even workloads.

Developed by Henry L. Gantt in 1917, the Gantt chart is the common tool for representing sequences of operations. In Gantt charts, time is listed across the top, and scarce resources, jobs, or both are listed down the side. The sequence of activities for individual jobs is marked on time lines for each job.

The jobs being done by each resource at each point in time are shown on time lines for each resource.

As an example, let us consider a Gantt chart for one week of planting (fig. 5.5). From this chart we can tell when a field will be planted and by whom. Waiting times and project completion times are easily visualized. For example, the north field will be planted starting on Monday by Frances. She should be finished there by midday on Wednesday when she will start on the Johnson farm. We can also tell an employee what field he or she should be planting and when. For example, Alan will be planting on the Watson farm starting on Tuesday and finishing on Thursday. Of course, nature and accidents may change this plan, but developing a plan will allow us to see where there are potential scheduling problems and to correct those problems before they become real problems.

The easiest way to compare alternative ways to schedule the same list of jobs is by comparing the total time required to complete all the work. The preferred plan is the one that completes all the work in the shortest time—provided that all work meets or exceeds specified quality standards. For the four fields to be planted in figure 5.5, the total time required is six days. In this simple example, it may be hard to come up with a better plan.

As another example of sequencing, let us consider how Robert Johns schedules his soybean harvest season in northern Indiana. Mr. Johns grows soybeans on his own land plus custom combines soybeans for two neighbors (table 5.6).

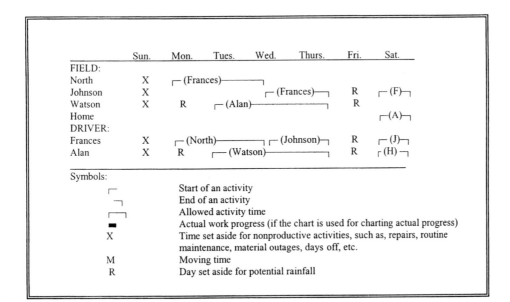

Figure 5.5. Gantt chart for one week of planting.

Table 5.6.   Mr. Johns' Soybean Harvest Commitment

Farm	Days needed	Estimated maturity date
Own: Home	6	September 15
Own: Smith	9	September 20
Custom: Johnson	10	September 16
Custom: Watson	2	September 25

He has two combines and one full-time employee; he also can hire a part-time combine driver. By planning his harvest season in August, Mr. Johns is able to (1) predict whether he can finish soybean harvesting by October 9, (2) know whether he has time to take on more custom work, and (3) decide whether it will be possible for him or his employees to have some time off between soybean and corn harvest.

On his own land, Mr. Johns needs to have a driver available to haul the grain to either his on-farm storage or to the elevator in town; this driver can be either the farmer or his full-time employee. The part-time employee will drive the combine only since the farmer and the full-time employee are familiar with the grain handling system and they do not plan to train the part-time employee in that system.

To keep his clients happy, Mr. Johns has decided that custom work has a priority when custom fields are ready. However, since he and his clients have agreed that while at least one combine will be in the customer's field if it is ready for harvest, they have also agreed that both combines do not need to be in the customer's field. He has also agreed that either he or his full-time employee, not the part-time employee, will be the driver. The contracting farmer will arrange for hauling the grain to storage.

Historical weather data show that, in seven years out of ten, he can expect to lose one day per week due to bad weather for field work during the harvest season. Mr. Johns, his employees, and his neighbors also do not care to work on Sundays unless it is an extremely difficult harvest season. He also estimates (on the high side) that it will probably take a half day of potential harvest time to move between his fields and his neighbor's farms. But it will not take any harvest time to change fields on his own land. Moving could take place on a rainy day but not Sundays. Let us suppose that September 13 is a Sunday this year and guess that the poor weather days during the soybean harvest season will be September 21 and 24 and October 5. Poor weather days affect all farms.

Developing a Schedule

The first step in developing a schedule is to decide what jobs and resources will be listed in the Gantt chart. In this scheduling plan, Mr. Johns wants to know when fields will be harvested, which combines will be used in each field, and which person will be driving the combine or hauling grain. So he wants to list the four fields: Home, Smith, Johnson, and Watson; the new and the old combines; the job of hauling grain; and the farmer and the two employees. This listing can be seen in the Gantt chart (fig. 5.6).

September	Sun. 13	Mon. 14	Tues. 15	Wed. 16	Thurs. 17	Fri. 18	Sat. 19
FIELD:							
Home	X	┌── (Part-time, & F½)────────────────					5½
Smith	X						
Johnson	X	M	┌── (Full-time)────────────				4
Watson	X						
COMBINE:							
Old	X	┌── (Home, Part-time)──────────────					
New	X	┌(H)– /M ┌── (Johnson, Full-time)───────					
HAULING:							
At home:	X	┌── (Farmer)────────────────					
DRIVER:							
Farmer	X	┌────────(Hauling at Home)──────────					
Full-time	X	┌(H) /M ┌──(Johnson, New)──────────					
Part-time	X	┌──(Home, Old)────────────────					

September	Sun. 20	Mon. 21	Tues. 22	Wed. 23	Thurs. 24	Fri. 25	Sat. 26
FIELD:							
Home	X	R	┌(P)6┐		R		
Smith	X	R	┌────(P)────1½		R		
Johnson	X	R	┌─(Full-time)────6		R	┌──(Full-time)──8	
Watson	X	R			R/M	┌──(Farmer)──2┐	
COMBINE:							
Old	X	R	┌(H, P)┐ ┌(S, P)───		R/M	┌(Watson, F)──┐	
New	X	R	┌─(Johnson, F)───		R	┌(Johnson, F)──	
HAULING:							
At home:	X	R	┌(F, H)┐ ┌(F, S)───		R		
DRIVER:							
Farmer	X	R	┌(H, h)┐ ┌(S, h)───		R/M	┌(Watson, O)──┐	
Full-time	X	R	┌─(Johnson, N)───		R	┌(Johnson, N)──	
Part-time	X	R	┌(H,O)┐ ┌(S, O)───				

Figure 5.6. Mr. Johns' Gantt chart for his soybean harvest scheduling. (*Continued on next page.*)

September	Sun. 27	Mon. 28	Tues. 29	Wed. 30	Thurs. Oct. 1	Fri. 2	Sat. 3

FIELD:

Home   X

Smith   X   M ⌐——(Part-time, 4½ & Full-time, 3)————————————9⌐

Johnson   X   ⌐(Full-time)——10⌐   M

Watson   X

COMBINE:

Old   X   M ⌐——(Part-time, Smith)——————————————————⌐

New   X   ⌐(F, Johnson)——⌐   M ⌐(F, Smith)————————————⌐

HAULING:

At home: X   ⌐——(Farmer)———————————————————————⌐

DRIVER:

Farmer   X   ⌐——(Hauling at home)————————————————⌐

Full-time   X   ⌐(Johnson)———10⌐   M ⌐(Smith, New)————————⌐

Part-time   X   ⌐——(Smith, Old)——————————————————⌐

Figure 5.6. (*Continued*).

The second step is to place any known down days and potential rain days on the schedule. This farmer has decided to not work on Sundays and has guessed which days may be rain days based on historical records.

The third step is to place the custom fields on the Gantt chart since they are a contractual obligation. They require at least one combine and driver in the field if it is ready for harvest; his own fields have second priority.

The fourth step is to place this farmer's own fields on the Gantt chart where possible. This shows that harvest on the Smith place will not start on the estimated day of maturity. The priority of custom work and the earlier maturity of the soybeans on the home place mean his soybeans on the Smith farm will be the last to be harvested.

Most of the time, the full-time employee is scheduled for the new combine and often on custom fields. This would allow the farmer to be either on the other combine for custom work or at home hauling grain as well as managing the part-time employee and anything else that may come up. This latter plan is changed only in the last few days when the full-time employee and the farmer are both custom combining but on different farms, and the full-time employee finishes first and comes home to harvest the Smith farm. Then the part-time employee drives the new combine and the full-time employee hauls grain. When the farmer finishes custom work, he continues to combine and the full-time employee continues to haul rather than trade jobs once the full-time employee has started hauling the grain. However, this could easily be changed if the farmer needed to be more available.

This chart shows that Mr. Johns can harvest all his soybeans before corn harvest is expected to start on October 9. It also shows that a part-time employee is needed for ten days in September.

Dispatching Rules

Although a manager may wish to have all activities planned well ahead of time, a manager also knows that reality does not always fit nicely into plans and things happen, so to say. So even though sequencing and Gantt charts are very useful for planing, say a few months or even longer, we still need a system to help decide which tasks need to be done this week, today, or this afternoon.

From general business management, we can find such a system called dispatching. As its name implies, dispatching comes from the work of a dispatcher, say a manager of a mobile repair business or the dispatcher for a taxi company. The job is to quickly assess which jobs need to be done and what their characteristics are. Then using a standard set of dispatching rules, the manager selects which jobs need to be done first, second, and so on. Some common dispatching rules are listed below. Each has its own logic and objective.

FCFS: first come, first serve. Who or what shows up first is served or worked on first. This is a very common rule. We see it being used in stores, restaurants, banks, and elsewhere. Most of us consider it to be very fair in those situations.

SPT: shortest processing time. The first task to be started is the one with the shortest processing time before being finished. Since this can lead to some long jobs never being started because shorter ones keep showing up, an adjusted version of SPT is often used. SPT with truncation uses the basic SPT rule of shortest processing time with a maximum "waiting" times. When a task has been waiting more then the maximum time, the task is started even if shorter tasks are also waiting.

EDD: earliest due date. The task that needs to be done the soonest is the first one to be started. Even though they may have short processing times, tasks that do not need to be done soon are not done soon.

LS: least slack time per operation. This rule compares the processing time needed with the amount of time before the task needs to be done. Slack time is an absolute measure of time. It is calculated as the time left before the due date minus the remaining processing time. In some instances, LS is standardized by dividing by the number of operations remaining for each job. This standardization provides a rough accounting for the time it can take to start, finish, and wait for different aspects of the same job.

MINCR: minimum critical ratio. This rule also compares the remaining processing time with the time to due date but does so in a relative sense. The critical ratio is calculated by dividing the time to due date by the remaining processing time.

ESD: earliest start date. The task that can be or has to be started the soonest is the one that is started first. This rule does not consider remaining processing time, due date, or economic importance of any rule. It just starts the task that has the earliest start date listed.

RANDOM: random selection from among jobs. This rule essentially has no rule. The manager or worker just starts whatever job happens along whether it is the first one thought of, the closest, or the easiest. This rule uses no objective standard by which to rank jobs. So it will contribute to the objectives of the farm only by random.

Economic importance. The dispatching rules just described do tend to ignore the economic importance of a task or job. The economic importance of a task includes its impact on the business: How important is it that the task be done, will the business suffer a major setback if it is not done? A manager may also trade off the direct benefits of accomplishing a task versus the potential regret of not doing the task soon. A task can also be evaluated for its ability to help the farm maximize profits, minimize costs, decrease risk, or meet other goals.

Many farmers may find the most beneficial dispatching rules to be SPT, EDD, and economic importance—especially when used together. However, each dispatching rule helps accomplish its own specific objective. So different situations and businesses may need a different set of rules. A study at Hughes aircraft indicated in manufacturing that SPT was the best for efficiency and flow rate while LS was the best for meeting due dates. The study found that FCFS and ESD did worse than RANDOM for most criteria (LeGrande 1963).

Using Dispatching Rules: Bjerke's Schedule

As an example of using dispatching rules, let us consider how Dennis Bjerke decides what tasks he should do tomorrow, October 24. Bjerke has made a list of jobs that need to be done or could be done tomorrow. He is willing to work 14–15 hours tomorrow since it is the middle of the harvest season (table 5.7).

As a first step in setting the priority of what to do, the jobs are ranked in terms of the processing time and due date. Bjerke estimates two jobs will take

Table 5.7. Mr. Bjerke's List of Jobs to Do and His Application of Dispatching Rules

Jobs to do	Dispatching rules					
	Processing time (hours)	Rank	Due date	Rank	Economic importance	Final priority rank
Go to FSA office to set LDP payment. It is predicted to start going down.	2	7	10/24	1	A	2
Replace fan belt on combine.	1	3	10/28	8	B	8
Go to coffee shop.	2	7	n/a	10	C	Don't go
Talk with custom harvester about cob pieces in grain.	½	1	10/25	3	B	6
Pay bills.	1	3	10/24	1	B	1
Talk to workers about conflict between them.	1	3	10/25	3	A	4
Clean out combine cab.	½	1	11/1	9	C	9
Lawyer says finish business plan by Friday or its too late.	3	9	10/26	6	A	5
Finish harvesting soybeans.	5	10	10/25	3	A	3
Weld reinforcement on wagon.	1	3	10/27	7	B	7

Note: For economic importance; A signifies that this job is very important economically, B signifies that this job has some economic importance, and C signifies that this job has little or no economic importance.

30 minutes each, so, using SPT, they are tied and both ranked #1. He estimates four jobs will take 1 hour each; these four are tied and third in rank. The other jobs are ranked in a similar fashion up to finishing harvest as #10, since he estimates it is the longest job to complete.

Two jobs have a due date of tomorrow, so, using EDD, they are both ranked first. Three jobs are due the day after tomorrow and are ranked third. The rest of the jobs are ranked similarly.

The final ranking is a result of subjectively balancing processing time, due dates, and economic importance. Paying bills is ranked first due to their due date being tomorrow. (Paying bills also is something that could possibly be done before the FSA office opens.) Due to its economic importance and early due date, going to the FSA office and setting the LDP payment is ranked #2. After this office work is done, the soybean harvest is finished because it is ranked A in economic importance.

Talking to the workers is considered economically important and Bjerke decides to do that next due to the problems that may happen if the conflict grows larger. Possibly this discussion could also take place, individually at least, during harvesting.

After the discussion is done, Bjerke heads back in to work on the business plan due to its importance to the farm and the need to finish it within two days.

This first set of jobs has the highest final priority due to its importance to the farm and the proximity of the due date. Since Bjerke is willing to work long days during harvest season, he ranks the next three jobs in descending priority: talk with the custom harvester about pieces of cob in the grain, weld reinforcement on a wagon, and replace the fan belt on the combine. If time permits, he may take time to clean out the combine cab.

Even though he would love to hear what is happening, Bjerke decides that going to the coffee shop is something he just cannot afford to do, due to the time it takes, its lack of economic importance, and the lack of a due date.

SUMMARY POINTS

- A process map is a description of a method or process of accomplishing a task.
- Process maps can help a farmer understand how the farm and its processes fit together and to see potential areas for improvement.
- Five general areas for improving processes are reducing input use, substituting inputs, increasing productivity, expanding the process, reorganizing the process.
- Enterprises can be competitive with each other, supplementary, or complementary.
- Gross margins can be used to select the best mix of enterprises for a farm.
- The gross margin equals gross income minus variable costs.
- The benefits of input supply management include taking advantage of price discounts, anticipating input needs, and more orderly and timely delivery of inputs.
- Scheduling is an allocation decision.
- The schedule shows what is to be done, when it will be done, who will do it, and what equipment will be needed.
- Scheduling has three distinct, competing objectives: cost, schedule, and performance.
- Two scheduling techniques can be useful for farming: sequencing and dispatching.
- Sequencing is a procedure for scheduling people, jobs, equipment, and other resources over several weeks or months. Gantt charts are used for sequencing.
- Dispatching rules help a manager decide which jobs have the highest priority and should be done first on a certain day.

- The most beneficial dispatching rules for farmers are SPT, EDD, and economic importance.

REVIEW QUESTIONS

1. Why develop a process map? Why and how is it useful to a farm manager?
2. What are the four broad types of products? How can this classification be useful in product selection?
3. What are gross margins? How can they be used for choosing enterprises?
4. Based on the projected budgets available from extension, a farmer has developed the following estimates of yields, prices, and costs for six crops (Swensen and Haugen 2001). This farm has 620 tillable crop acres. She does not plant more than 50% of her farm to one crop. Given just this information, what is the best crop mix for this farm?

Table 5.8. Estimated Crop Budgets for a Farmer in North Dakota in 2002.

	Spring wheat	Durum wheat	Malting barley	Corn	Soybeans	Lentils
Yield	28 bu	28 bu	50 bu	64 bu	22 bu	1,150 lb
Price ($ per bu or lb)	3.10	3.36	2.16	2.00	4.70	.10
Other income ($/ac)*	9	9	12	12	3	0
Variable costs ($/ac)	51	52	50	90	63	52
Fixed costs ($/ac)	56	56	58	73	58	55

*Other income includes government payments and straw sales.

5. How can enterprise budgets and the chosen enterprise mix be used for input supply management? How could this be beneficial to a farmer? When is it needed? When is it not needed?
6. How can operations be scheduled using sequencing (with Gantt charts)? In what situations would schedule planning be needed for farming?
7. Describe the main dispatching rules and how they can be used by farmers.

6

Quality Management and Control

Quality has not been talked about very explicitly in farm management. We have established product quality standards, such as USDA grades for livestock and eggs, USDA standards for grains, and USDA inspections in processing plants. We have established and encouraged the use of procedures to reduce pesticide spills, bruised meat, safe operation of equipment, and so on. But quality on the farm has not been emphasized in the way it has in general business. The pressures and changes that are taking place now in the food and agriculture industry require farmers to look at quality from a new perspective—beyond product quality to process quality and the needs and expectations of the consumer.

This chapter focuses on quality management and quality control. *Quality management* involves a holistic view of the farm. It aims to develop a philosophy of improving quality in all functions: marketing, production, finance, and personnel. It develops policies and procedures to instill a commitment to quality throughout the farm. Quality management should include the initial product choice and design, input purchases, choice of production processes, production and marketing, and the warranty and repair service supplied after production. The responsibility of producing quality can be carried out only by the application of proper management and control in all phases and levels of the business. A few of the concepts being adopted by businesses today are total quality management (TQM), zero defects, and continuous improvement.

Quality control applies to the already selected production process for the products already chosen and designed. Two major parts of quality control discussed in this chapter are process control and process improvement. Process control consists of the procedures to monitor production for compliance with the original plan and development of corrective actions designed to bring the process back into compliance. Process improvement includes a set of tools used to understand the current process better and to look for potential ways to improve both the process and the product.

This chapter starts with a definition of quality and the costs of quality. It then addresses quality management, process control, and process improvement.

QUALITY DEFINED

Quality is commonly defined as "meeting or exceeding customer requirements." This definition points to the first and most important lesson in quality management: *The customer, not the producer, defines quality.*

Farmers themselves (and manufacturers, professors, and even hamburger flippers) may want to define quality, because they "know" what quality is! But, whether we like it or not, the customer decides whether a product is what they need.

The phrase "fitness for use" can also help us understand that quality is related to value received by the customer and to customer satisfaction. In both ideas we see that it is the customer and not the producer who defines quality.

A customer defines quality using several dimensions. These include performance, reliability, durability, service, reputation, conformance, and safety. Price is included in some lists of quality dimensions. In one sense, price does not affect the performance of a product or service. Yet in another sense, price can affect a customer's ability to meet cost and profit goals, and thus, price does affect the customer's view of a product and its "fitness for use."

To understand better what customers are looking for, their dimensions of quality can be defined in four ways or into four determinants of **product quality**: quality of design, quality of conformance, the "abilities," and service after delivery. In other words, a customer is looking for how well the product or service is designed, how well it meets the design specifications, how available it is, and how well it is taken care of by the producer.

Quality of Design

Quality of design is determined before production takes place. How well does the product design meet the customer's needs and wants? For farmers, design quality has been present for a long time in the price breaks defined by such things as grain moisture standards, market animal weights, milk fat, and protein levels. For a farmer, quality of design involves crop variety choice, the weight and type of finished animal, the fat content of the product, and the protein content. Design quality also involves the choice of production methods to meet food quality and safety concerns, desire for organically produced foods, and animal welfare concerns. Design quality is a common part of agricultural production contracts: choice of crop varieties, planting dates, delivery dates, and such. Design quality is also important in a service product such as custom harvesting where the planning of the schedule is important for harvesting the customer's crop at the optimal time.

Quality of Conformance

Quality of conformance is about how well the production process has produced the product or service to the design specifications. Was the crop fertilized correctly? Were the proper crop protection methods used? Were chemical labels followed? Were organic practices followed if dictated by the product? Was the machinery calibrated and set correctly? Were the animals fed the correct rations to produce the fat levels requested? Were animals with the right genetics used? Were sick animals treated appropriately? Was the crop harvested at the optimal time?

The "Abilities"

The "abilities" are availability, reliability, and maintainability. All these have a time element that reflects the continued satisfaction of the customer.

Availability. A product is available if it is in an operational state and not down for repairs or maintenance. For a machine, availability can be expressed as the percentage of time that it is up and ready for work out of the total of uptime and downtime. For agricultural products, availability can also be viewed as the time that the product is available to the consumer in a certain price range. Seasonality of production may affect availability.

Reliability is the length of time that a product can be used before it fails. It can also be described as the probability that a product will function for a specified period without failure. For a machine, reliability is related to the mean time between failures (MTBF). For food products, reliability can be viewed as shelf life in a grocery store and at home. For fiber products (cotton, for example), reliability can be measured by tensile strength of the cloth.

Maintainability is the restoration of a product to service once it has failed. It can be measured as the mean time to repair (MTTR). So, availability can be defined also as: availability = MTBF / (MTBF + MTTR).

Service after Delivery

Service after delivery is the warranty, repair, and replacement after the product has been sold. In farming, service examples include the replacement of breeding animals if they do not perform as expected, the hiring of replacement custom operators if other circumstances do not allow a contract to be fulfilled, the buyback or replacement of the product that does not meet the expectations of the customer, and so on.

Process quality has an external and an internal component. Externally, process quality ensures that the product will be made so that the product conforms to what the customer wants. Internally, process quality is concerned with the efficiency of the production process, that is, how well the farm is operating. This view of process quality involves physical and economic measures of efficiency and productivity. Process quality is not the same as product quality and is measured differently. Process quality is discussed later in this chapter in the sections on process control and process improvement.

THE COSTS OF QUALITY

The costs of quality include both the costs of meeting customers' requirements and the costs of *not* meeting customers' requirements. These costs are divided into control costs and failure costs. Some of these costs, such as higher costs for better sanitation, are obvious and easily seen and measured. Other costs of quality, such as lost sales, cannot be seen so easily.

Control costs are the costs of removing defects by prevention or by appraisal or inspection. An example of cost prevention is choosing and following procedures that have higher costs but provide higher quality (e.g., higher sanitation procedures in the milking parlor, taking the time to test the accuracy of the sprayer). Another example is maintaining equipment to meet higher standards, not just to get a job done (e.g., operating a dryer to provide a lower percentage of burnt kernels, not just dry corn). Examples of appraisal or inspection costs include checking for correct seed placement, testing for correct nutrient levels in feedstuffs, monitoring the moisture level of grain in storage, sorting and selling livestock only within desired market weight ranges, and testing for chemical residues.

Failure costs are the costs of the product failing to meet quality standards. Failure costs can be divided into internal and external costs. Internal failure costs occur during production. Examples of internal costs include scrap, rework, quality downgrading, and downtime. Animal disease causing poor productivity or death is an example of the cost of internal failures on farms. Equipment breakdowns, repair costs, and lost timeliness due to poor maintenance or replacement are other examples. External failure costs occur after production and shipment. Examples of external costs include warranty charges, returned goods, and allowances. Bruised meat being rejected by the processor is an external failure cost. Products rejected due to chemical residues found after they left the farm are an example of external failures. One external failure cost that may be hard to measure is receiving a lower price due to lower quality and the resulting lack of or lower consumer demand. Lost contracts and lack of repeat sales are also external costs of poor quality, but they too are hard to measure.

The *total cost of quality* is the sum of control costs and failure costs, that is, the sum of prevention costs, appraisal costs, internal failure costs, and external failure costs. The manager should strive to lower the total cost of quality. The hidden costs (e.g., lower prices due to external failure) may be higher than the internal control costs of prevention. While seeing the costs of prevention in the expenses is very easy, the manager must evaluate that against the unseen losses of external failure.

The *benefits of quality* include an enhanced reputation, increased business, greater customer loyalty, and fewer production and service problems. All of this means that the business will have higher productivity, fewer complaints, lower production costs, and higher profits. In general business, the companies who have won the national Baldridge quality award outperform other companies financially.

QUALITY MANAGEMENT

Every worker is responsible for quality; every aspect of a farm needs to be scrutinized for quality. The concern for quality has to be present at all levels including top management, product and process design, input purchasing, production and operations, storage and shipping, marketing and sales, and customer service. Even on a one-person farm, the operator has to consider the need for high quality and consequences of low quality in all decisions and operations. This all-encompassing need to pay attention to quality is why we start talking about "quality management."

In this section, we first look at the ideas of a group of people—the quality gurus—who started talking about quality management early, who left lasting contributions to quality management, and whose ideas are still relevant for farmers today. Then we briefly consider three approaches for improving quality that we hear about often and that will affect farmers in the future. First, we look at the overall concept of total quality management (TQM) and how it can be used by farmers. Second, we consider how farmers can use the ISO standards from the International Organization for Standards to improve the consistency of conforming to product design and assuring customers that the product advertised will be provided. The third approach we will study briefly is Hazard Analysis and Critical Control Point (HACCP), which is used more and more in the food industry.

Lessons from the Quality Gurus

A small group of people has made a tremendous impact on how business views and improves quality. The most famous of this group includes these gurus: Deming, Juran, Feigenbaum, Crosby, and Ishikawa. Let us look at the ideas that made each one famous and how these ideas can be used by farmers.

Deming

W. Edwards Deming is the best known of this elite group. One of his main messages was that the cause of inefficiency and poor quality is the system, not the workers. Deming felt that management was responsible for correcting the system. He expressed his "requirements for a business whose management plans to remain competitive in providing goods and services that will have a market" in 14 management principles (fig. 6.1). The key elements of his 14 points are constancy of purpose, continual improvement, and profound knowledge. Profound knowledge means understanding (1) the system and the impact of everyone in the system, (2) the causes of variation, (3) the need to learn from theory, and (4) the importance of psychology and how to motivate workers to contribute their collective efforts to a common goal.

Deming defined quality as continuous improvement of a stable system. Two points can be seen in this definition. First, all systems (administration, design, production, sales, etc.) must be stable in a statistical sense (i.e., a constant variance around a constant average). Second, continuous improvement is needed to reduce variation and better meet the customer's needs. To reduce variation, Deming saw the need to distinguish between special causes of variation (i.e., correctable or assignable causes) and common causes of variation (i.e., random or uncontrollable).

Briefly, Deming wants top management to (1) look at the long-run not short-run profits, (2) cease dependence on mass inspection of products, and (3) stress prevention of defects. However, Deming is also a strong advocate of quantitative, statistical tests to find sources of variation. He started working in statistical quality control but saw that more was needed than simple acceptance or rejection of the final product. He saw that the systems needed change starting with top management.

Juran

Joseph M. Juran described quality as "fitness for use" as defined by the customer. Juran described quality management as a trilogy of quality planning, quality control, and quality improvement. Quality planning is needed to develop processes that can meet quality standards. Quality control is needed to know when corrective actions are needed to bring the process back in control. Quality improvement is needed to help find better ways of doing things. Juran

1. Create constancy of purpose toward improvement of product and service with a plan to become competitive and to stay in business. Decide to whom top management is responsible.
2. Adopt the new philosophy. We are in a new economic age. We can no longer live with commonly accepted levels of delays, mistakes, defective materials, and defective workmanship.
3. Cease dependence on mass inspection. Require, instead, statistical evidence that quality is built in. (Prevent defects rather than detect defects.)
4. End the practice of awarding business on the basis of price tag. Instead, depend on meaningful measures of quality along with price. Eliminate suppliers that cannot qualify with statistical evidence of Quality.
5. Find problems. It is management's job to work continually on the system (design, incoming materials, composition of material, maintenance, improvement of machine, training, supervision, retraining).
6. Institute modern methods of training on the job.
7. The responsibility of foremen must be changed from sheer numbers to quality ... [which] will automatically improve productivity. Management must prepare to take immediate action on reports from foremen concerning barriers such as inherent defects, machines not maintained, poor tools, and fuzzy operational definitions.
8. Drive out fear, so that everyone may work effectively for the company.
9. Break down barriers between departments. People in research, design, sales, and production must work as a team to foresee problems of production that may be encountered with various materials and specifications.
10. Eliminate numerical goals, posters, and slogans for the workforce, asking for new levels of productivity without providing methods.
11. Eliminate work standards that prescribe numerical quotas.
12. Remove barriers that stand between the hourly worker and his right to pride of workmanship.
13. Institute a vigorous program of education and retraining.
14. Create a structure in top management that will push every day on the above 13 points.

Figure 6.1. Deming's 14 principles for quality management (From Deming 1982 as quoted in Stevenson 2002, 403).

showed the potential for increased profits if the costs of poor quality could be reduced.

Feigenbaum

Armand Feigenbaum used the "cost of nonconformance" to obtain management's commitment to quality. He also saw that by improving the system in one part of a business, other areas also enjoyed improvements. With this understanding of systems, Feigenbaum encouraged (1) cross-functional work (i.e., teams of people from production, finance, sales, and other parts of a business) and (2) the development of environments in which workers could learn from each other's successes. Feigenbaum argued that improving human factors were more important to improved quality than technological factors. He also argued that it is the customer who defines quality and that it is important to control quality at the source rather than at the end of the system.

Crosby

Philip B. Crosby coined the phrases: "zero defects" and "do it right the first time." He argued against the idea of acceptable levels of mistakes and defects. Like the other quality gurus, Crosby was convinced that quality improvement must start with top management's commitment to, support of, persistency toward, and communication for good quality. He argued that "quality is free" because the costs of poor quality are much greater than the costs of improving quality.

Ishikawa

Kaoru Ishikawa followed both Deming and Juran but also provided the quality tools of the cause-and-effect diagram and quality circles. Cause-and-effect diagrams (also called fishbone diagrams) help solve problems by helping to identify potential causes. Quality circles involve workers in quality improvement. Ishikawa was the first to talk about the internal customer, that is, the next person in the process. By providing internal customers with better quality inputs to their part of the process, the overall quality of the final product would be improved. This description of the system allowed workers to see potential ways to improve the part of the operation that was in their control. For most farmers, the next person is the elevator or processor. By improving the quality of the product supplied to them, farmers can help improve the quality of the products supplied to the final consumer.

In summary, twelve points to remember from the quality gurus and what they mean to a farmer are:

1. Quality is defined by the customer. Farmers need to listen to customers. They are the ones who define a quality product.
2. Quality improvement must start with management's commitment. If a manager does not have quality as a top concern neither will his or her employees.
3. Problems in the system, not the workers, cause inefficiency and poor quality. Employees will respond to how a farm is organized, how buildings are designed, how incentives are described, and so on. If these systems and structures are not designed well, employees will not work well.
4. Continual improvement is needed. Due to the world always changing and our ability to learn how to do things better, we need to keep changing our products, processes, systems, and ourselves.

5. Prevent defects, don't just throw them out. Preventing problems by changing methods is usually more profitable and provides a better product for the consumer.

6. Quality planning, quality control, and quality improvement must go together. A manager needs to work on the whole package to make the best progress toward higher quality product and process.

7. Improving human factors is more important than improving technological factors. Simply buying new machines and switching to new inputs will not make large improvements in product and process quality if workers do not fully understand how to use either the new or old technologies and systems.

8. Do it right the first time. It is better to take the time to plan and learn how a job needs to be done before the job is started than to start the job and waste time and inputs, ruin products, and/or destroy customer faith.

9. Quality is free. The expenses for improving processes and products will be more than paid back through improved efficiencies and other benefits received due to the improvements.

10. Cause and effect diagrams can help us see potential solutions. Understanding all the possible factors affecting the desired outcome can help a manager see potential areas for improvement. Cause and effect diagrams are especially useful for helping managers see more than the "usual suspects" when trying to solve problems.

11. Quality circles involve workers in improving the business. Listening to workers and customers is beneficial in many ways including: Workers are closer to the work and know how it can be improved, workers feel respected when management asks them for information, customers can provide a better understanding of what they want, costs go down, and efficiencies go up.

12. Helping internal customers perform better will help the whole business. Whether one person or several persons are involved, the whole farm can see improvements when each person looks for ways to help the next person perform his or her task easier and better. For example, how can the mechanic help the planter plant? How can the corn producer provide a better feed to the hog feeder? How can the worker provide better records to the financial analyst?

Total Quality Management (TQM) for Farmers

Total quality management (TQM) is a management philosophy that strives to involve everyone in a continual effort to improve quality and achieve customer satisfaction. Three points should be noted in this description of TQM: *every-*

one, *continual*, and *customer satisfaction*. Everyone should be involved in improving quality on a farm: from the owners and managers to the part-time workers. A farm practicing TQM to its fullest extent will even involve suppliers in the quest for quality improvement. TQM also involves a continual process of checking all aspects of the farm that affect quality: inputs, processes, and products. TQM also has customer satisfaction as its primary goal which means, as discussed earlier, meeting or exceeding customer expectations.

For farmers, TQM involves the following steps:

1. *Find out what the customers want.*

   For commodities such as #2 yellow corn, this could start with checking the standards for moisture, foreign material, and so on. This could also mean talking with the elevator about its preferences for delivery method and timing. For specialty products, such as high-oil corn or contract vegetables, this step starts with checking the terms of the contract.

   For services such as custom harvesting, this step involves talking with current and potential customers and understanding what they want concerning the work done: speed, timeliness, daily start and end times, product left in the field, and clean up afterward.

2. *Design a product or service that will meet or exceed what customers want.*

   Most farmers will not participate in all the details of product design since they produce basic raw commodities (e.g., milk, pork, corn) that have long-established specifications that are part of their product description. However, farmers do control the product that leaves their farm. For example, they set the moisture percentage they will dry their corn to, the temperature of milk in the bulk cooler, the weight at which hogs are sold, the genetic potential of their livestock, and so on. They do need to consider what their customers want.

3. *Design processes that will facilitate doing the job right the first time.*

   Designing, or choosing, how to produce a product, should involve not just picking the steps, inputs, equipment, and workers needed. Designing the process should also look at the risks and uncertainties of production and include steps, procedures, and other aspects to ensure that the desired product will be produced with as little physical and economic waste as possible.

   Process design and product design need to be done simultaneously although they are listed as two steps here. For example, in producing #2 yellow corn, the decision of what the delivered moisture level should be is a joint process-product decision. The market defines a certain moisture level for pricing, but if the cost of on-farm drying is too great compared with the benefit, corn can be delivered at a higher moisture level and, for a price, dried by the local elevator.

4. *Keep track of results and use them to improve the system.*

Keeping the books for quality management involves more than financial data. Livestock and field records can be used to identify both good and poor producers, to spot areas that need improvement, and to eliminate waste.

5. *Strive to have suppliers and processors adopt the ideas of TQM.*

Improving quality on the farm has to involve improving the quality of inputs and discussing the standards requested by the buyers of the products. For example, after improving the feeding equipment and processes on a hog farm, the next step is to be sure that purchased feedstuffs are delivered at the desired and expected nutrient levels. This may involve more specific definitions of nutrient requirements, testing or requiring testing for nutrient levels in feedstuffs, and discussion with suppliers about the nutritional needs. This may also involve farmers serving on boards of directors.

Besides the five steps above, TQM includes other ideas and concepts that are crucial to its use by farmers. They are:

Continuous improvement. Always look for ways to improve the process and results of the farm, never being satisfied with the status quo, not accepting "normal" mistake levels.

Competitive benchmarking. Compare the processes and results of your own operation with the competition. This may involve visiting other farms, joining farm business associations, reading government and popular media that present the performance levels of other farms, and so on. At times, this benchmarking may not always be done with farms in the same business. For example, feed grain producers may learn how to maintain identity preservation in a commodity environment by visiting rice and bean farmers and their processors.

Employee empowerment. Giving employees the responsibility for quality improvement and the authority to make changes allows access to their intimate knowledge of the process they are using and motivates them to be more involved in the success of the farm.

Team approach. Teams provide benefits in many ways. Teams can provide synergies, better understanding of problems and potential solutions, and increased motivation and cooperation. Independent farmers can benefit from discussions with family members and from professional discussions with neighbors and others close to their business.

Data-based decisions. Instead of relying on opinions and ideas of what may be the causes of problems, TQM stresses the need to gather and understand the data relevant to a situation before decisions are made.

Knowledge of quality management tools. Understanding the basic tools will help farmers understand and analyze their production processes, what aspects need to be improved first, how to improve that aspect, and how to keep a process under control and producing the desired product.

Supplier quality. At some point, the on-farm processes are at a level of quality where suppliers need to become involved in understanding the farmer's needs and the implications for input quality in order to improve the results of the total system.

Quality at the source. Quality at the source is a philosophy that makes every worker responsible for the quality of his or her own work.

ISO 9000 Standards

Since 1947, the International Organization for Standards (ISO) has promoted worldwide standards to improve operating efficiency, improve productivity, and reduce costs. Most of the ISO standards have been highly specific for a particular product, material, or process. Engineers have used ISO standards to ensure that a certain kind of bolt, for example, is the same in all corners of the world. In 1987, the ISO deviated from this engineering specificity to develop the ISO 9000 standards for quality management. The ISO 9000 standards are generic management system standards versus specific product, material, and process standards. Before we discuss how these standards are being used by farmers, let us first look at the standards themselves.

The ISO 9000 standards are generic management system standards concerned with quality management. ISO's definition of quality is conforming to the requirements of the customer, and quality management is ensuring that the products conform to the customer's requirements. The ISO 9000 standards are generic in that any size and type of business can use them. A "management system" is what an organization, company, or business does to manage its processes, activities, and services. The ISO 9000 standards are a set of guidelines on how to set up and operate a management system to ensure that its products conform to the customer's requirements. ISO 9000 standards are not concerned with what the customer's requirements are; the standards are to be used by an organization to show a customer how the organization is ensuring that the customer's requirements are being met.

There are three standards within the ISO 9000 family (ISO 2000).

1. *ISO 9000:2000, Quality Management Systems—Fundamentals and Vocabulary*

    These establish a starting point for understanding the standards and define fundamental terms and definitions.
2. *ISO 9001:2000, Quality Management Systems—Requirements*

    This is the requirement standard used to assess an organization's ability to meet customer and applicable regulatory requirements and address customer satisfaction. This is now the only standard in the ISO 9000 family against which third-party certification can be carried out.

3. *ISO 9004:2000, Quality Management Systems—Guidelines for Performance Improvements*

   This guideline standard provides guidance for continual improvement of an organization's quality management system to benefit all parties through sustained customer satisfaction.

The ISO 9000 family of standards and specifically the ISO 9001:2000 standards focus on assuring the quality of the process being used in the organization. This quality assurance system must include procedures, policies, and training. Extensive documentation is required, which includes a quality manual, process flow charts, operator instructions, inspection and testing methods, job descriptions, and organization charts. To be certified, an organization must review, refine, and map functions such as process control, inspection, purchasing, and training. ISO registrars look for a well-documented quality system, completed training, and whether the actual process being used conforms with the process described in the documentation. The organization must be reregistered every three years.

Once it is registered as ISO 9000 certified, an organization can advertise and promote its products as being made consistently in the same way. It can show customers that standard processes and inputs are used in the production of its products. The ISO 9000 family is especially important for companies involved in international trade—particularly in Europe, but certification can be important for domestic customers also.

As rigorous and time-consuming as becoming certified sounds, ISO 9001:2000 standards do not address all aspects of quality, namely, product. An organization may be ISO 9001:2000 certified and follow the process flawlessly but be producing a product that nobody wants to buy. These standards also do not address organization leadership, strategic planning, customer satisfaction, continuous improvement, and business results.

Farmers can benefit from ISO 9000 certification in two ways. First, working through guidelines can help farmers take advantage of new ideas for decreasing costs and increasing profits by changing their process of production and service. Second, ISO certification can be a marketing advantage to show far-flung buyers that production and identity preservation processes are designed to ensure that the resulting product will conform to the customer's requirements.

Certification in the ISO standards allows potential buyers to understand and trust that they are buying the product described. Developing the processes and becoming ISO certified can allow farmers to produce products that meet the increasing concerns of food quality, food safety, and identity preservation. These concerns are present both domestically and internationally.

ISO certification is not a theoretical idea for farmers. In September 2000, a group of Iowa farmers became the first U.S. farmers to be ISO 9000 certified in grain production (Swoboda 2001). Their aim was to improve their competitive advantage in the export market. They needed to provide a higher valued product to their customers. The higher value comes from being able to certify and document that their grain is produced in a certain way using certain identifiable inputs.

Farmers in Minnesota can participate in quality programs designed to follow the ideas involved in ISO 9000:2000, although they do not become officially ISO certified as the programs are currently set up. These two programs—Minnesota Certified Pork (MNCEP) and Minnesota Certified (MinnCERT)—are a result of discussions with the Minnesota Pork Producers Association, the University of Minnesota, and the Minnesota Department of Agriculture (Starner 2001). MNCEP was started to help pork producers find a market for a standard product inspected and audited to ensure that a set of standard operating procedures was followed. To broaden the potential use of these ideas, MinnCERT was started to help farmers who want to become part of a value-added chain to find markets, to implement market-tailored quality standards, and to certify the defined production procedures.

Hazard Analysis and Critical Control Point (HACCP)

Hazard Analysis and Critical Control Point, or HACCP (pronounced *hassip*), started 30 years ago to ensure safe foods for astronauts. Now it is a food safety program that does and will affect almost all farmers because it is used by both of the federal agencies that monitor and regulate the U.S. food system. The Food Safety and Inspection Service (FSIS) of the USDA inspects and regulates all federally inspected meat and poultry processing plants. The Food and Drug Administration (FDA) inspects and regulates the rest of the U.S. food system.

In brief, the goal of HACCP is to design systems to monitor and reduce contamination through preventive and corrective measures instituted at each stage of the food production process where food safety hazards could occur (FSIS 1999). It strives to do this by applying science-based controls through the entire system from raw materials to finished products. This goal is in contrast to the traditional system of spot-checks of manufacturing conditions and random sampling of final products to ensure safe food.

In 1996, USDA published a final rule on Pathogen Reduction; Hazard Analysis and Critical Control Point (HACCP) Systems (PR/HACCP) that "requires meat and poultry plants under Federal inspection to take responsibility for, among other things, reducing the contamination of meat and poultry products with disease-causing (pathogenic) bacteria" (FSIS 1999, 3). FDA now applies HACCP to seafood and juice and intends to eventually use it for much of the U.S. food supply (FDA 2001).

HACCP has seven principles:

1. *Analyze hazards.* Potential hazards associated with a food are identified and the measures that can be used to control these hazards are identified. Potential hazards include biological (bacterial, parasitical, or viral), chemical (toxins, drugs, cleaners, etc.), and physical (glass, metal, etc.). These hazards may be in the raw material or introduced at points in the processing and transportation process. Since farmers deliver raw materials, they are not immune from HACCP.

2. *Identify critical control points.* Critical control points are points at which a potential hazard (identified in the first principle or step) can be controlled or eliminated. Examples of these critical points include grinding, mixing, cooking, cooling, metal detection, and so on. The entire production process for a specific food is evaluated starting with its raw ingredients and continuing through processing, storage, and transportation to consumption by the consumer. Delivery of raw products is usually a critical control point for potential entry of hazards. Upon delivery, the temperature of milk may be measured as an indicator of safety, health of animals is checked, grain is inspected for mold, and so on.

3. *Establish preventive measures with critical limits for each control point.* These are exact, specific measures with both upper and lower limits that cannot be exceeded. Examples include cooking temperatures, cool down time, equipment washing temperatures, and dimensions of metal fragments. Specific milk temperatures or time above critical temperature are identified and maximum mold infection levels are identified.

4. *Establish procedures to monitor the critical control points.* Monitoring procedures are those things which are done routinely, either by an employee or by mechanical means, which measure the process at a given critical control point and create a record for future use. These procedures describe, for example, how cooking time and temperatures are to be monitored, or how mixing of ingredients is to be monitored. Raw products could be evaluated before being unloaded. Records of how and where raw products have been transported and stored may be required. Analytical tests could be identified that will provide an accurate measurement of the potential hazard.

5. *Establish corrective actions.* HACCP requires that corrective actions be identified for those occasions when monitoring shows that a critical limit has been violated. Examples include intercepting affected food, reprocess-

ing, disposing of adulterated food, cleaning equipment, rejecting contami-nated raw material, changing procedures, retraining of employees, and so on. FSIS describes four points that its regulators will be checking: (1) iden-tification of the cause of the deviation, (2) verification that the system will be in control after the corrective action is taken, (3) measures taken to pre-vent recurrence of the deviation, and (4) steps taken to ensure that no adul-terated product enters the food chain.

6. *Establish record-keeping procedures.* These are needed to document the HACCP system including records of hazards, control methods, monitoring records, corrective actions taken.

7. *Establish verification procedures.* Verification involves testing the system to be sure it is working as intended. This means checking testing equipment such as thermometers and other analytical equipment as well as observing that monitoring is taking place as intended, corrective actions are being taken when needed, and records are being kept as designed.

PROCESS CONTROL

In the first part of this chapter, we saw how farmers could benefit from the ideas developed by the quality gurus and become involved in overall quality management programs such as TQM and ISO 9001:2000. However, even though a farmer may develop a great TQM system and be certified in the most rigorous ISO 9001:2000 program, uncertainty and risk still creep in and cre-ate problems due to products not meeting customer expectations and process-es being out of control. In this section we will look at how farmers can moni-tor and control their farm processes to be sure that all are working as they were designed to do.

Three-Part System

A process control system consists of three simple parts: standards, measure-ment, and corrective actions. (1) Standards or expected performance levels come from the plans developed as part of strategic planning and short-term planning. Enterprise budgets are a main source of standards for yields, pro-duction levels, prices, input use, and so on. (2) The second part of a process control system is deciding how, when, where, and which results should be measured. These measurements show whether the process is operating according to the standards expected in the plans. If the system is not in com-pliance, corrective actions are needed to bring the process back into compli-ance. (3) Corrective actions may be specified before common problems occur or they may be the result of problem-solving sessions held when new prob-lems occur.

Three Types of Control

Three types of control are available for the manager to use to keep the production process in control: preliminary control, concurrent control, and feedback control.

Preliminary controls are anticipatory actions that take place before problems are noticed. Preliminary control is closely related to planning. An example of preliminary control is the use of pre-emergence herbicides even though no weed problems are seen or may even occur. Another example is the development of so-called fail-safe plans to be sure all steps are done at a crucial point in a process, such as caring for newborn pigs or calves.

Concurrent control occurs during the production process. Two examples on farms are disease treatment for livestock and plant tissue testing to determine the plant's fertilizer needs.

Feedback control starts after production has occurred. Changes in disease control and sanitation methods for the next production period are examples of feedback. Taking classes to understand new marketing options better is another example. Feedback control often starts with problems that were noted on trend charts or check lists during the process. Cause-and-effect diagrams (discussed later in this section and in the next) can be used to identify the causes of the problems and, thus, potential solutions and ways to improve the process. Feedback control uses many of the same tools used in process improvement, but they are different. In feedback control, we are trying to solve a problem to bring the process back into compliance with the plan. In process improvement, the process is working as currently designed, but we are looking for ways to make it better.

Developing Process Control Systems

Process control is the inspection of the product or service while it is being produced. It is the process of determining and implementing the necessary actions to make certain that plans are transferred into desired results. Through effective control, we are better able to achieve the goals and objectives that have been established.

The number of critical points, and thus the detail and formality of the control system, depends upon the benefits versus costs of inspection, operator differences and needs, and relative importance of the enterprise to the whole farm.

How to Develop a Process Control System

To develop a process control system, a manager needs to evaluate the system or operation and determine what opportunities exist for control. This can be done in five steps.

1. *Identify critical points where inspection and testing are needed and can be done.*

    These critical points can be at any place or time in the process. Process maps, which were described in chapter 5, can be extremely useful. These critical points may be to ensure incoming raw materials or purchased services meet specifications, test work in process or the service while it is being delivered, or inspect the finished product or service. This step should result in a list of critical points in order of importance.

2. *Choose the type of measurement to be used at each inspection point.*

    Measurement can be by variables or by attributes. Testing by variables uses a continuous scale for such factors as length, height, weight, time, and dimensions. Acceptance is based on how close the product or service meets these actual measurements. Testing by attributes uses a discrete scale by counting the number of defective items or defects per unit. These are qualitative or discrete measurements (good or bad, broken or not, etc.). Acceptance is based on the percentage of items that meet the standards.

3. *Identify appropriate methods and frequency of measuring.*

    The plan for measuring performance involves deciding the timing of measurement, the units to measure in, and the recording system. How the performance is measured and what sensors will be used need to be specified. The reliability of the alternative measurement methods needs to be evaluated.

    The measurement plan also needs to specify how much inspection to use: 100% of the production or a sample of the production. If the testing destroys the product (protein tests in feedstuffs, for example), a sample is the obvious choice. If the product is not destroyed, the choice revolves around the cost of the testing process, the underlying variability of the measure, and the need for information.

    Another part of developing the measurement plan is to decide whether the workers or someone else should do the inspection. By having the workers inspect their own work, they can see problems directly. Since they are closer to the process, they can quickly offer ideas for improving the process and the resulting quality. The traditional method of having someone else inspect the work may be needed if a third-party inspection is needed or if worker inspection needs to be verified as accurate.

4. *Specify standards and "in-control" range.*

    Standards need to be quantitative in both physical and monetary terms. Even if qualitative attributes are used to measure quality, the standard needs to be defined in quantitative terms (e.g., percent acceptable). They need to be measurable and recorded as part of the plan. Standards can be developed from historical records for the farm, records from other farms, information from suppliers, and private and public budgets.

5. *Establish rules of action to bring the system back into control.*

   The final part of a process control system is the corrective actions to be taken if the measurements show the process to be out of control. These corrective actions take one of three forms: change the plan, change the implementation of the plan, or change the goals or standards by which production is evaluated.

For examples, consider the illustrative components of a process control for soybeans (table 6.1) and for a beef-finishing operation (table 6.2). These examples are only illustrative of how a process control system could be developed and are not meant to be the entire system that a farmer may need to control his or her own process.

Fail-Safe Plans

A fail-safe plan is a process and product design tool that looks for ways to eliminate the possibility of problems or mistakes occurring. One example is task lists that include spaces for the worker's initials and the date the task was completed. Fail-safe plans are not always lists on paper. Other examples include tying red flags on equipment parts that need repair; designing the planter supply truck to be sure all needed materials, parts, and tools are included; and developing forms that check for completeness and consistency in the activities. These are especially useful for preliminary control. A seed monitor mounted on the tractor is a fail-safe tool for planting because it provides an immediate signal of seeds not being fed through the planter at the appropriate rate.

   As another example of a fail-safe method, consider the jobs that need to be done to care for feeder yearlings in the first hours after they arrive in the feedlot. The tasks that need to be done are very crucial for the health of the cattle and the success of the farmer. A fail-safe plan for arriving feeders would include a checklist of the tasks that need to be done, the timing of those tasks, the expectations or standards for each task, corrective actions needed if deviations are noted, and a place for the initials of who completed the task and when (table 6.3).

Check Sheets

Check sheets are forms that merely require a check to be made in the appropriate spot to show the frequency of certain events that relate to quality or process characteristics. They can be used to show many things: the frequencies of error or problem sources, the variation of time required to do the same job, and productivity differences. No statistics are used with this tool. Examples of check sheets include causes of baby pig deaths (table 6.4), frequency of corn header plugging by row (table 6.5), and so on. These can help pinpoint

Table 6.1. Illustrative Components of a Soybean Production Control System

Critical point	Type of measurement	Sensor	Monitoring schedule	Control standards	Corrective actions
Soil moisture for planting	Attribute	Visual, feel, weather reports	Daily during planting season	Dry enough	Wait for soil to dry
Soil texture for planting	Attribute	Visual, feel	Before planting	Proper soil texture	Till to break up clods or level seed bed
Seed placement: distance and depth	Variable	Tape measure	Test before planting begins & then once a day	¾–1¼" between seeds & 1½–2" deep	Adjust planter to attain proper distance & depth
Seeding rate	Attribute	Planter monitor	Continuous	Lights flashing in expected pattern for desired population & the same for each row	Inspect planter for plugs, broken parts, & incorrect settings
Planting speed	Variable	Acres per day	Daily	65–75 acres per day	If weather is not the problem, check for correct speed & other problems
Weed population	Variable	Visual	Daily during May & June	Weed count below thresholds for treatment	Apply postemergence herbicide
Grain moisture at harvest	Variable	Moisture tester	Daily shortly before harvest begins	Moisture at or below critical level	Wait for soybeans to dry before harvesting
Harvest wastage	Variable	Visual count	Daily	Number of soybeans on ground behind combine below critical level	Inspect combine for proper settings / Verify proper speed of combine

174

Table 6.2. Illustrative Components of a Beef-Finishing Control System

Critical point	Type of measurement	Sensor	Monitoring schedule	Control standards	Corrective actions
Feeding silage: 12 lb./day	Variable	Scale	Mondays	11.5 to 12.5 lb/ head/day	Adjust amount delivered to desired level
Average daily gain	Variable	Scale	Every other Monday	1.5 to 1.7 lb/day	Inspect animal health, feed quality; calibrate scale
Water	Attribute	Visual	Daily	Waterers clean & working, fresh water	Clean waterers, fix waterers, & supply lines
Clean lot & sleeping area	Attribute	Visual & smell	Daily	No excessive manure buildup	Clean when manure reaches excessive level
Animal health	Attribute	Visual	Daily	No coughing, no slowness, no down animals, no sores, etc.	Inspect closer, call the vet

175

Table 6.3. Illustrative Fail-Safe Plan for the First 36 Hours after Receiving Feeder Yearlings

Item to be monitored	Method	Schedule	Control standard	Corrective actions, if needed		Done by: initials	day	time
Initial health	Visual, auditory	On arrival and at 8 hour intervals	Walking properly, active, no coughing	Call vet	at arrival / 8 hours / 16 hours / 24 hours / 36 hours	___ ___ ___ ___ ___	___ ___ ___ ___ ___	___ ___ ___ ___ ___
Water	Visual, float	Before arrival, at arrival, and at 2 hour intervals	Fresh water is available, waterer operating properly	Repair and replace parts as needed to restore water supply	before arrival / at arrival / 2 hours / 4 hours / (and so on)	___ ___ ___ ___	___ ___ ___ ___	___ ___ ___ ___
Old hay only for first 24 hours	Visual	Before and on arrival	No feed in bunks except old hay	Clean bunks & replace with hay	before arrival / at arrival	___ ___	___ ___	___ ___
Vet inspection	Visual	Within 6 hours	Healthy, active	Administer care as needed	within 6 hours	___	___	___
Feed silage	Visual	After 24 hours	Fresh silage	Feed if not done, repair equipment as needed	after 24 hours	___	___	___

176

Table 6.4. Check Sheet for Baby Pig Deaths

Cause of death	Week of: *April 3–9*	Total deaths
Stillborn	√√√√√√√	7
Diarrhea	√√√√√√√√√√√√√√√√√√√	19
Laid on	√√√	3
Unknown	√√√√√	5
	Total	34

Comments: *Need to use more complete cleaning methods after this farrowing.*   Initials:
                                                                                      RW

Table 6.5. Check Sheet for Corn Header Plugging

Row (from left as seen by combine driver)	Week of: *October 15–17*	Total plugs
1	√√	2
2	√	1
3		0
4		0
5	√√√√√√	6
6	√√	2
7	√	1
8	√√	2
	Total	14

Comments: *The right chain in row 5 seems looser than the others. What else could be wrong?*   Initials:
                                                                                      KD

causes and sources of problems. Having a written record helps focus resources on the problem rather than missing or trying to recall information. The written record is especially useful when multiple workers operate the machinery or work with livestock.

Trend or Run Charts

A trend or run chart is simply a running plot of measured quality characteristics: stored grain moisture, pigs born per litter, milk per cow per day, acres per day, bacteria or somatic cell counts, and so on. These characteristics can be good, bad, or neutral. These are characteristics that the manager decides need to be watched for trends that may suggest the need for corrective actions. We are not deciding if the process is in control, we are just watching how the process is working and if anything needs attention. Two examples of trend or run charts are stored grain temperatures (fig. 6.2) and market prices (fig. 6.3).

Trend charts can provide the data for data-based decisions rather than trying to remember what something has looked like before or how the measurement has changed over time. These real data can be especially useful for a multiperson situation.

Cause-and-Effect (CE) Diagrams

In process control, if measurement and inspection find that the process is out of control (i.e., a problem is present), the process is stopped and an assignable cause is searched for. Assignable causes are specific factors or things that can be identified as the cause of a problem and thus controllable. Assignable causes are different from common causes, which are random fluctuations and not controllable. For example, if the seed monitor indicates a problem in one row, planting should stop, the assignable cause searched for, and the problem solved. Searching for an assignable cause can be done with the help of cause-and-effect (CE) diagrams. Some causes, such as minor movements of the planter, may be deemed random and uncontrollable even though they can cause actual seed placement to differ slightly from planned placement.

CE diagrams are drawn with the problem or opportunity typically on the right and potential causes along the spine of the diagram. When developing a CE diagram, any potential cause can be listed. For organization and readability, the causes should be grouped into major categories. Categories can be added and deleted as appropriate for the business and situation being analyzed.

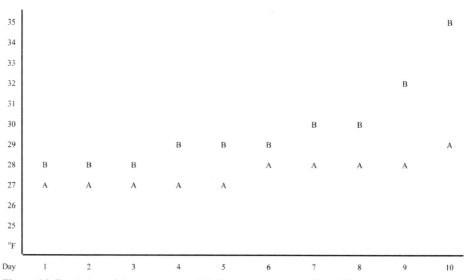

Figure 6.2. Trend chart of the temperature of shelled corn in two bins (A and B).

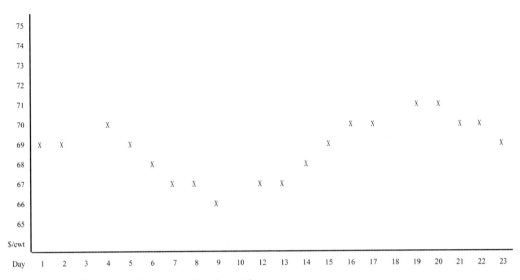

Figure 6.3. Trend chart of the price of market cattle.

For farming, five main categories to start with are workers, material, inspection, tools, and weather.

In agriculture, finding the assignable cause could lead to one of many potential changes. Labor may need training, replacement, a rest period, or a vacation. A machine may need fixing, adjusting, tuning, or replacing. The inputs or raw material may need to be replaced, reordered, or found from a new supplier. Animals may need to be treated for illnesses or even replaced. Even the inspection process may need adjustment in its frequency, standards, and methods. For farmers, weather may be the assignable cause of a problem. Here, since we cannot change the weather, the change would be an adjustment to the current weather conditions. For example, when applying chemicals, choosing a different pressure and nozzle may be needed to compensate for different wind conditions. Adjusting the ration mix for different seasons during the year to maintain the daily gain at the desired level is another example.

As a first example of a CE diagram, diagnosing a farm-earning problem can be drawn as a CE diagram (fig. 6.4). If the return to labor and management is lower than desired, a farmer would sort through the potential causes until the current causes are found and corrected. The major categories of causes are listed as workers, equipment, enterprises, weather, inspection, marketing, and resources. Some examples shown in the diagram are workers who have poor training, a wrong mix of enterprises, and being undercapitalized.

In another example, the potential causes of seeding rate and placement problems are drawn as a CE diagram (fig. 6.5). The planter is a tool that may

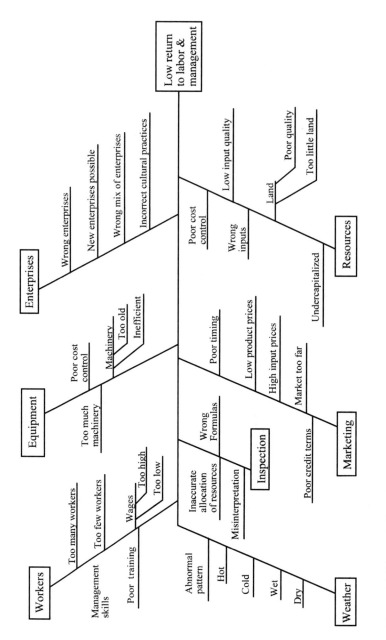

Figure 6.4. Diagnostic financial analysis drawn as a cause-and-effect diagram.

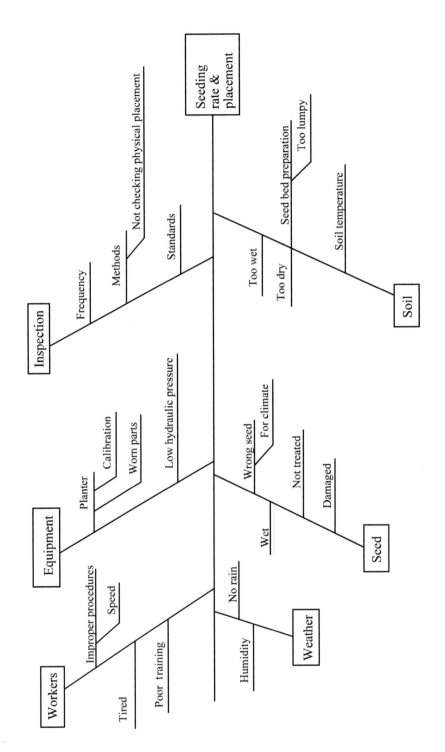

Figure 6.5. A cause-and-effect diagram for seeding rate and placement problems.

181

cause problems due to improper calibration or worn parts. The soil may be too wet, too dry, or not prepared properly. The seed itself may be too wet or damaged; it may be the wrong seed. The workers may be poorly trained in both operation of the planter and in inspection procedures. They may be tired and thus not watching the process properly. With some equipment and material, high humidity can cause problems.

At times, an obvious, major cause needs to be solved first. If no obvious causes can be seen quickly, potential causes can be ranked by frequency of occurrence, if known or inferred, or evaluated one by one to find the true cause of the problem (or opportunity for improvement). A check sheet, as described earlier, may be useful for choosing which causes should be addressed first.

While it may seem too obvious or too time-consuming to draw a CE diagram when a farmer encounters a problem, the CE diagram is a useful concept to learn for three reasons. First, a CE diagram can be used to step back from an incorrect fixation on a potential solution, review the symptoms of a problem, and find the true problem elsewhere in the CE diagram. By having this CE model in their mental toolkit, farmers may not even have to draw the physical diagram.

The second reason for using a CE diagram is the increasing complexity of the systems that farmers manage. Complex systems require a better working knowledge of the system to diagnose symptoms and problems. Farmers have to deal with several complex systems—even if the farmer is specializing in only one or two products. For example, the concepts involved with sustainable agriculture are complex and the concept of the CE diagram may help a farmer find alternatives to chemical practices.

The third reason for learning the concepts in a CE diagram is that it can be helpful in reducing variability. This is part of process improvement discussed in the next section. Reducing variability generally requires problem solving or changes in the design of the product or process itself. Reducing variability can mean lower costs, higher quality, higher prices, or any combination of these improvements.

TOOLS FOR PROCESS IMPROVEMENT

Process control, as discussed in the last section, is concerned with making sure the process (growing corn or hogs, for example) works as planned or is corrected during the process. Process improvement looks at a process in a different way: it is concerned with making a process better. A process does not have to be broken for improvement to be possible.

We may meet or even exceed customer expectations and the standards we have developed. Nevertheless, we sense we could do better. Through benchmarking against other farms or by watching our own farm by using some of the other tools listed below, we can see possible ways to improve both averages and variability of critical measures. This can include physical performance, cost of production, marketing results, and profitability.

Process improvement is implemented by following a set of six steps that are similar to those used for other aspects of quality management:

1. Define quality characteristics based on customer needs.
2. Decide how to measure each attribute.
3. Set quality standards.
4. Establish appropriate tests for each standard.
5. Find and correct causes of poor quality or ways to improve the process.
6. Continue to evaluate and make improvements.

A very useful classification of process-improvement tools uses these groupings: general tools, coarse-grained tools, and fine-grained tools (fig. 6.6). This classification can help a person choose which tool is the appropriate one to use in a specific situation.

General Tools

Team building and group interaction tools can improve how a group (of even two) can learn more about each other to improve trust, communication, and

GENERAL TOOLS
     1. Team building and group interaction tools
     2. Specific process and technology tools

COARSE-GRAINED TOOLS
     3. Process maps
     4. Benchmarking
     5. Check sheets and histograms
     6. Pareto analysis (and Pareto charts)
     7. Cause-and-effect charts (fishbone charts)

FINE-GRAINED TOOLS
     8. Fail-safe plans
     9. Trend (or run) charts
     10. Scatter diagram and correlation
     11. Experiments and trials

Figure 6.6. Process improvement tools for farmers.

thus quality. Quality circles are groups of workers or, say, neighboring farmers, who meet to discuss ways to improve products and processes.

Specific process and technology tools are anything specific that helps improve process output. An example of this idea is buying specialized equipment to improve crop yields such as one planter specifically designed for corn and a drill specifically designed for soybeans.

Coarse-Grained Tools

Process maps involve describing the process and asking "Why this way?" As described in chapter 5, process maps can increase our understanding of our current process and make us see obvious areas to improve and potential areas that require further questioning.

Benchmarking is the process of measuring a farm's performance on a key customer requirement against the best in the industry (or even wider than a specific industry). By benchmarking, a farmer can establish standards against which performance is judged and to identify models for learning how to improve. When benchmarking, we have four basic questions to answer: (1) What farm or group of farms does it best? (That is, meets or exceed customers' expectations best.) (2) How do they do it? (3) How do we do it now? (4) How can we change to match or exceed the best?

Check sheets are forms that merely require a check to be made in the appropriate spot to show the frequency of certain events that relate to quality or process characteristics. They can be used to show many things: the frequencies of error or problem sources, the variation of time required to do the same job, productivity differences, and so on. Two examples are causes of baby pig deaths (table 6.4) and frequency of corn header plugging by row (table 6.5).

*Pareto analysis (*and *Pareto charts)* organize the frequencies of causes from most to least frequent and thus show which problem or source to try to solve first. The data for these can come from a check sheet—pig deaths, for example (table 6.4).

*Cause-and-effect (CE) diagrams (*or *fishbone charts)* identify the major groupings of causes and then the specific causes and subcauses of observed and potential problems or areas that could be improved. CE diagrams can be extremely beneficial by showing the connections between causes and results and, thus, helping a farmer identify a potentially broader list of ways to improve the process. They are discussed under "Developing Process Control Systems" above.

Fine-Grained Tools

Fail-safe plans are best used on those parts of an operation that are very crucial to successful completion of the process and the business. This quality tool is a process and product design tool that looks for ways to eliminate the possibility of problems or mistakes occurring. One example of a fail-safe plan is the checklist and accompanying instructions concerning the jobs that need to be done to care for feeder yearlings in the first 36 hours after they arrive in the feedlot (table 6.3). The successful completion of these jobs is very crucial for the health of the cattle and the success of the farmer.

Trend (or run) charts are, as described in an earlier section, plots of measured quality characteristics. These charts provide a quantitative record on which better decisions can be made.

Experiments and trials are used when the sources of problems are unknown or not obvious; designing and using experiments in a controlled environment can help us understand how a process works; how, where, when, and why problems can occur; and how improvements could be made. On-farm experiments can be used to develop farm-specific data rather than using information from experiments that may have been done in different soil and climate environments.

Scatter diagrams and correlations are used to help decide whether events, things, and conditions are related. Data from experiments and checklists can be used to check on relationships and potential cause and effect relationships. Simple plots of pairs of data can show whether one event or result is related to another. Correlation is a statistical measure of two events' interrelatedness. The decision of whether this is cause and effect or simply correlation caused by other events depends on our knowledge of the process.

SUMMARY POINTS

- Quality management involves a holistic view of the entire farm and aims to develop a philosophy of improving quality in all functions.
- Quality control refers to controlling already selected processes for the products already chosen and designed.
- Process control consists of the procedures to monitor production for compliance with the original plan and development of corrective actions designed to bring the process back into compliance.
- Process improvement includes a set of tools used to understand the current process better and to look for potential ways to improve both the process and the product.
- Quality is meeting or exceeding customer requirements.
- The customer, not the producer, defines quality.
- Quality has many dimensions depending upon the customer. Price may be a determinant of quality for some customers.
- Product quality can be defined in four dimensions: quality of design, quality of conformance, the "abilities," and service after delivery.

- Process quality has external and internal dimensions. Externally, process quality ensures that the product will be made so that the product conforms to what the customer wants. Internally, process quality is concerned with the efficiency of the production process.
- The costs of quality include control costs (prevention and appraisal) and failure costs (internal and external).
- The "quality gurus" have provided new views on quality and many tools for improving quality.
- Total quality management (TQM) is a management philosophy that strives to involve everyone in a continual effort to improve quality and achieve customer satisfaction.
- TQM involves a set of steps and several ideas and concepts: continuous improvement, competitive benchmarking, employee empowerment, team approach, data-based decisions, knowledge of quality management tools, supplier quality, and quality at the source.
- The ISO 9000 family of standards is concerned with assuring consistent application of a business' chosen production process.
- A process control system consists of three parts: standards, measurement, and corrective actions.
- Preliminary controls are anticipatory actions that take place before problems are detected.
- Concurrent control occurs during the production process as problems are detected.
- Feedback control evaluates production and makes recommended changes after production is finished.
- Process control systems can be developed by following a series of five steps.
- A fail-safe plan is a process and product design tool that looks for ways to eliminate the possibility of problems or mistakes occurring.
- Check sheets are forms that merely require a check to be made in the appropriate spot to show the frequency of certain events that relate to quality or process characteristics.
- A trend or run chart is a running plot of measured quality characteristics.
- Cause-and-effect (CE) diagrams show the factors that affect a desired result and are used to search for assignable causes of detected problems.
- Benchmarking is the process of measuring performance against the best in the industry.
- On-farm experiments and trials can be used to develop local data to help solve problems and improve processes.

REVIEW QUESTIONS

1. What makes quality management different from quality control?
2. What is product quality?
3. Who defines product quality?
4. How is product quality different from process quality?
5. How can quality management be incorporated into the management of a farm?
6. Summarize Deming's 14 points for implementation on a farm.
7. How can a process map be useful in quality control and management?
8. What are the three types of process control?
9. What are the five steps in developing an enterprise control system?
10. Describe a corrective action.
11. Develop the control plan for one input in the planting subsystem for corn production. Start with the input to be monitored and end with the corrective action if control standards are violated.
12. Develop the control plan for one output in the milking subsystem for dairy production. Start with the output to be monitored and end with the corrective action if control standards are violated.
13. Develop a fail-safe plan that could be used by a farmer for a specific operation of your choice.
14. Develop a check sheet for potential sources of problems and errors for a specific operation on a farm of your choice.
15. Describe how a farmer could use a trend or run chart.
16. What is a cause-and-effect (CE) diagram and how can it be used in farm management?
17. How could a farmer benefit from benchmarking?
18. Why would a farmer decide to perform an on-farm experiment or trial?

7

Financial Analysis

How do we know what time it is? If a clock does not have any hands (or digits), we cannot tell what time it is even if the rest of the clock is running perfectly. If all we have is a clock with only the minute hand or digits, we do not know what time it is. If our clock has only the hour hand or digits, we may have a better idea of the time but we still do not know for sure. To be sure we know the correct time of day, we need both hands or all the digits on our clock and some "outside information" on whether it is day or night.

Similarly, we need some "hands" to tell what the "financial time" is on a farm. Even if the business is running well, we do not know how well it is running without a set of financial statements. Also, as with our clock, if we do not know how to interpret the statements (much less build them), we cannot analyze the financial position or performance of the farm. ***Financial position*** refers to the total resources controlled by a business and the claims against those resources. ***Financial performance*** refers to the results of decisions over time.

The purpose of any financial analysis is to find problems and opportunities, to develop potential solutions to those problems, and to develop plans to take advantage of those opportunities. However, what we often see first are symptoms and not actual problems or opportunities. For example, while low income is a big problem, it is not a problem that can be corrected directly. Low income is a symptom of a problem elsewhere in the farm business, such as improper animal nutrition, poor crop yields, poor marketing, or the wrong strategy. The first step is to find the symptoms. Then we explore and analyze the business in depth to find the problems that need to be solved.

Several points need to be remembered in analysis. First, the initial analysis should be used to find areas that need more analysis. We should not try to analyze everything and every part of the business in the initial analysis. Second, calculating every possible measure and ratio will only confuse the analyst and the manager. In this paper, the list of measures and ratios is kept short on pur-pose. In subsequent analyses, other measures and ratios may be calculated and used to help ferret out the problem whose symptom has been found in the first step of the analysis. Third, financial statements should be interpreted in the light of the history of a farm, the history of its environment, its current envi-ronment, its expected future, and its goals and objectives.

This chapter has four main sections. The financial statements are described in the first section. In the second section, the main measures of profitability, solvency, liquidity, repayment capacity, and financial efficiency are described. The last two sections contain procedures for an initial analysis of a farm busi-ness and for diagnostic analysis of problems and opportunities facing the farm business. The three sections of Appendix B at the back of this book relate to this chapter. The first describes several financial measures that are important either regionally, in some institutions, or historically but are not in the primary list of 16 measures. The second appendix explains the steps for taking an inventory of the farm's resources. The third appendix describes how to check the accuracy of farm records.

UNDERSTANDING FINANCIAL STATEMENTS

The three main financial statements are the income statement, balance sheet, and cash flow statement. These financial statements allow us to calculate a business' profitability, solvency, and liquidity. They record (or project) the progress toward a person's and business' goals.

For general businesses, certified public accountants (CPAs) are required to follow a very structured set of accounting procedures known as the Generally Accepted Accounting Principles (GAAP). Historically, the U.S. Congress has exempted agricultural producers below a certain sales level from following GAAP. Consequently, in agriculture, the format of these statements and how certain elements are calculated has varied between institutions and regions of the United States and over time. Due to the concerns caused by the farm debt crisis of the 1980s, the Agricultural Bankers Division of the American Bankers Association initiated a discussion to standardize financial reporting and finan-cial analysis for agricultural producers. This discussion has evolved into the Farm Financial Standards Council (FFSC) and its financial guidelines. Anoth-er educational effort used extensively in the United States and in other coun-tries involves the software program, FINPACK, produced by the Center for Farm Financial Management (CFFM) at the University of Minnesota. This section describes the financial statements developed by both CFFM and FFSC, since they are both widely used and known.

Income Statement

The income statement (or profit/loss statement) shows the difference between the gross income and the costs incurred to produce that income. The income statement measures the *flow* of income for a certain *period*. It is used to measure profitability. It includes both cash and noncash income and expenses and usually covers a period of one year. Since it includes depreciation and amortization of capital assets, the income statement does not include transactions such as machinery purchases or sales and loan principal payments or receipts.

Farmers have a choice of two ways to keep their accounts: cash accounting and accrual accounting. With a cash accounting system, income and expenses are recorded when received or paid (or considered being received or paid). The exception to this is the money paid for machinery, buildings, and other capital assets depreciated over a period of years. Although all income and expenses are considered and counted eventually, the cash system does not accurately measure income attributable to a specific year or production period because it does not include changes in inventory. With an accrual accounting system, transactions are recorded as production occurs or expenses are committed— regardless of whether it is received or paid (or considered being received or paid). Because the accrual accounting system includes changes in inventory, the accrual system more accurately measures income generated within one year. The accrual system is better for evaluating business results and plans and is the system used in the examples in this section.

In the example income statement using the CFFM format (fig. 7.1), gross cash farm income is $563,461 from the sales of products and other income sources and total cash farm expenses are $417,181. Net cash farm income is the difference between gross cash farm income and total cash farm expenses or $146,280. The net inventory change is an increase of $72,273. Although it is shown as a negative number on the income statement, depreciation and other capital adjustments are usually referred to as a positive $62,836 (without accounting for changes in real estate). Net farm income (NFI) before real estate changes is $155,717. This is an accrual estimate of NFI since the net cash income is adjusted for changes in inventories plus for depreciation and other capital adjustments. Net nonfarm income is $0 for this farm. Proprietor withdrawals (or family living expense) are $68,016, so the addition to retained earnings is $87,701. If the change in real estate value is included, net farm income (with real estate) is $155,717 and retained earnings are $87,701. However, since the real estate values are cost basis values (more on that in the next section), these figures are not adjusted for any taxes that the owner may owe for income or when the real estate is sold.

FARM OPERATING RECEIPTS		FARM OPERATING EXPENSES	
Crop Sales:		Seed	$38,164
Corn sales	$133,063	Fertilizer	43,794
Soybean sales	88,143	Crop chemicals	42,007
		Crop insurance	6,556
		Drying fuel	2,123
		Crop miscellaneous	894
		Purchased feed	45,133
Livestock & Livestock Products:		Feeder livestock bought	12,303
Market hog sales	213,121	Veterinary	9,378
Cull breeding stock	0	Livestock supplies	14,452
		Interest	26,967
Other Cash Income:		Fuel and oil	9,503
Government program payments	52,111	Repairs	36,549
Contract hog income	27,689	Custom hire	12,934
Insurance income	0	Hired Labor	16,953
Miscellaneous	49,334	Land rent	59,690
		Real estate taxes	16,521
		Farm insurance	3,508
		Utilities	6,622
GROSS CASH FARM INCOME	563,461	Trucking and Marketing	228
		Miscellaneous	12,902
		TOTAL CASH FARM EXPENSE	417,181
		NET CASH FARM INCOME	146,280

ADJUSTMENTS FOR CHANGES IN INVENTORY

	Crop & Feed	Market Livestock	Accounts Rec'ble	Supplies & Prepaid Expenses		Accounts Payable	
Ending Inv.	229,274	158,012	12,037	59,154	Begin	27,485	
Beginning Inv.	214,117	118,527	7,932	51,846	End	21,267	
Net Adjustment	15,157	39,485	4,105	7,308		6,218	
						TOTAL:	72,273

DEPRECIATION AND OTHER CAPITAL ADJUSTMENTS

		Breeding Livestock	Mach. & Equip.	Bldgs & Imprv'ts	Other Assets	Real Estate	
Ending Inventory	(+)	0	94,621	199,423	9,816	406,570	
Capital sales	(+)	0	2,127	0	0	0	
Beginning inventory	(-)	0	99,200	199,089	10,109	406,570	
Capital purchases	(-)	0	37,167	9,908	13,350	0	
Depr. & cap. adj.	(=)	0	-39,619	-9,574	-13,643	0	
				TOTAL (without Real Estate)			-62,836

NET FARM INCOME	155,717
NET NONFARM INCOME	0
TOTAL NET INCOME	155,717
PROPRIETOR WITHDRAWAL	68,016
ADDITION TO RETAINED EARNINGS	87,701
NET FARM INCOME (with real estate changes)	155,717
ADDITION TO RETAINED EARNINGS (with real estate)	87,701

Figure 7.1. Income statement using CFFM's FINPACK format.

To move agricultural producers to a standard format closer to GAAP, the FFSC recommends using one of two formats: the gross revenue format (fig. 7.2) or the value of farm production (VFP) format (fig. 7.3). For this example farm, both procedures and formats recommended by FFSC calculate the same accrual net farm income ($155,717) as the procedures followed by CFFM.

CROP CASH SALES	221,206	
+/- Change in Crop Inventories	15,157	
Gross Revenues from Crops		236,363
MARKET LIVESTOCK/POULTRY CASH SALES	213,121	
+/- Change in Inventories	39,485	
Gross Revenues from Market Livestock/Poultry		252,606
Livestock Products		0
Government Programs		52,111
Gain/Loss from Sale of Culled Breeding Livestock		0
+/- Change in Value Due to Change in Quantity of Raised Breeding Livestock		0
Crop Insurance Proceeds		0
Change in Accounts Receivable		4,105
Other Farm Income		77,023
GROSS REVENUES		622,208
FARM OPERATING EXPENSES		
Purchased Feed/Grain	45,133	
Purchased Market Livestock	12,303	
Other Cash Operating Expenses	332,778	
+/- Accrual Adjustments	-13,526	
Depreciation/Amortization Expense	62,836	
TOTAL OPERATING EXPENSES		439,524
Cash Interest Paid	26,967	
+/- Change in Interest Payable	0	
Total Interest Expense		26,967
TOTAL EXPENSES		466,491
NET FARM INCOME FROM OPERATIONS		155,717
Gain/Loss on Sale of Farm Capital Assets		0
Gain/Loss Due to Changes in General Base Value of Raised Breeding Livestock		0
NET FARM INCOME, ACCRUAL ADJUSTED		155,717
Miscellaneous Revenue	0	
Miscellaneous Expense	0	
Total Miscellaneous Revenue and Expense		0
INCOME BEFORE TAXES AND EXTRAORDINARY ITEMS		155,717
Cash Income Tax Expense	0	
+/- Change in Income Tax Accruals	0	
+/- Change in Current Portion of Deferred Taxes	0	
+/- Change in Non-Current Portion of Deferred Taxes	0	
Total Income Tax Expense (Farm Business Only)		0
INCOME BEFORE EXTRAORDINARY ITEMS		155,717
Extraordinary Items (Net of Tax)		0
NET INCOME, ACCRUAL ADJUSTED		155,717

Figure 7.2. Statement using FFSC's gross revenue format.

However, several items are grouped differently in the FFSC formats compared with the CFFM format (fig. 7.1). The differences between the two FFSC formats are in the treatment of items purchased and held for sale later—feeder cattle, for example. In the gross revenue format, the costs of these items are listed with all other operating costs. In the VFP format, the costs of these items

CROP CASH SALES	221,206	
+/- Change in Crop Inventories	15,157	
Gross Revenues from Crops		236,363
MARKET LIVESTOCK/POULTRY CASH SALES	213,121	
+/- Change in Inventories	39,485	
Gross Revenues from Market Livestock/Poultry		252,606
Livestock Products		0
Government Programs		52,111
Gain/Loss from Sale of Culled Breeding Livestock		0
+/- Change in Value Due to Change in Quantity of Raised Breeding Livestock		0
Crop Insurance Proceeds		0
Change in Accounts Receivable		4,105
Other Farm Income		77,023
GROSS REVENUES		622,208
Less:		
Purchased Feed/Grain		45,133
Purchased Market Livestock		12,303
VALUE OF FARM PRODUCTION		564,772
Other Cash Operating Expenses	332,778	
+/- Accrual Adjustments	-13,526	
Depreciation/Amortization Expense	62,836	
TOTAL OPERATING EXPENSES		382,088
Cash Interest Paid	26,967	
+/- Change in Interest Payable	0	
Total Interest Expense		26,967
TOTAL EXPENSES		409,055
NET FARM INCOME FROM OPERATIONS		155,717
Gain/Loss on Sale of Farm Capital Assets		0
Gain/Loss Due to Changes in General Base Value of Raised Breeding Livestock		0
NET FARM INCOME, ACCRUAL ADJUSTED		155,717
Miscellaneous Revenue	0	
Miscellaneous Expense	0	
Total Miscellaneous Revenue and Expense		0
INCOME BEFORE TAXES AND EXTRAORDINARY ITEMS		155,717
Cash Income Tax Expense	0	
+/- Change in Income Tax Accruals	0	
+/- Change in Current Portion of Deferred Taxes	0	
+/- Change in Non-Current Portion of Deferred Taxes	0	
Total Income Tax Expense (Farm Business Only)		0
INCOME BEFORE EXTRAORDINARY ITEMS		155,717
Extraordinary Items (Net of Tax)		0
NET INCOME, ACCRUAL ADJUSTED		155,717

Figure 7.3. Income statement using FFSC's value of production format.

are subtracted from gross revenue (before other costs are subtracted) to estimate the value added by the farm after accounting for purchase price of these items.

Both FFSC formats estimate gross revenue to be $622,208. This is CFFM's gross income adjusted for changes in the inventories of crops and livestock (except for changes in the base value of raised livestock). In the VFP format, the value of farm production ($564,772) shows the value added by the farm after accounting for the purchases of feeder livestock and feed and grain for feed. The VFP format is preferred by people who want to see the contribution of a farm versus the sales of a farm.

Besides the obvious difference in their visual format, FFSC procedures are different from CFFM procedures. The first difference is the treatment of inventory change. CFFM calculates the net cash farm income and then adjusts for inventory changes. FFSC adjusts for inventory changes at the point that sales and expenses occur. For example, changes in crop inventories are listed right after crop sales in the FFSC income statements. Thus, the FFSC gross revenue accounts for the value of sales from the period in the statement; that is, the beginning value of crops produced last year and carried into the current year is subtracted from this year's sales to provide a more accurate figure for the current year. Change in expense inventories is treated similarly within the list of expenses. Also, several individual items in the CFFM format are listed as totals in the two FFSC statements.

Another difference between FFSC and CFFM is in the treatment of changes in asset values. FFSC recommends that valuation changes due to changes in the number of animals be treated as either positive or negative revenue. This treatment is recommended because a change in the number of animals is not a capital gain or loss. A change in the base value (price per animal, for example) creates a capital gain or loss and FFSC lists that separately for accounting reasons and for taxation reasons.

The FFSC income statements also include current and deferred income taxes and the value of extraordinary items after taxes. Extraordinary items are those events or transactions that are both unusual in nature (i.e., not part of the usual and customary activity of the business) and not reasonably expected to happen again in the foreseeable future. FFSC's 1997 report contains more information on how these statements are developed and how they differ from GAAP.

A farmer's choice of format depends on the preferences of the farmer and his or her creditor and accountant. The FFSC statements more closely follow GAAP while the CFFM statements do not. For farms with simpler information and reporting requirements, the CFFM statements may be sufficient.

Balance Sheet

The balance sheet (also called the financial or net worth statement) shows the assets and the liabilities of a business on a specific date. The balance sheet measures the *stock* of assets and liabilities of a business at a certain *point* of time. An *asset* is an item owned or controlled by a person or business for holding wealth or producing income. An asset can be a tangible item, such as land, machinery, or cash, or it can be an intangible item such as a patent or knowledge.

The balance sheet is used to measure solvency and solvency-based liquidity. Solvency is the long-run ability to meet financial commitments. The most common measure of solvency is net worth. Liquidity is the short-run ability to meet financial commitments and to cope with unexpected financial needs. Solvency-based liquidity is liquidity calculated from the balance sheet; it does not consider income flows, just short-term assets versus short-term liabilities.

The difference between the net worth statement and the balance sheet is only slight and comes from the accounting procedure used to develop the information. Decades ago, before computers and even before electronic calculators, accountants kept their handwritten tallies of income, expenses, assets, and liabilities in separate ledgers. Since values were entered twice, once in two different ledgers, the system was called double-entry accounting. When financial statements were prepared, the accountants would compare their hand-calculated total asset values with the sum of separately calculated totals of liabilities and equities. If the comparison showed that the totals were equal, the accounts were said to be "in balance," hence, the name "balance sheet" for the statement showing assets, liabilities, and equity. Today, computers allow us to keep our tallies and totals differently, but the name has stayed.

Assets

In the CFFM format, assets are classified traditionally as current, intermediate, or long-term. Current assets are unrestricted cash, assets that can be quickly converted to cash, or are normally sold by the business within 12 months. Examples of current assets include cash, checking accounts, grain in storage, livestock held for sale, and accounts receivable. Intermediate assets are items held and used for one to ten years. Examples of intermediate assets include machinery, breeding livestock, equipment, and retirement accounts. Long-term assets are held and used for more than ten years. Common long-term assets are land and buildings.

The FFSC recommends that assets be classified as either current or noncurrent. Current assets are unrestricted cash or assets that will be converted into cash or consumed in the normal course of business. All other assets have a useful life greater than one year, are usually not purchased for resale, and are used in the production of salable products or services. The FFSC does not see a need to differentiate between intermediate and long-term assets since most financial questions involve only the distinction between current and noncurrent classes.

Methods of Asset Valuation

Since different asset valuation methods can result in very different asset values, the analyst has to know which methods were used in order to interpret the balance sheet information and, thus, the true financial position of a farm correctly. The biggest distinction between methods is the choice between using a cost-less-depreciation basis or a current market value basis. These are commonly called "cost basis" and "market value." The names are very literal. The **cost basis** value is the original basis of the asset minus any depreciation taken since it was purchased. The original basis is the cash paid for the asset adjusted for the cost basis of any assets traded as part of the purchase. The cost basis is also called the "book value." GAAP prefers the cost less depreciation since the cost and depreciation can be obtained objectively. The **market value** is the value of the asset on the open market minus any selling commissions. The differences between cost basis and market basis values are typically larger for assets purchased several years before the preparation of the balance sheet. For example, the market value of land will be closer to the purchase price of land bought last year than to the purchase price of land bought 15 years ago.

Both cost basis and market values are required for a proper analysis of a business' borrowing capacity and financial position. GAAP allows the presentation of market values as supplemental information, but its accuracy cannot be commented on. The FFSC recommends that when both market value and cost information are available, one of two formats can be used: either a balance sheet with market values in the figure and cost information supplied as supplemental information or a double-column balance sheet. To avoid potential confusion and interpretation problems, a double-column balance sheet that includes both cost and market values is used in this book.

The basic methods of asset valuation are:

- Net selling price is market price less transaction costs. This method is recommended for assets normally sold within one year (slaughter cattle, grain, etc.) or assets that could be easily sold (machines, breeding stock, etc.).
- Cost is the amount paid for an asset. This is recommended for supplies such as feed, fuel, and fertilizer. Growing crops can be valued based on the costs incurred with any needed adjustments if the crop were damaged. To be valid, specific identification of the asset is needed; otherwise, an average price should be used. For supplies stored and replenished throughout the year, an inventory control system needs to be chosen. FIFO (first in–first out) assumes items purchased first are used first; thus, items purchased most recently are in inventory and the most recent prices are used to value the inventory. LIFO (last in–first out) assumes items purchased most recently are used first; thus, the inventory consists of items purchased earlier so use

earlier prices. In most situations, FIFO is the recommended method to keep values close to the current market conditions.

- Purchase cost less depreciation is the new or original cost less accumulated depreciation. This cost basis method is recommended for long-lived assets during stable prices, such as machinery or breeding stock.
- Replacement cost less depreciation is used to estimate owner equity. It is recommended for machinery and breeding stock during periods of inflation or deflation.
- Replacement cost for equivalent function less depreciation also is used to estimate owner equity in assets that are now used for other than the original purchase. It considers the product value of assets. This method is recommended for old buildings remodeled for new uses, an old barn converted to hog finishing, for example.
- Income capitalization is used to convert a stream of income into the value of an asset at a particular point in time. This is used for land or buildings. It can be compared with the estimated market value to determine the proportion of the market value that is coming from appreciation versus income generation. It can also be used to evaluate whether the market price is incorrect.

Estimating Asset Value

The recommended methods of valuation for estimating asset value and equity for different types of assets are

- Current assets to be sold within one year should be valued at their expected selling price minus selling costs.
- Supplies should be valued at their original cost.
- Growing annual crops should be valued at the incurred cash costs adjusted for any damage.
- Machinery, breeding stock, and other intermediate assets should be valued using either the *cost basis method* (their original basis at purchase less any depreciation taken) or the *market value method* (their expected selling price minus selling costs).
- Real estate improvements and other long-term depreciable assets should be valued using either the *cost basis method* (their original basis at purchase less any depreciation taken) or the *market value method* (their expected selling price minus selling costs).
Another seldom-used method is *replacement cost less a depreciation amount* reflecting the age of the asset or replacement cost for equivalent function less depreciation.
- Farmland should be valued using (1) the *cost basis method* (the original basis less any depreciation taken on associated improvements), (2) the *market value method* (the expected selling price minus selling costs), or (3) the

income capitalization value. The last two methods are described in chapter 10, "Land Purchase and Rental."

Liabilities

Liabilities are considered claims on the assets by outside entities (e.g., banks). Liabilities are classified in the same way as assets. If a debt is due within 12 months, it is a current liability. Operating notes is a common name for loans held less than 12 months and used to finance the current year's operations. Accounts payable, accrued interest, and government crop loans are current liabilities. Any loan payment due within the next 12 months is classified as a current liability even if the loan is a multiyear or term loan. In figure 7.4 the "current portion of term debt" of $37,090 is listed as a current debt and the remaining portion of the total term debt is included as the "Noncurrent portion— notes payable" in the intermediate liability category ($83,629) and "Real estate debt deferred" in the long-term liability category ($131,866). In CFFM's format, loans for intermediate assets such as machinery are classified as intermediate liabilities, and loans for long-term assets, such as real estate, as long-term liabilities. In FFSC's format, loans for noncurrent assets are noncurrent liabilities.

When developing a "market value" balance sheet, the potential taxes that would have to be paid when the asset was sold are listed as deferred liabilities. These are called capital gains taxes when they actually do occur. Since the assets have not been sold, these potential taxes are thought of as being deferred to a future year and thus called "deferred taxes" on market value balance sheets. They are classified as intermediate or long-term liabilities based on whether the assets are intermediate or long-term assets. Deferred taxes are calculated as a percentage of the difference between the market basis and cost basis values of the assets. The percentage is the effective tax rate that the asset owner is estimated to incur if the asset were sold.

Net Worth or Equity

A business' *net worth* (or equity) is the difference between total assets and liabilities. In contrast to liabilities, equity is considered a claim on the assets by the business itself (e.g., the owners). The cost basis net worth is the difference between the original cost of the assets (less any depreciation) and the debts held against those assets. The market value net worth is the difference between what the assets could be sold for on the market and the debts held against those assets and deferred taxes due to the sale of those assets. In other words, the market value net worth is how much money the owner would have if all assets were sold and all debts and deferred taxes were paid.

Changes in the cost basis net worth over time are due to accumulated earnings. These earnings may be used to pay off debt, buy assets, or just accrue as savings. Increases in the cost basis net worth are used to show the earning power of a farm business and the ability of the manager to accumulate those earnings.

Changes in the market value net worth over time are due both to accumulated earnings and to changes in the asset market. An increase in the market value of a farm is good for the farmer-investor, but the market value increase does not signal a profitable operation. Increases in market value net worth may disguise operating problems. Increases in land prices may increase the market value of a farm if the farm has not suffered an operating loss that offsets the market gain. If the market values are based on good valuations done by trusted appraisers, the market value balance sheet can be very useful in obtaining additional loans for expansion.

FARM ASSETS			FARM LIABILITIES		
CURRENT ASSETS	Cost Basis	Market Value	CURRENT LIABILITIES	Cost Basis	Market Value
Cash and checking account	5,000	5,000	Accrued interest	0	0
Hedging accounts	0	0	Accounts Payable	21,267	21,267
Crops held for sale or feed	229,274	229,274	Notes Due Within One Year	44,518	44,518
Market livestock held for sale	158,012	158,012	Income Taxes Payable	0	0
Crops under government loan	0	0	Current Portion of Term Debt	37,090	37,090
Accounts receivable	12,037	12,037	Other Accrued Expenses	0	0
Prepaid expenses and supplies	59,154	59,154	Government crop loans	0	0
Other current assets	0	0	Other Current Liabilities	0	0
TOTAL CURRENT ASSETS	463,477	463,477	TOTAL CURRENT LIABILITIES	102,875	102,875
INTERMEDIATE ASSETS			INTERMEDIATE LIABILITIES		
Breeding livestock	0	0	Non-Current Portion - Notes Payable	83,629	83,629
Machinery, equip., & vehicles	94,621	322,992			
Other intermediate assets	9,816	27,215	Intermediate Portion - Deferred Taxes	xx	68,816
TOTAL INTERM. ASSETS	104,437	350,207	TOTAL INTERM. LIABILITIES	83,629	152,445
LONG TERM ASSETS			LONG TERM LIABILITIES		
Real Estate	406,570	665,570	Real Estate Debt deferred	131,866	131,866
Buildings & Improvements	199,423	281,742			
Other long-term assets	0	0	Long Term Portion - Deferred Taxes -----	xx	95,569
TOTAL LONG-TERM ASSETS	605,993	947,312	TOTAL LONG-TERM LIABILITIES	131,866	227,435
TOTAL BUSINESS ASSETS	1,173,907	1,760,996	TOTAL BUSINESS LIABILITIES	318,370	482,755
			FARM NET WORTH	855,537	1,278,241

Figure 7.4. Ending balance sheet using CFFM's format. (*Continued on next page*).

In the example balance sheets shown here, current farm assets are estimated to be $463,477 and current farm liabilities are estimated to be $102,875 using both CFFM's format (fig. 7.4) and FFSC's format (fig. 7.5). Current assets and current liabilities will have the same values with both cost basis and market value methods because the methods of estimating values are the same in both systems. In either format, the total farm assets have a total value of $1,173,907 using a cost basis and a value of $1,760,996 using a market value. The two values are different because, for noncurrent assets, cost basis valuation methods are different from the methods used in determining market values. Total farm liabilities are $318,370 using the cost basis method. After the deferred taxes are added in the market value column, total farm liabilities are $482,755. From these estimates, farm net worth is estimated to be $855,537 using the cost basis method and $1,278,241 using market values.

PERSONAL ASSETS			PERSONAL LIABILITIES		
CURRENT PERSONAL ASSETS	Cost Basis	Market Value	CURRENT PERSONAL LIABILITIES	Cost Basis	Market Value
Cash & checking account	4,484	4,484	Accrued interest	0	0
Savings accounts	0	0	Accounts payable	0	0
Readily marketable securities	7,001	7,001	Current notes	3,572	3,572
Other current assets	5,900	5,900	Income tax due	0	0
			Principal due on term debt	0	0
TOTAL CURRENT ASSETS	17,385	17,385	TOTAL CURRENT LIABILITIES	3,572	3,572
INTERMEDIATE ASSETS			INTERMEDIATE LIABILITIES		
Retirement Accounts	19,597	19,597	Interm. Term Debt- Deferred Portion	0	0
Cash Value of Life Insurance	10,784	10,784	Other Intermediate Liabilities	0	0
Nonfarm Equipment	5,487	5,487			
Household Goods	7,615	7,615			
Other Intermediate Assets	7,470	7,470	Intermediate Portion - Deferred Taxes	xx	0
TOTAL INTERMEDIATE ASSETS	50,953	50,953	TOTAL INTERMEDIATE LIABILITIES	0	0
LONG TERM ASSETS			LONG TERM LIABILITIES		
Nonfarm Real Estate	36,023	40,562	Long Term Debt - Deferred Portion	0	0
Other Long Term Assets	15,683	17,926	Long Term Portion - Deferred Taxes	xx	1,899
TOTAL LONG-TERM ASSETS	51,706	58,488	TOTAL LONG-TERM LIABILITIES	0	1,899
TOTAL PERSONAL ASSETS	120,044	126,826	TOTAL PERSONAL LIABILITIES	3,572	5,471
			PERSONAL NET WORTH	116,472	121,355

TOTAL FARM & PERSONAL ASSETS			TOTAL FARM & PERSONAL LIABILITIES		
	Cost Basis	Market Value		Cost Basis	Market Value
TOTAL ASSETS	1,293,951	1,887,822	TOTAL LIABILITIES	321,942	488,226
			TOTAL NET WORTH	972,009	1,399,596

Figure 7.4. (*Continued.*)

FARM ASSETS			FARM LIABILITIES		
	Cost Basis	Market Value		Cost Basis	Market Value
Cash and checking account	5,000	5,000	Accounts Payable	21,267	21,267
Hedging accounts	0	0	Notes Due Within One Year	44,518	44,518
Crops held for sale or feed	229,274	229,274	Current Portion of Term Debt	37,090	37,090
Market livestock held for sale	158,012	158,012	Accrued interest	0	0
Crops under government loan	0	0	Income Taxes Payable	0	0
Accounts receivable	12,037	12,037	Current Portion -- Deferred Taxes	0	0
Prepaid expenses and supplies	59,154	59,154	Other Accrued Expenses	0	0
Cash Investment in Growing crops	0	0	Government crop loans	0	0
Other current assets	0	0	Other Current Liabilities	0	0
TOTAL CURRENT ASSETS	463,477	463,477	TOTAL CURRENT LIABILITIES	102,875	102,875
Breeding livestock	0	0	Non-Current Portion - Notes Payable	83,629	83,629
Machinery, equip., & vehicles	94,621	322,992	Non-Current Portion - Real Estate Debt	131,866	131,866
Investments in Capital Leases	0	0	Non-Current Portion - Deferred Taxes	xx	164,385
Investments in Other Entities	0	0	Other Non-Current Liabilities	0	0
Investments in Cooperatives	0	0			
Real Estate	406,570	665,570			
Buildings & Improvements	199,423	281,742			
Other Assets	9,816	27,215			
TOTAL NON-CURRENT ASSETS	710,430	1,297,519	TOTAL NON-CURRENT LIABILITIES	215,495	379,880
			TOTAL BUSINESS LIABILITIES	318,370	482,755
			FARM BUSINESS EQUITY (Retained Capital + Valuation Equity)	855,537	1,278,241
TOTAL BUSINESS ASSETS	1,173,907	1,760,996	TOTAL BUSINESS LIABILITIES AND OWNER EQUITY	1,173,907	1,760,996

Figure 7.5. Balance sheet using FFSC's format. (*Continued on next page.*)

Since personal and business finances are closely related on many farms, the balance sheet is often split between farm and personal areas. This split allows the manager to keep a better distinction between business and family assets and liabilities and still see the entire financial position of the family. Personal net worth for this example farm is $116,472 on a cost basis and $121,355 on a market value basis. Adding the farm business and personal information together, the total net worth of this example farm owner is $972,009 on a cost basis and $1,399,596 on a market value basis.

Cash Flow Statement

The cash flow statement shows the annual flow and timing of cash in and out of a business. It is used to estimate cash flow liquidity and to show credit needs and loan payment ability. A cash flow statement using the CFFM format is

PERSONAL ASSETS			PERSONAL LIABILITIES		
	Cost Basis	Market Value		Cost Basis	Market Value
Cash & checking account	4,484	4,484	Accrued interest	0	0
Savings accounts	0	0	Accounts payable	0	0
Readily marketable securities	7,001	7,001	Current notes	3,572	3,572
Other current assets	5,900	5,900	Income tax due	0	0
	0	0	Principal due on term debt	0	0
TOTAL CURRENT ASSETS	17,385	17,385	TOTAL CURRENT LIABILITIES	3,572	3,572
Retirement Accounts	19,597	19,597	Non-Current Portion - Notes Payable	0	0
Cash Value of Life Insurance	10,784	10,784	Non-Current Portion - Real Estate Debt	0	0
Nonfarm Equipment	5,487	5,487	Non-Current Portion - Deferred Taxes	xx	1,899
Household Goods	7,615	7,615	Other Non-Current Liabilities	0	0
Nonfarm Real Estate	36,023	40,562		0	0
Other Non-Current Assets	23,153	25,396		0	0
TOTAL NON-CURRENT ASSETS	102,659	109,441	TOTAL NON-CURRENT LIABILITIES	0	1,899
TOTAL PERSONAL ASSETS	120,044	126,826	TOTAL PERSONAL LIABILITIES	3,572	5,471
			PERSONAL NET WORTH	116,472	121,355

TOTAL FARM & PERSONAL ASSETS			TOTAL FARM & PERSONAL LIABILITIES		
	Cost Basis	Market Value		Cost Basis	Market Value
TOTAL ASSETS	1,293,951	1,887,822	TOTAL LIABILITIES	321,942	488,226
			TOTAL NET WORTH	972,009	1,399,596

Figure 7.5. (*Continued.*)

shown in figure 7.6 for the example farm used earlier. Using the same example, FFSC's statement of farm business cash flows is shown in figure 7.7.

A cash flow statement includes only and any cash transactions for the year. Cash transactions are included even if they are due to the activity in another year. For example, a cash transaction may be the sale of grain produced in the previous year and prepaid expenses for the next year besides income and expenses for the current year. It includes cash transactions such as the receipt of loans and the payment of loan principal. These are not included in the income statement. The cash flow statement does not include noncash items such as depreciation, milk from the farm's dairy cows fed to the farm's calves, or corn grown on the farm and fed to livestock on the farm.

A typical cash flow statement (such as fig. 7.6) has four major sections: sources of cash, uses of cash, a flow of funds summary, and the loan balances. Sources of cash, or cash receipts, include cash receipts from crops and livestock; sale of capital items; reduction in savings, stocks, and so on; and other

	Jan.-Mar.	Apr.-Jun.	Jul.-Sept.	Oct.-Dec.	ANNUAL
CASH RECEIPTS					
Corn sales	58,032	14,532	0	60,499	133,063
Soybean sales	35,820	0	0	52,323	88,143
Market hog sales	51,692	48,360	51,537	61,532	213,121
Government payments	28,537	5,069	2,561	15,944	52,111
Contract hog income	8,059	5,327	9,564	4,739	27,689
Other farm income	0	22,826	0	26,508	49,334
Capital Sales:					
Breeding Stock	0	0	0	0	0
Machinery	0	2,127	0	0	2,127
Total cash receipts	182,140	98,241	63,662	221,545	565,588
CASH OUTFLOW	Jan.-Mar.	Apr.-Jun.	Jul.-Sept.	Oct.-Dec.	ANNUAL
Operating Expenses					
Seed	19,582	1,690	0	16,892	38,164
Fertilizer	25,106	9,630	0	9,058	43,794
Crop chemicals	20,387	0	0	21,620	42,007
Crop insurance	6,556	0	0	0	6,556
Drying fuel	0	0	0	2,123	2,123
Crop miscellaneous	0	894	0	0	894
Purchased feed	12,680	13,422	10,954	8,077	45,133
Feeder livestock bought	0	0	12,303	0	12,303
Veterinary	846	472	5,806	2,254	9,378
Livestock supplies	938	2,599	6,335	4,580	14,452
Fuel and oil	1,154	2,865	1,860	3,624	9,503
Repairs	17,608	6,433	2,585	9,923	36,549
Custom hire	0	0	0	12,934	12,934
Hired Labor	3,960	3,960	3,960	5,073	16,953
Land rent	35,846	0	0	23,844	59,690
Real estate taxes	8,260	0	0	8,261	16,521
Farm insurance	0	0	0	3,508	3,508
Utilities	1,935	1,438	1,118	2,131	6,622
Trucking and Marketing	0	0	0	228	228
Miscellaneous	4,526	2,080	3,058	3,238	12,902

Figure 7.6. Quarterly cash flow statement (CFFM). (*Continued on next page.*)

sources. Nonfarm income is included as a source of cash if it is counted as part of the cash flow supporting the farm. Uses of cash, or cash outflow, include farm operating expenses; capital purchases; proprietor withdrawals or dividends for corporations; principal and interest payments; and the ending cash. The first section of the flow of funds summary shows the calculation of the cash difference, which is the beginning cash balance plus the total cash receipts minus the total cash outflow. If the cash difference is not sufficient to meet a minimum desired ending cash balance and pay any scheduled payments for principal and interest on operating loans, new borrowings are needed and entered as appropriate. For example, the cash difference of $5,591 in the second quarter in figure 7.6 is not sufficient to meet the desired ending

	Jan.-Mar.	Apr.-Jun.	Jul.-Sept.	Oct.-Dec.	ANNUAL
Capital purchases					
Breeding livestock	0	0	0	0	0
Machinery & equipment	0	37,167	0	0	37,167
Buildings & implements	9,908	0	0	0	9,908
Other assets	0	0	0	13,350	13,350
Proprietor withdrawals	15,000	15,000	15,000	23,016	68,016
Term Loan Payments					0
Principal	0	0	0	34,350	34,350
Interest	0	0	0	22,890	22,890
Total cash outflow	184,292	97,650	62,979	230,974	575,895

FLOW-OF-FUNDS SUMMARY

	Jan.-Mar.	Apr.-Jun.	Jul.-Sept.	Oct.-Dec.	ANNUAL
Beginning cash balance	13,520	5,000	5,000	5,000	13,520
Total cash receipts	182,140	98,241	63,662	221,545	565,588
Total cash outflow	184,292	97,650	62,979	230,974	575,895
Cash difference	11,368	5,591	5,683	(4,429)	3,213
Borrowing this period	0	399	315	10,432	11,146
Payment on operating loan					
Principal	5,282	0	0	0	5,282
Interest	1,086	990	998	1,003	4,077
Ending cash balance	5,000	5,000	5,000	5,000	5,000

LOAN BALANCES AT END OF PERIOD

	Beginning Balance					End Balance
Term debt	286,935	286,935	286,935	286,935	252,585	252,585
Operating	59,921	54,639	55,038	55,353	65,785	65,785

Operating loan interest rate: 7.5%
Desired minimum balance: 5,000

Figure 7.6. (*Continued.*)

cash balance ($5,000) and the scheduled interest payment of $990 on the operating loan. So new borrowings of $399 are needed in the second quarter. New borrowings are also needed in the third and fourth quarters because cash differences are not sufficient to cover the desired ending cash balance and pay scheduled interest payments on the operating loan.

An astute reader may already see a red flag. The new borrowings in the fourth quarter ($10,432) appear to be partly used to pay for scheduled term loan payments (listed in the cash outflow). Either the farmer or a financial advisor should spend more time trying to understand why performance appears to have taken a turn for the worse in the fourth quarter. Is this a sign of future trouble or something that will be corrected in the next year? Further questioning this year and a projection for next year can help answer the question.

	ANNUAL
Cash Received from Operations	563,461
Cash Paid for Feeder Livestock, Purchased Feed, and Other Items for Resale	57,436
Cash Paid for Operating Expenses	332,778
Cash Paid for Interest	26,967
Net Cash -- Income and Social Security Taxes	0
Net Cash -- Other Operating Activities	0
Net Cash -- Other Miscellaneous Revenue	0
Net Cash Provided by Operating Activities	146,280
Cash Received from Sale of Breeding Livestock (other than normal culling)	0
Cash Received from Sale of Machinery and Equipment	2,127
Cash Received from Sale of Real Estate and Buildings	0
Cash Received from Sale of Marketable Securities	0
Cash Paid to Purchase Breeding Livestock	0
Cash Paid to Purchase Machinery and Equipment	37,167
Cash Paid to Purchase Real Estate and Buildings	9,908
Cash Paid to Purchase Marketable Securities	13,350
Net Cash Provided by Investing Activities	-58,298
Proceeds from Operating Loans	11,146
Proceeds from Term Debt Financing	0
Cash Received -- Capital Contributions, Gifts, Inheritances	0
Scheduled Principal Payments -- Term Debt	34,350
Unscheduled Principal Payments -- Term Debt	0
Principal Paid on Capital Lease Obligations	0
Repayment of Operating and CCC Loans	5,282
Owner Withdrawals (net)	68,016
Dividends and Capital Distributions	0
Net Cash Provided by Financing Activities	-96,502
Net Increase (Decrease) in Cash/Cash Equivalents	-8,520
Cash/Equivalents at Beginning of Year	13,520
Cash/Equivalents at End of Year	5,000

Figure 7.7. Statement of farm business cash flows using FFSC format.

MEASURES OF FINANCIAL POSITION AND PERFORMANCE

To assess financial position and performance adequately, profitability, solvency, liquidity, repayment capacity, and financial efficiency need to be assessed. While many measures and ratios could be calculated and analyzed, this section defines and explains a limited set of measures including the 16 measures recommended by the Farm Financial Standards Council (FFSC). Each measure is estimated using the example farm's financial statements presented in figures 7.1 through 7.7. These measures provide a comprehensive view of the financial position and performance of a farm. Several other measures are important historically, used for specific questions, or common in some financial discussions but are not part of the ones listed by the FFSC. These are discussed in appendix B.

The choice of which measures to use for financial analysis depends on the farmer's preferences, the creditor's preferences, and the local customs. Several measures provide the same information because they are merely a rearrangement of numbers. For instance, as will be discussed below, net worth equals total assets minus total liabilities; so a ratio of total liabilities to total assets provides essentially the same information as a ratio of net worth to total assets.

What is important is the choice of both an absolute measure and a relative measure. The two kinds of measures can show different kinds of problems. For example, net farm income provides an absolute measure of income that can be compared with the income needs of the farmer. The rate of return on assets (ROA) is a relative measure that shows how well the assets or resources were used to generate income. A small set of assets may have been used very efficiently (i.e., with a high ROA) but the resulting income may still be too low to feed the family. Alternatively, a high income may provide for a family very well, but the resources were not used efficiently because the ROA is so low. Each problem calls for a different kind of solution.

Measures of Profitability

Several measures and ratios help analyze the profitability of a farm. With these measures, a higher value is desired. The nonfarm and family portion of a set of records is excluded from the analysis unless explicitly stated otherwise.

Gross Cash Farm Income

Gross cash farm income (GCFI) is the total of cash sales before expenses are deducted. In figure 7.1, the CFFM format of the income statement shows the example farm has a gross cash farm income of $563,461. In figure 7.2, FFSC's *gross revenue* of $622,208 is equivalent to gross cash farm income adjusted for changes in the inventories of crops, market livestock, raised breeding livestock, and accounts receivable (figs 7.2 and 7.3). FFSC's *value of farm production* of $564,772 is calculated by subtracting the cost of inputs such as feeder livestock and purchased feed and grain from gross revenue (fig. 7.3). The U.S. tax forms use a different method where gross cash income is the sales total minus the original cost of items purchased for resale even if the original cost was incurred in a previous tax year.

Net Cash Farm Income

Net cash farm income (NCFI) is equal to the gross cash farm income less cash expenses. For the example farm, NCFI is $146,280. Sometimes, NCFI is incorrectly called the net operating profit. The more accurate estimate of the operating profit is found on the FFSC statements as net farm income from operations, which accounts for inventory changes and depreciation.

Net Farm Income from Operations

FFSC's *net farm income from operations* is essentially the same as CFFM's NFI except that gains and losses on the sale of farm capital assets and due to changes in the general base value of raised breeding livestock are not subtracted yet. Using the FFSC's definition, NFI from operations is $155,717.

Net Farm Income

Net farm income (NFI) is an absolute measure of profitability. It represents the returns to unpaid labor, management, and owner equity. (The return to taking on risk could be stated explicitly, but is usually considered part of the return to management, equity, and labor.) NFI is the GCFI minus cash expenses and depreciation and adjusted for inventory change. Inventory change is ending inventory minus beginning inventory (except the change in accounts payable, which needs to be defined as beginning minus ending). In both the CFFM and FFSC formats, NFI is $155,717, however, the official FFSC name is *net farm income, accrual adjusted.* Using the CFFM format, NFI (without real estate) is calculated as:

$$NFI = NCFI + \text{inventory change} - \text{depreciation}$$
$$= 146,280 + 72,273 - 62,836 = 155,717$$

Net farm income using the cash basis is net cash farm income minus depreciation without adjusting for inventory change. Cash-basis net farm income is used in income tax preparation but not for analyzing financial position. Net farm income as defined above is used for financial analysis since it gives a more accurate picture of the profitability of a farm within a specific year.

Labor and Management Earnings

Labor and management earnings (LME) are the estimated return to the farm operator's labor and management after subtracting a cost for the owner's equity in the farm business. If the farm uses a large amount of unpaid family labor, a charge for that may also be subtracted to provide a more accurate comparison of labor and management earnings between farms. The cost for using equity in the farm should be the return that equity could have received in another investment of similar risk; this is also called the opportunity cost of that equity. The opportunity cost often assigned to equity capital is the yield received on government bonds or the savings rate at a bank, although these violate the similar risk rule. The opportunity cost could also be estimated using the average rate of return in the stock market, which accounts for more risk. The opportunity cost of equity is the owner's equity multiplied by the chosen rate. The value of unpaid family labor is determined from the labor rates paid locally. Thus, labor and management earnings measure the return to the operator's investment of time and management ability. Using an opportunity cost of 5% on the farm's equity of $1,278,241 (market value basis) and assuming no unpaid family labor, LME is $91,805.

$$LME = NFI - \text{opportunity cost of equity} - \text{unpaid family labor}$$
$$= 155,717 - (1,278,241 \times 0.05) - 0 = \$91,805$$

LME should not be confused with the opportunity cost or value of operator labor and management. LME is the estimate of what the operator receives by farming. The opportunity cost or *value of operator labor and management* is what the operator could receive in a nonfarm job that requires similar labor and management skills. Thus, the comparison between LME and the value of labor and management will show how well a farmer is doing compared with his or her opportunities elsewhere.

Rate of Return on Farm Equity

The *rate of return on farm equity (ROE)* is the first of three relative measures of profitability recommended by the FFSC. ROE represents the interest rate being earned on the farm's average net worth. The farm's average net worth is the average of the net worth at the beginning of the year and at the end of the year. If assets are valued at market value, ROE can be compared with returns available if the assets were liquidated and invested in alternate investments. If assets are valued at cost value, ROE closely represents the actual return on the funds invested or retained in the business. Using FFSC's NFI from operations avoids any distortions that may result from the gain or loss from the sale of capital assets. This gain or loss is included in CFFM's NFI so FFSC's NFI from operations is preferred for calculating ROE.

The example farmer estimates he could receive a salary of $65,000 working off the farm, so the value of his labor and management is set at $65,000.

The balance sheet at the beginning of the year shows the example farm's net worth is $798,146 (market value basis). So, using FFSC's definition, ROE for the example farm is 8.7%.

$$ROE = \frac{NFI \text{ from operations} - \text{value of operator's labor and management}}{\text{average farm net worth}}$$

$$= \frac{155,717 - 65,000}{(798,146 + 1,278,241) \,/\, 2} = 8.7\%$$

While a higher ROE is preferred to a lower ROE, the ROE also should be compared with the opportunity cost of equity capital, that is, the rate of returns available in alternative investments, such as savings or stocks. ROE is essentially the same measure as general business' ROI where I is the owner's or stockholder's equity investment.

The value or opportunity cost of unpaid labor and management includes both the value of unpaid family labor and the opportunity cost or value of the operator's labor and management. For partnerships and corporations that pay wages to the operator(s), this value can be calculated as the owners' withdrawals for unpaid labor and management. In sole proprietorships, this value can be estimated from local, non-farm-labor rates and the time spent working on the farm. A typical value range for the operator's labor and management used in recent years is from $25,000 to $30,000 per farmer. However, as the example farmer did, this value can and should change over time and due to a farmer's abilities and the local job market. When comparing across farms, the same value of operators' labor and management is typically given to each farmer so that the farms will be compared on a consistent basis. When analyzing multiple operator farms, the value of operator labor and management is increased by the number of operators. The value of a part-time operator should be a proportion of that for a full-time operator. A common proportion is one-half of the typical value for an operator who also has a full-time, nonfarm job. Similarly, instead of estimating the number of hours worked by family members, their value could be expressed as a proportion of a full-time operator. For example, a teenager who does some work during the school year and works full-time on the farm in the summer could be valued at one-fourth to one-third of a full-time operator.

If any part of the ROE equation is negative or close to zero, special care is needed when interpreting the calculated ROE. If equity is positive and ROE is negative, it can be interpreted as a very negative event. If a farm's equity is negative, ROE cannot be used very well for analysis and other measures may be more important to analyze than ROE. If equity is positive but very close to zero, ROE may be overinflated and comparison with other farms is impossible with ROE. In these cases of negative equity or low equity, the rate of return on assets (ROA) should be used for comparisons instead of ROE.

Rate of Return on Farm Assets

The *rate of return on farm assets (ROA)* is the second relative measure of profitability recommended by FFSC. ROA represents the average interest rate being earned on all investments in the farm or ranch business. If assets are valued at market value, the rate of return on assets can be looked at as the opportunity cost of farming versus alternate investments. If assets are valued at cost value, the rate of return on assets more closely represents the actual return on the average dollar invested in the farm. ROA includes both equity capital and debt capital so the interest paid on debts has to be added in the equation since it has already been subtracted in the calculation of NFI. Again from the balance sheet at the beginning of the year, the market value of all assets is $1,258,830. Using the FFSC definition, ROA for the example farm is 7.8%.

$$\text{ROA} = \frac{\begin{array}{c}\text{NFI from operations + farm interest paid} - \\ \text{value of operator's labor and management}\end{array}}{\text{average farm investment}}$$

$$= \frac{155,717 + 26,967 - 65,000}{(1,258,830 + 1,760,996) / 2} = 7.8\%$$

Financial Leverage Ratio

The *financial leverage ratio* evaluates how well the business used debt capital versus equity capital. The financial leverage ratio is defined as:

Financial Leverage Ratio = ROE / ROA = 8.7 / 7.8 = 1.1

If debt capital is used effectively, the return to that capital will be greater than its cost, ROE will be greater than ROA, and the financial leverage ratio will be greater than one. However, if debt capital is not used effectively, its cost will be greater than its return and the ratio will be less than one. Therefore, the financial leverage ratio identifies potential problems by signaling when debt capital is not being used effectively. For the example farm, the financial leverage ratio is 1.1 showing that debt capital contributed more than it cost in this year.

Operating Profit Margin Ratio

The *operating profit margin ratio* (OPM) is the FFSC's third relative measure of profitability. OPM is a measure of the operating efficiency of the business. If expenses are held in line relative to the value of output produced, the farm will have a healthy net profit margin. A low net profit margin may be caused by low prices, high operating expenses, or inefficient production. Using the FFSC definition with gross revenue, OPM is calculated to be 18.9%.

$$OPM = \frac{\text{NFI from operations + farm interest paid −}}{\text{value of operator's labor and management}}{\text{gross revenue}}$$

$$= \frac{155,717 + 26,967 − 65,000}{622,208} = 18.9\%$$

Measures of Solvency

The first solvency measure is the *net worth (NW)* or *owner's equity* of the farm. Net worth is the difference between total assets (TA) and total liabilities (TL). It is the absolute measure of the cushion between asset values and liabilities. This cushion can be used for increased borrowing and (or) enduring loss in asset value. For the example farm, net worth using cost basis is $855,537 and net worth using market values is $1,278,241.

NW = TA − TL
NW (cost basis) = 1,173,907 − 318,370 = 855,537
NW (market value) = 1,760,996 − 482,755 = 1,278,241

The FFSC chose three relative measures of solvency: the debt-to-asset ratio, the equity to asset ratio, and the leverage ratio. All three show the farm's overall financial risk of the farm to the farmer and the lender. They all show the relationship between a farm's total assets, debt level, and equity. Since these ratios contain the same information, they are very similar and the choice among the three depends upon personal preference and experience.

The first FFSC solvency measure, *farm debt/asset ratio (D/A)*, measures the size of the farm's debt load compared with the total asset value. This ratio became popular during the agricultural financial crisis in the early 1980s. It is usually expressed as a percentage. The debt-to-asset ratio shows the extent to which the farm's assets are financed by debt capital versus equity capital. A lower value is preferred. A higher value shows that creditors have a larger share of farm assets and the farmer and, thus, any creditor faces a higher level of financial risk. Using the FFSC definition and market values, the debt/asset ratio for the example farm is 27%.

D/A = (total farm liabilities / total farm assets) × 100%
  = (482,755 / 1,760,996) × 100% = 27%

The *farm equity/asset ratio* is also recommended by the FFSC. The equity-asset ratio measures the extent to which the farm's assets are financed by equity capital versus debt capital. A higher value shows that the farmer has a larger share of farm assets and the farmer and, thus, any creditor faces a lower level of financial risk. The equity-asset ratio is the opposite, or mirror-image, of the debt-to-asset ratio. These two measures always add up to 100% because they describe how total farm assets are financed. Increases in the equity-value ratio improve the financial position of the farm: a smaller proportion of assets is owed to creditors. Using the FFSC definition and market values, the equity/asset ratio for the example farm is 73%.

Equity/asset ratio = (total farm equity / total farm assets) × 100%
  = (1,278,241 / 1,760,996) × 100% = 73%

The *farm debt/equity ratio* measures farm debt relative to farm net worth or owner's equity. It shows the debt held by a farm for every dollar of equity. Historically, the debt/equity ratio was called the leverage ratio. While the debt-to-asset ratio is used extensively in the press, the leverage ratio has historical importance and is still used by many people in the financial sector. Using the FFSC definition and market values, the debt/equity ratio for the example farm is 38%.

Debt/equity ratio = (total farm liabilities / total farm equity) × 100%
  = (482,755 / 1,278,241) × 100% = 38%

Measures of Liquidity

Liquidity is the ability of the firm to meet its financial obligations during the next 12 months. Liquidity can be thought of in two ways. Balance sheet liquidity evaluates the ability to meet obligations at a point in time. Balance sheet liquidity calculates working capital and the current ratio from the net worth statement. Cash flow liquidity measures the ability to meet obligations as a flow of money over time. Cash flow liquidity is also called repayment capacity and is discussed in the next subsection.

Working capital is the money that a farmer has available for use in the current and near future for production. The FFSC calculates working capital by subtracting total current farm liabilities from total current farm assets. While some definitions include intermediate assets and intermediate liabilities in the calculation of working capital, FFSC does not. This measure is called working capital because it is capital that is available for use (or work) in the near term within the business. Using FFSC's definition, working capital for the example farm is $360,602.

$$\text{Working capital} = \text{total current farm assets} - \text{total current farm liabilities}$$
$$= 463,477 - 102,875 = 360,602$$

The *current ratio* shows the value of the current assets relative to the current liabilities. It shows the farm's ability to meet debt obligations in the next 12 months from current assets. Using FFSC's definition, the current ratio for the example farm is 4.5.

$$\text{Current ratio} = \frac{\text{total current farm assets}}{\text{total current farm liabilities}}$$

$$= \frac{463,477}{102,875} = 4.5$$

While interpreting the current ratio depends upon the type of business in which the farm is involved, a current ratio value equal to or greater than two generally is considered healthy for farms. This value shows that the farm could cover its current obligations and still have some current working capital. A value of less than one obviously should cause concern because it says that the farm could not cover its current obligations from current assets. A value between one and two is of concern, but requires further analysis before actions are recommended.

Measures of Repayment Capacity

The *term debt and capital lease coverage ratio* (or term debt coverage ratio) measures whether the business generated (or is projected to generate) enough cash to cover term debt payments. Total term debt payments are the annual scheduled principal and interest payments on intermediate and long-term debt. Note that this ratio requires the scheduled payments; these may or may not be the same as the actual payments. It is usually written as a percentage. A ratio less than 100% indicates the business did not generate sufficient cash to meet scheduled payments in the past year. Other sources of cash are needed. A ratio greater than 100 indicates the business generated enough cash to pay all term debt payments. Using FFSC's definition and the term debt payment information from figure 7.7, the term debt coverage ratio for the example farm is 3.03 or 303%.

Term debt coverage ratio = (NFI from operations +/– total miscellaneous revenue/expense + total nonfarm income + depreciation/amortization expense + interest on term debt + interest on capital leases – total income tax expense – total owner withdrawals) / (Annual scheduled principal and interest payments on term debt + annual scheduled principal payments on capital leases)

$$= \frac{155,717 + 0 + 0 + 62,836 + 22,890 + 0 - 0 - 68,016}{(57,240 + 0)} = 3.03 \text{ or } 303\%$$

The *capital replacement and term debt repayment margin* is the money remaining after all operating expenses, taxes, family living costs, and scheduled debt payments have been made. It is the cash generated by the farm business that is available for financing capital replacement such as machinery and equipment. Using the FFSC definition, the capital replacement and term debt repayment margin for the example farm is $110,905.

Capital replacement and term debt repayment margin =

NFI from operations +/– total miscellaneous revenue/expense
+ total nonfarm income + depreciation/amortization expense
– total income tax expense – total owner withdrawals
– payment on unpaid operating debt from a prior period
– principal payments on current portions of term debt
– principal payments on current portions capital leases
– total annual payments on personal liabilities (if not included in withdrawals)
= 155,717 + 0 + 0 + 62,836 - 0 - 68,016 - 5,282 - 34,350 - 0 - 0
= 110,905

Measures of Efficiency

The *asset turnover ratio* is a measure of how efficiently farm assets are being used to generate revenue. It measures how fast the assets of the farm are turned over annually. Typical asset turnover ratios in farming are between 40% and 50% with some variation between types of farms. For example, a dairy farm typically has a lower asset turnover ratio than a fresh vegetable farm because the dairy farm has a more stable income flow and can operate with a thinner margin. The vegetable farm operates at a higher risk level so requires a faster rate of asset payment. Most cash crop and crop-livestock farms have asset turnover ratios between these two. Using the FFSC definition and assets valued at market, the asset turnover ratio for the example farm is 41%.

$$\text{Asset turnover ratio} = \frac{\text{gross revenues}}{\text{average total farm assets}}$$

$$= \frac{622{,}208}{(1{,}258{,}830 + 1{,}760{,}996) / 2} = 41\%$$

The value of production can be used in place of gross revenues when calculating the asset turnover ratio. However, as the FFSC points out, the analysis should consistently use either gross revenues or value of production to avoid misinterpretation. Multiplying the asset turnover ratio by the operating profit margin ratio should result in the rate of return on farm assets. If the value of production or gross revenue is not used consistently, this relationship will not exist.

This ratio can cause problems when comparing farms that have different mixes of owned and rented assets even though other measures of size are similar. Without adjusting the equation, the farms with more rented assets will have a falsely higher asset turnover rate because their value of assets owned is less than a farm that owns all of the assets it uses. So, although this is neither mentioned nor recommended by the FFSC, an analyst may find it appropriate to use the value of controlled assets (that is, both rented assets and owned farm assets) to provide a more accurate comparison between farms. Then the asset turnover ratio is defined as:

$$\text{Asset turnover ratio} = \frac{\text{value of production}}{\text{average total controlled assets}}$$

The FFSC defines four *operational ratios* that the distribution of gross revenues to cover total farm expenses and generate farm income. The sum of the operating expense ratio, the depreciation expense ratio, and the interest expense ratio equals the percent of gross revenues used to pay farm expenses. The amount remaining is net farm income from operations. In all four of these ratios, the FFSC allows the use of the value of farm production rather than gross revenues if one of these is used consistently.

The *operating expense ratio* shows the percentage of the gross farm income used to pay operating expenses. FFSC's recommended definition of operating expense is on a pretax basis that includes inventory changes and depreciation/amortization expense but does not include interest expense. Using the FFSC definition, the operating expense ratio for the example farm is 61%.

Operating expense ratio

$$= \frac{(\text{total farm operating expense} - \text{depreciation/amortization})}{\text{gross revenues}}$$

$$= \frac{439{,}524 - 62{,}836}{622{,}208} = 61\%$$

The *depreciation expense ratio* shows the percentage of the gross revenues used to cover depreciation/amortization. Using the FFSC definition, the depreciation expense ratio for the example farm is 10%.

Depreciation/amortization expense ratio

$$= \frac{\text{depreciation/amortization expense}}{\text{gross revenues}}$$

$$= \frac{62,836}{622,208} = 10\%$$

The *interest expense ratio* shows the percentage of the gross revenues used for farm interest expenses. Using the FFSC definition, the interest expense ratio for the example farm is 4%.

$$\text{Interest expense ratio} = \frac{\text{total farm interest expense}}{\text{gross revenues}}$$

$$= \frac{26,967}{622,208} = 4\%$$

The *net farm income from operations ratio* shows the percentage of the gross revenues that remains after all expenses. Using the FFSC definition, the net farm income from operations ratio for the example farm is 25%.

Net farm income from operations ratio

$$= \frac{\text{net farm income from operations}}{\text{gross revenues}}$$

$$= \frac{155,717}{622,208} = 25\%$$

Cost measures show the efficiency with which costs are controlled relative to production levels. The total cost per unit (bushel, head, cwt., etc.) can be compared with other farmers and, if adjusted for inflation, with other years. The total costs per unit can be divided between direct operational costs per unit (i.e., variable) and overhead costs per unit (i.e., fixed). Feed costs per livestock production unit are especially important for dairy operations. Comparing the investment per acre or per livestock unit can explain differences between farms and changes over time. For more detailed analysis, the total investment could be divided into building investment (especially for livestock) and machinery investment (especially for crops). Cost measures can be used for (1) efficiency comparisons, (2) comparison of operational differences between farms (e.g., labor-intensive versus capital-intensive farms), (3) determination of whether a cash flow problem is due to a production problem or a marketing problem (or both), and (4) break-even and marketing analysis.

INITIAL ANALYSIS

We have many ways to analyze the financial position of a farm. The procedure described in this section is for a first-time analysis of a farm and assumes no previous knowledge of the farm. This procedure is not the only procedure that will work. It is offered as a guide to be used as described or to help the development of the reader's own analysis procedures. This procedure is designed to find the areas that have problems or opportunities so that further analysis can better focus on those areas.

The steps to analyze the overall financial position and performance of a farm are:

1. Gain an overview of the farm.
2. Identify the vision, goals, objectives, and standards of the farm.
3. Check the accuracy of the records.
4. Prepare the financial statements.
5. Evaluate profitability.
6. Evaluate solvency.
7. Evaluate liquidity.
8. Evaluate repayment capacity.
9. Evaluate efficiencies.
10. Examine marketing performance.
11. Analyze livestock and crop enterprises.
12. Analyze internally, historically, vertically, and comparatively.
13. Evaluate managerial ability and personal characteristics.

Gain an Overview of the Farm

What is the location of the farm and its major enterprises? How large is the farm? What is its history? What are the characteristics of the farm family or families: size, participation, and interests? What are the management abilities and risk attitudes of the principals involved in the farm business?

One step that could be taken at this time, especially if the analyst has no experience with the farm, is to complete an inventory of resources for the farm. This step is also necessary if the financial statements are not available. The process of taking an inventory is described in appendix B2, "Inventory of Resources."

Identify the Vision, Goals, Objectives, and Standards of the Farm

Vision, goals, objectives, and standards are compared against actual performance and plans. They tell us what the owners, operators, managers, and family members want to do and where they want to go. If we do not know the goals, objectives, and standards, we will not know how to evaluate or grade the performance and position of the farm.

Vision, goals, objectives, and standards are similar yet different. A vision is what the farmer and others involved in the business want the farm to be or become in the future. Goals are usually thought of as long-run targets, such as retirement by 60 or the farm paid for in ten years. Objectives are the specific items that a person wants to accomplish in a shorter time (say a year). Objectives may be items like a dairy herd average of 25,000 pounds of milk per cow, a corn yield of 180 bushels per acre, feed costs of $35 per feeder pig (given the assumptions and practices of a farm). Standards are very similar to objectives but standards are usually more realistic in expectations. Objectives are what we strive for, knowing that we may not be able to reach them. Standards are what we wish to be measured against because we feel that we probably could grasp or attain those goals.

The development of goals, objectives, and standards should involve all the principals of the farm: the operator(s), other partners and equity holders, and family members. They can be developed from past performance; comparable farms; benchmarking with the best farms; experiments and recommendations; requirements by a bank, the government, or other institution; dreams and expectations; normal prices, yields, efficiencies; and what the experts say.

Check the Accuracy of the Records

To trust the results of the analysis, we must trust the initial information used for the analysis. Various consistency checks can be done to assess the accuracy of the records. These are explained in appendix B3, "Checking Record Accuracy."

Prepare the Financial Statements

Once the records are deemed accurate (that is, accurate enough), the three main financial statements should be prepared. These are the (1) income statement (or profit/loss statement), (2) balance sheet (or net worth statement), and (3) cash flow statement.

The income statement shows net income left after expenses are paid. The balance sheet shows the balance between the value of the assets and the debt owed by the farm. The cash flow statement shows, as its name implies, when cash comes into the business and when cash goes out of the business. The development and interpretation of these three statements appear at the beginning of this chapter.

Evaluate Profitability

Profitability is usually the first goal of farmers, so it is listed first in the analysis procedures.[1] When analyzing profitability, we are concerned not only with the absolute amount of profit but also with the relative amount. Indirectly, it also measures the managerial ability of the farmer. We are interested in (1) whether the actual level of net farm income provides enough for family living and other commitments and (2) whether that net farm income is adequate compared with a farmer's equity capital investment, the unpaid operator and family labor used in business, and the contribution of management ability.

Using the information in the financial statements and the FFSC's definitions, calculate the net farm income, the rates of return to farm assets and to farm equity, and the operating profit margin. To gain a better understanding of this farm's financial condition and performance, compare the actual measures to widely accepted expectations by using a form as shown in table 7.1 for the example farm.

Evaluate Solvency

Solvency evaluates the asset-liability balance at a particular point in time. It measures both the ability to repay all debts and the financial ability of the firm to withstand periods of poor profitability and, if measured with assets at market value, decreases in asset values. Solvency does not consider past or future profitability or liquidity. It only considers the farm's position on a certain date.

Using the information in the financial statements and the FFSC definitions, calculate the farm's net worth, debt-to-asset ratio, equity-to-asset ratio, and debt-to-equity ratio, and compare the actual results with accepted expectations as shown in table 7.1 for the example farm.

Evaluate Liquidity

Liquidity is the ability of the firm to meet its financial obligations in the next 12 months. Liquidity is measured by the working capital and the current ratio. Sometimes these measures are referred to as measuring balance sheet liquidity because they require information from the balance sheet.

Using the information in the financial statements and the FFSC definitions, calculate the current ratio and the working capital. The actual current ratio can be compared with accepted standards as shown in table 7.1 for the example farm, but there is no standard for working capital since farms differ widely in size and type and thus the need for working capital.

Table 7.1. Evaluation Form for the Example Farm's Financial Position and Performance

Category and Measure	Actual Results	Relative Rating of Actual Measure**				
		Vulnerable				Strong
Liquidity						
Current ratio	4.5		1.0		2.0	✓
Working capital	$360,602					
Solvency (market)						
Net worth (equity)	$1,278,241					
Farm debt-to-asset ratio	27%		70%		40%	✓
Farm equity-to-asset ratio	73%		40%		70%	✓
Farm debt-to-equity ratio	38%		150%		43%	✓
Profitability (market)						
Net farm income (from oper.)	$155,717					
Rate of return on farm assets	7.8%		1%		5%	✓
Rate of return on farm equity	8.7%		5%	✓	10%	
Operating profit margin	18.9%	✓	20%		35%	
Repayment Capacity						
Term-debt coverage ratio	303%		110%		135%	✓
Capital-replacement margin	$110,905					
Financial Efficiency						
Asset-turnover rate (mkt)	41%		20%		40%	✓
Operating-expense ratio	61%		80%		60%	✓
Depreciation-expense ratio	10%		20%		10%	✓
Interest-expense ratio	4%		20%		10%	✓
Net farm income ratio	25%		10%		20%	✓

*These 17 measures are taken from and defined by the Farm Financial Standards Council (1997).
**While the borders between vulnerable and strong are not well defined, these guidelines serve to help describe the farm relative to widely accepted expectations for financial condition and performance.

Evaluate Repayment Capacity

The ability to repay loans is critical both to the farmer and the creditor. The traditional measures of liquidity ignore a farm's cash inflow and outflow and, thus, are unable to predict adequately the future ability to service debt and other commitments. To look at a business' ability to generate cash to meet obligations, term debt coverage ratio and capital replacement margin are suggested. Sometimes these measures are referred to as measuring cash flow liquidity because they require information from both the balance sheet and the cash flow statement.

Using the information in the financial statements and the FFSC definitions, calculate the term-debt coverage ratio and the capital-replacement margin. The actual term-debt coverage ratio can be compared with accepted standards as shown in table 7.1 for the example farm, but there is no accepted standard for the capital-replacement margin since farms differ widely in size, type, and capitalization and, thus, in need for capital replacement margin.

Examine Efficiencies

Financial measures of efficiency include the asset turnover rate, the distribution of gross income, and costs and investment per unit. Using the information in the financial statements, calculate the five financial measures of financial efficiency defined by the FFSC: asset-turnover rate (market), operating-expense ratio, depreciation-expense ratio, interest-expense ratio, and the net farm income ratio. The actual values for these five measures can be compared with standard expected values for these measures as shown in table 7.1 for the example farm.

Physical measures of efficiency include crop yields, livestock production per animal, feed efficiency, calving and weaning rates, death loss, and output per labor hour. These and other measures are used to compare actual performance with the farm's internal standards and goals, the farm's historical performance trends, and the performance of comparable farms. When these physical measures are calculated and compared, the same rules and procedures should be used. For instance, the weaning rate (number of young weaned per adult female) can be altered drastically by different methods of counting the numbers of breeding adults. Sometimes, only the adults that have young are counted; in other places those bred but never kept for birthing are counted also. The question is not really which way is right, but whether the same method is being used for all calculations.

Examine Marketing Performance

The prices that a farmer receives or pays can suggest marketing ability, market timing, storage, discounts, and other factors. A farmer's marketing performance in economic terms can be measured by the price-received rate and the price-paid rate. Physical measures of marketing performance include storage and handling losses, insect and mold damage levels, and so on. The critical question when evaluating marketing performance is not whether the best price was obtained but does the marketing plan need to be changed.

When evaluating *prices received*, several items need to be checked. Products marketed in a timely fashion will usually command a higher price. To be timely, a farmer needs to monitor the markets and watch for signals that suggest changes in market patterns and seasonal differences. Market strategies such as hedging, options, and contracting need to be used to their best advantage and just not always used or never used. Also, inattentiveness to quality can cause price penalties at marketing time. Quality problems can be caused by many problems such as improper ventilation in grain storage or feeding livestock to weights that receive price penalties (unless market movements suggest that prices are going up faster than the price penalty). A good indicator of marketing performance is the price-received rate.

$$\text{Price received rate} = \frac{\text{farmer's average price}}{\text{average market price}} \times 100\%$$

The farmer's average price is that weighted average of what the farmer actually received for his/her product. The average market price is calculated from market information for the general market area and marketing year where the farm is located.

The analysis of *prices paid* is done similarly to the analysis of prices received. Supplies purchased in a timely and orderly fashion should obtain more favorable prices for the farmer. The price-paid rate is a good indicator of input purchasing ability.

$$\text{Price paid rate} = \frac{\text{farmer's average price}}{\text{average market price}} \times 100\%$$

Any analysis of prices and marketing performance should be done with a good knowledge of the farmer's marketing plan. For instance, a farmer without input storage cannot take advantage of large purchase discounts or preseason sales; the farmer may have already decided that the cost of owning storage is greater than any potential price discounts. The same reasoning needs to be applied when analyzing prices received.

Analyze Livestock and Crop Enterprises

So far, the analysis has focused on the whole-farm level without explicitly considering the parts of the business, that is, its crop and livestock enterprises. At this point in the analysis, the enterprises need to be analyzed individually. Enterprises need to be evaluated to discover problems and opportunities hidden at the whole-farm level and to answer questions generated during previous analysis, such as why prices or yields are low. Enterprise analysis can also help answer questions, such as "Should I stop growing crops and specialize in dairy?"

The basic questions to be answered in enterprise analysis are these:

1. Is each enterprise being operated efficiently from an economic view? Does it produce the correct products? Does it produce them with the proper mix of inputs? Is it the proper size?
2. Is the mix of enterprises appropriate for this farm, its management, and its economic environment? Is each enterprise of the correct size? What is the strategic position, or desired position, of the farm?

Enterprise analysis is very similar to whole-farm analysis, but limited to one enterprise. However, the process of getting enterprise information can be involved. The steps for developing enterprise budgets from whole-farm information are discussed in chapter 4, "Budgeting," in the section "Developing Enterprise Budgets from Whole-Farm Records."

The analysis of enterprises can be done in three ways.

1. *Gross margins (or contribution to overhead expenses).* The gross margin (gross income minus variable expenses) is used for short-run analysis. On a per-unit basis (e.g., per acre or per dollar invested), gross margins can be used to decide what to produce and how much of each product to produce. Gross margin analysis ignores fixed and sunk costs in the short run.
2. *Allocation of direct and indirect expenses and income.* This type of enterprise analysis includes both variable and fixed costs to estimate the enterprise's contribution to whole-farm profit. The allocation of indirect expenses can be an arbitrary process, but to evaluate the enterprises on a longer-run basis, the allocation needs to be done.
3. *Allocate assets and liabilities to enterprises.* This level of analysis involves more arbitrary decisions on what assets and liabilities belong to each enterprise. With this allocation, a financial analysis can be done for each enterprise including profitability, solvency, liquidity, financial stress, and efficiency analysis. While some people argue that this type of analysis cannot be done because enterprises do not own assets or take on debt by themselves, how else can long-run decisions be made concerning starting and quitting major enterprises?

Analyze Internally, Historically, Vertically, and Comparatively

Once the measures and ratios are calculated, there needs to be some point of comparison for analyzing these numbers. Four methods of comparison are internal, historical, vertical, and horizontal.

Internal. The vision, goals, objectives, and standards are the first points of comparison. Are they being met? Is there sufficient income to meet the family's needs and run the business? Does the business perform at expected levels? Are the performance standards (such as crop yields, weaning rates, or debt repayment) being met or exceeded?

Historical. How has the business performed over time? What are the trends for the same farm business? Are the trends meeting plans and objectives? Is this year better or worse than previous years?

Vertical. Vertical analysis gets its name from its comparison of a column of numbers, say, income and expense items, and calculating each item as a percentage of the total income or expenses. Then the farmer or analyst can quickly see which item(s) dominates the other income or expense items. This provides information on which items need attention because they are a large component of the total. Vertical analysis can be used when comparing with other farms to see how the mix of income and expense items differ or are the same.

Horizontal. How does the business compare with other farms? One group of other farms should be similar in size and type to compare performance and efficiency. Another group of farms could be different to show the effect of different resource allocation and enterprise and strategy selection. These other farms should have their financial analysis done with similar methods so that comparisons are possible.

Evaluate Managerial Ability and Personal Characteristics

While evaluating people is a hard and subjective process, a financial analysis is not complete without this evaluation. Without managerial evaluation, the probability of success cannot be determined. Factors that need to be considered are skills, knowledge, experience, capacity (to learn and to work), honesty, integrity, risk attitudes, goals, and drive. Other factors may be evaluated also depending on the geographic location, type of farm, and purpose of the financial analysis.

Final Points for Analysis

Remember that one primary purpose of analysis is to identify portions of the business that need further study for improvement or cessation. After completing a financial analysis of a farm, the decisions that need to be made are:

1. Do we continue the farm business largely as it is?
2. Do we make adjustments or changes in the farming business that will result in a higher net farm income in the future or solve other problems? This requires a more thorough analysis of the business.
3. Do we liquidate the business, invest the capital elsewhere, and seek off-farm work?

When problems or symptoms of problems (or opportunities) are discovered in the initial analysis or are known to exist, the source of the problems needs to be found. Usually this problem can be found with a more thorough analysis of the business, but having a diagnostic structure to follow can be helpful in the search for the solution. This diagnostic structure or system is described in the last section of this chapter.

DIAGNOSTIC ANALYSIS

Once a problem or symptom of a problem becomes evident, a diagnosis of the problem and its solution is needed. This is similar to the process used by a medical doctor or a veterinarian who, when confronted with symptoms, diagnoses the disease and its treatment.

Problem diagnosis and solution discovery start with asking what is the problem or the symptom of the problem? Problems and their symptoms are usually tied to the lack of performance toward goals and objectives. Goals and objectives can include farm earnings level, wealth or equity growth, leisure, high yields, high milk production, new pickups, and many others. So problems and symptoms can be such things as low income, insufficient money to cover debt payments, no time for vacation, no cash for a new pickup, and second place in the yield contest. The problem can also be competing goals, so farmers and their families may need to decide which goals have higher priorities than other goals. Priorities may also change as different levels of goals are attained. For example, income may be quite important until a certain level is reached at which point, vacation time becomes more important than higher income.

Once a problem or symptom is found, a structured diagnosis procedure follows a set of steps similar to those identified below. The procedures follow the discovery of an earnings or income problem. Other problems such as lack of wealth accumulation or lack of vacation time could follow similar procedures for finding solutions.

1. Is there an earnings problem?
   a. Calculate these measures: net farm income, rates of return on equity and assets, residual return to operator's labor and management. Do these meet expectations? If not, continue.

    b. Is the problem absolute or relative?
      (1) It is absolute if there is just not enough money or cash inflow to meet expenses, debt obligations, family living, and other commitments.
      (2) It is relative if there is enough money and cash inflow, but earnings do not meet internal goals and standards or other farms have better earnings.
      (3) It might be both absolute and relative.
    c. An absolute problem suggests an immediate cash flow problem while a relative problem does not.
    d. Both relative and absolute earning problems can be either permanent or temporary or either organizational or operational.
2. Is the problem temporary or permanent?
    a. A temporary problem is caused by an event that is not controlled by the farmer and not expected again. Temporary problems can be caused by events such as drought, flood, or market collapse. If a problem is temporary, its solution can be developed knowing that it will not likely occur again. However, the short-run problems caused by temporary problems still can be very serious and dangerous to business survival.
    b. A permanent problem is due to a situation that can be expected to occur again. The situation may not last forever, but it is expected to last for more than one or two years.
      (1) Has a structural change occurred? This may not be a sudden change but a realization that a change has occurred or is occurring. Examples of these changes are closing of slaughter plants at nearby locations, new processing plants built near the farm, changes in consumer demand (e.g., decreasing red meat consumption), the entry of other countries into the world trade market.
      (2) Is it a management problem? Some problems occur due to mismanagement of the farm's resources. If these problems persist, severe financial damage could occur. These problems could occur in many areas of the business: organization, operation, marketing, finance, personnel, and timeliness.
    c. The distinction between permanent and temporary problems may be found by comparing the current prices, yields, and inputs with their normal values. These normal values come from historical data for the farm and for the local area and national trends. With these norms, the current year can be put in perspective concerning trends, changes, and other geographical areas. If a farm's current year's figures differ from its own normals, the problem is most likely temporary (unless it is the beginning of a permanent change). If a farm's current year's figures and its own normals differ from the local or national normals, the problem is most likely permanent.

3. Is it an operational or organizational problem?
  a. Operational problems are due to inefficient use of resources.
    (1) What are the practices followed in carrying out the production, mar-keting, and financial plans? Are they specified correctly and done in a timely manner?
    (2) What are the production costs and efficiencies?
    (3) How do the actual results compare with standards, expected values, normal values, and comparable farms?
    (4) Each enterprise and other phases of the business (e.g., financial management) need to be analyzed separately. The basic question is whether each enterprise or phase is performing well and as expect-ed or is improvement needed?
  b. Organizational problems are due to poor choice of enterprises, poor allocation of resources, insufficient size, or lack of resources. If the enterprises met their standards and normals as described in the previous section on operational problems and there is still a problem, it is an organizational problem.
    (1) To solve this problem, we have a few options:
      (a) select a different enterprise combination
      (b) move resources from one enterprise to another enterprise
      (c) add more resources (land, labor, and capital)
      (d) move resources from farming to other investments and uses
    (2) To analyze the enterprise combination, calculate the enterprise budgets and analyze them for gross margins, profit contribution, asset requirements, and debt load. Analyze alternative mixes of cur-rent and new enterprises to find a mix that increases earnings or solves other problems. If alternative enterprise mixes do not solve the problem, the problem is most likely a size problem.
    (3) Size problems
      (a) Is the farming unit large enough to take advantage of modern technology? Or large enough to produce enough sales to pro-vide sufficient income? Is the farm using current technology but just is not large enough? How, where, and when can it be expanded? What are the options for owning, renting, leasing, or hiring more resources and would they help solve the problem?
      (b) Is the farm too large given its management level or is one area too large (e.g., acreage) compared with another area (e.g., labor and machinery resources)? What are the alternatives to solve these problems? Decreasing the size of the farm may help solve problems if the smaller resource base can be managed more efficiently and still meet other objectives. Adding labor resources may be all that is needed to improve a farm that has expanded faster in land and livestock than in labor supply.

4. If the options considered so far have not solved the problem(s), a manager must consider the option of putting the resources into other uses. This includes other employment for the operator. While this drastic option may not meet the goal of being a farmer, it may help meet other goals such as living standards and leisure time.

SUMMARY POINTS

- Financial statements are needed to analyze financial position and performance.
- Financial position refers to the total resources controlled by a business compared to the claims against those resources.
- Financial performance refers to the results of decisions over time.
- The three main financial statements are the income statement, balance sheet, and cash flow statement.
- The income statement shows the difference between the gross income and the costs incurred to produce that income.
- The balance sheet shows the assets and liabilities on a specific date.
- The cash flow statement shows the annual flow and timing of cash in and out.
- To assess financial position and performance adequately, profitability, solvency, liquidity, repayment capacity, and financial efficiency need to be assessed.
- Both absolute and relative measures are needed.
- Profitability is measured by net farm income, rate of return on farm equity (ROE), and rate of return on farm assets (ROA).
- Solvency is measured by net worth (or equity), debt/asset ratio, equity/asset ratio, and debt/equity ratio.
- Liquidity is measured by working capital and current ratio.
- Repayment capacity is measured by the term debt and capital lease coverage ratio and the capital replacement and term debt repayment margin.
- Financial efficiency is measured by the asset turnover ratio, four operational ratios (the operating expense ratio, depreciation expense ratio, interest expense ratio, and net farm income from operations ratio), and cost measures appropriate to the farm.
- An initial analysis is performed to gain an understanding of the farm and to find areas that have problems or opportunities so that further analysis can better focus on those areas.
- Diagnostic analysis is used once a problem or symptom of a problem becomes evident.

REVIEW QUESTIONS

1. Why do we need to know a farmer's goals, objectives, and standards for financial analysis?
2. What is the difference between a symptom and a problem?
3. Define and contrast the income statement, the balance sheet, and the cash flow statement.
4. Identify the appropriate financial statement(s) for each of the following and where on that statement the value should be placed.

hog sales, $45,000	purchased feed, $15,000
hired labor, $18,000	unpaid family labor, $25,000
corn sales, $55,000	corn in storage, $125,000
fertilizer expense, $35,000	prepaid fertilizer expense, $25,000
unpaid feed bill, $2,500	farm truck, book value, $5,500
depreciation, $33,000	farm land loan payment, $29,000
new tractor, $85,000	new operating note from bank, $250,000
new tractor loan, $45,000	cash sale of old tractor, $20,000

5. When assessing the financial position of a farm, why do we need both absolute measures and relative measures?
6. Describe and define profitability, solvency, liquidity, and repayment capacity. Name both an absolute and a relative measure for each.
7. Define and compare balance sheet liquidity and cash flow liquidity.
8. Using the data shown in table 7.2, what are the total assets and net worth for this farm? What is the equity/asset ratio for this farm?

Table 7.2.   Assets and Liabilities for an Examle Farm

Assets (market values)	Jan. 1 ($)	Liabilities	Jan. 1 ($)
Cash and checking accounts	−2,173	Accounts payable	0
Feed and grain in storage	42,028	Current farm loans	19,000
Market livestock	46,346	Intermediate farm liabilities	75,149
Breeding livestock	78,400	Long-term farm liabilities	231,703
Accounts receivable	0		
Supplies and prepaid expenses	0		
Machinery, equipment, & buildings	54,469		
Land & buildings	227,168		

9. Last year, another farm had a farm profit of $41,015, an average farm asset value of $487,924, and an average farm debt of $326,772.
   a. If she did not farm, the farmer thinks she could get an off-farm job that pays $20,000 per year. Using these figures, calculate the residual rate of return on equity (ROE) for this farm.
   b. If she could get 7% on her equity, what is her residual return to her labor and management?
   c. If $20,000 and 7% were her goals, did she achieve her goals?
10. Define and contrast an initial analysis and a diagnostic analysis.

NOTE

1. Under different circumstances, say when liquidity is known to be the main problem, other aspects of the farm's financial position may be diagnosed first.

8

Financial Management

Financial management is the process of obtaining, using, and controlling capital: both cash and credit. It uses some of the same tools as financial analysis. However, unlike financial analysis, which looks very much in a feedback mode of management, financial management works mostly on a planning and concurrent basis for managing the farm. In this chapter, we look at alternative sources of capital, estimating the costs of credit, the steps for projecting a cash flow, and developing and using financial control systems.

SOURCES AND USES OF CAPITAL

Capital, that is, money, is a major input for today's farmer. Capital is needed to own the resources required to farm, such as land, buildings, machinery, and other capital assets that are not used up during production. Capital is also needed to pay for operating expenses such as fertilizer, feed, market animals, and other current assets that need to be replaced for each production cycle.

Capital comes in two basic forms: equity capital and debt capital. Equity capital is the money of the owner(s), partners, and other investors. It is the cash and other financial resources they own directly. Equity capital can be used for any purpose in the business: buy capital assets, pay operating expenses, and make loan payments. Equity owners expect a return to their money (much like expecting interest on a savings account). However, equity capital on a farm is a residual claimant on income, that is, paid after other expenses and claims are paid. Debt capital is a liability or other financial obligation on which interest and perhaps other fees have to be paid. Credit is debt capital. Debt capital has to be paid before equity capital receives a return. The return to debt capital is fixed by the terms of the loan or other financial contract. Except for bankruptcy by the business owner, debt capital will receive the same fixed return whether the business has a good or poor year. The owners of equity capital can enjoy the higher returns during high-income years but also incur the risk of very low returns or even loss of equity during poor years.

Owner-operator equity is the largest source of capital for U.S. farmers. This equity comes from savings or reinvestment from farm business income, decreased family consumption, gifts and inheritances, off-farm jobs, other investment income, and other sources. When asset markets, such as land markets, are appreciating, equity can come from increases in the value of the farm's assets, but if assets decrease in value, say land prices fall, equity can also shrink quickly.

Outside equity is another source of equity capital for farmers. Nonfarm partners and other investors may be willing to invest in farms and, thus, increase the capital base for the operator of the farm. While the increased size may be of benefit to the owner-operator, this investment does not directly increase the farmer's equity, and the farmer is also obligated to share farm income with the outside investors. Usually, outside investors are more likely to be involved in the ownership of real estate and other capital assets rather than in supplying operating capital for the farm.

The use of credit (that is, taking on debt) is the second largest source of capital for U.S. farmers. Debt or loans are used for real estate, machinery, and other assets as well as for operating expenses. Sources of loans include commercial banks, government credit sources, insurance companies, machinery manufacturers, merchants, dealer credit, and individuals.

Taking on debt is a risky venture and is avoided by some people. Debt creates future fixed financial obligations and commits a farm's financial resources to a fixed set of assets. Yet, debt can improve farm income, wealth, and progress toward goals in several ways as listed below. Debt can

- create and maintain an adequate business size—to take advantage of economies of size, for example.
- increase the efficiency of the farm business—to change production systems or methods to improve process and product quality by reducing costs, improving timeliness, and so on.
- adjust the business to changing economic conditions—to purchase new technology, adapt to new markets, for example.
- meet seasonal and annual fluctuations in income and expenditures—to allow for purchasing sufficient resources (fertilizer and seed at planting, for example) before products are sold (at harvest, for example).
- protect the business against adverse conditions—to allow for payment of expenses and other obligations when revenue falls short due to adverse weather, disease, price fluctuations, for example.
- provide continuity of the farm business—from one generation to the next.

Leasing and contracting are methods for obtaining the use of capital assets without needing to use equity capital or take on the debt required to purchase those assets. However, leases and contracts are financial instruments and thus create financial obligations for the farm. Through leasing and contracting, a farmer can increase the capital base of the farm. Since they are usually shorter in duration than ownership, leases and contracts can preserve flexibility for adapting a farm to changing conditions in the future. Leasing and contracting do increase the risk of losing control of those specific assets if the leases or contracts are not renewed, but they reduce the risk involved in increasing debt in order to purchase the asset.

CALCULATING LOAN PAYMENTS

Using credit can be beneficial to a farmer, but loans and their associated interest need to be paid according to the terms of the loan. For simple loans such as operating notes taken out and paid back within one year, the payment is a simple calculation of the loan principal times the stated interest rate. For example, suppose a farmer borrows $50,000 at 10% to pay for feeder pigs purchased in April and paid back six months later in October when the pigs reach market weight and are sold. Since the loan is paid back after half a year, the total interest charge is $2,500, which is half of 10% of $50,000. The payment in October for this farmer is $52,500. If there are any, additional noninterest charges would be added to this total payment.

Amortized Loans

For loans that have multiple payments per year or which last longer than one year with annual payments, the original loan principal is amortized (that is, spread out) over the length of the loan and interest is included in the periodic payment. In this situation, we know the loan principal, the quoted interest rate, the length of the loan, and the number of payments per year. What we want to calculate is the periodic payment necessary to pay back the loan. This is usually called the principal and interest (PI) payment. The PI is calculated by the following formula:

$$PI_{(n,i)} = \text{loan principal} \times \text{amortization factor}_{(n,i)}$$

where PI is the periodic principal and interest payment, loan principal is the face value of the loan, n is the number of periods (that is, number of years times the number of payments per year), and i is the interest rate specified in the terms of the loan. The amortization factors can be found in appendix D, table D4, for the specified loan terms or by using the following formula:

Amortization factor$_{(n,i)}$ = $\{i/[1 - (1 + i)^{-n}]\}$

As an example, suppose a farmer wants to buy a used truck that costs $24,000. The banker will finance 75% of the cost at 9% interest for three years with monthly payments and a $100 loan fee. Under these terms, the stated loan is $18,000, which has to be paid back in 36 equal monthly payments over three years. Since the payments will be monthly, the annual rate of 9% is converted to an estimated monthly rate of 0.75% by dividing 9% by 12. The monthly amortization factor is calculated by inserting the monthly rate of 0.0075 and 36 periods into the equation above as shown here:

Amortization factor$_{(36,.0075)}$ = $0.0075/[1 - (1 + 0.0075)^{-36}]$ = 0.0318

Thus, each monthly payment is the loan amount (or principal) multiplied by the amortization factor:

PI$_{(36,.09)}$ = $18,000 \times 0.0318$ = $572.40

As another example, suppose a farmer needs a loan of $250,000 to expand her operation and could obtain financing from the local bank with these terms: annual payments, 20 years, 8.5% interest, 1% loan origination fee, and a $250 appraisal fee. Under these terms, each annual principal and interest payment is:

PI$_{(20,.085)}$ = $250,000 \times 0.1057$ = $26,425

where the amortization factor$_{(20,.085)}$ = $0.085/[1 - (1 + 0.085)^{-20}]$ = 0.1057. This amortization factor can be estimated using the amortization factors for 8% and 9% in the row for the 20th period in appendix table D4. Since 8.5% is average of 8% and 9%, its amortization factor is the average of the factors for 8% and 9%.

Suppose this farmer finds she could also obtain financing from an insurance company at these terms: annual payments, 20 years, 7% interest, 2.5% loan origination fee, and no other costs. Under these terms, each annual principal and interest payment is:

PI$_{(20,.07)}$ = $250,000 \times 0.0944$ = $23,600

where the amortization factor$_{(20,.07)}$ = $0.07/[1 - (1 + 0.07)^{-20}]$ = 0.0944. This amortization factor can be found in appendix table D4 at the intersection of the row for the 20th period and 7%.

Now this farmer is faced with a decision of whether to take a loan with a lower interest rate and thus a lower annual payment but a higher fee or to accept a higher interest rate with a lower fee. This question is addressed in the next section.

ESTIMATING THE COST OF CREDIT

Due to the many different sources of credit that can be used by farmers and the very different terms that those sources may use, the true cost of the alternative sources can be difficult to see immediately. The federal Truth-in-Lending legislation requires lenders to disclose the dollar amount of the interest charges and the annual percentage rate being charged on loan transactions, but many farm loans are exempt from this reporting requirement under Truth-in-Lending. However, if it is not presented, the effective actual percentage rate (APR) can be calculated in two ways. First, using the procedures for estimating the internal rate of return (IRR) shown in chapter 9, "Investment Analysis," the true effective interest rate can be estimated. We will leave this method for chapter 9. Second, the APR can be approximated by solving the following formula for the amortization factor and using the amortization factors in appendix D, table D4, to estimate the effective APR. These estimates of the effective APR can be used to compare alternative sources of credit.

Periodic loan payment = net amount borrowed $\times$ amortization factor$_{(n,i)}$

The net amount borrowed is the original or stated loan amount minus any noninterest charges and fees such as loan points, origination and other signing fees, appraisal fees, stock purchases, and so on. The amortization factor$_{(n,i)}$ is, thus, equal to:

$$\text{Amortization factor}_{(n,i)} = \frac{\text{periodic loan payment}}{\text{net amount borrowed}}$$

This estimated amortization factor is used in the amortization tables (appendix D) to find the interest rate related to the number of periods for the loan being analyzed. If the loan requires more than one payment per year, then the resulting interest rate estimate from the table is multiplied by the number of payments per year to estimate the annual rate.

As an example, let us look again at the two loans the farmer was considering in the last section. This farmer needs a loan of $250,000 to expand her operation. The local bank was willing to provide a loan with these terms: annual payments, 20 years, 8.5% interest, 1% loan origination fee, and a $250 appraisal fee. Under these terms, as we saw in the last section, each annual principal and interest payment is:

$PI_{(20,.085)} = \$250,000 \times 0.1057 = \$26,425$

This farmer could also obtain financing from an insurance company at these terms: annual payments, 20 years, 7% interest, 2.5% loan origination fee, and no other costs. Under these terms, as we saw in the last section, each annual principal and interest payment is:

$PI_{(20,.07)} = \$250,000 \times 0.0944 = \$23,600$

To decide whether she should accept a loan with a lower interest rate but higher fees or one with a higher interest rate and lower fees, this farmer can estimate the effective APR to compare the two options.

The bank will require the farmer to pay a 1% loan origination fee ($2,500) plus a $250 appraisal fee. Thus, the net amount borrowed is $250,000 minus $2,750 or $247,250. Coupling this net amount with the annual payment of $26,425, the amortization factor is calculated to be 0.1069 for this loan (26425/247250). In appendix D, table D4, this factor lies between the factors for 8% and 9% in the row for 20 years. A more accurate estimate of the effective APR can be found by linear interpolation. For example, the factor 0.1069 is 66% of the way from 0.1019 (the factor for 8%) to 0.1095 (the factor for 9%), so the effective APR is 66% of the way from 8% to 9% or 8.66%.

The insurance company will require the farmer to pay a 2.5% loan origination fee ($6,250) but it has no other fees. Thus, the net amount borrowed is $243,750. Coupling this net amount with the annual payment of $23,600, the amortization factor is calculated to be 0.0968 for this loan (23600/243750). In table D4, this factor lies between the factors for 7% and 8% in the row for 20 years. Again using linear interpolation, the factor 0.0968 is 32% of the way from 0.0944 (the factor for 7%) to 0.1019 (the factor for 8%), so the effective APR is 32% of the way from 7% to 8% or 7.32%.

Based on the effective APRs, the loan from the insurance company is a lower cost loan even though it has a higher fee associated with it. However, the final choice between these two loans also has to consider the availability of funds to pay the higher fee versus the benefits of lower annual payments. This part of the decision needs the preparation of a cash flow of the potential expansion and the impact on the balance sheet of the farmer.

CASH FLOW MANAGEMENT

Cash flow management is critical to all farms. Financially strapped farms need to watch their cash flow. Even profitable and growing farms can experience periods of tight cash flow. Farm growth and wise use of cash reserves depend on knowing when cash will flow into the farm and when cash needs to flow out to cover expenses, loan payments, and living expenses, for example. The cash flow statement is the main tool of cash flow management.

As described in the previous chapter, the cash flow statement shows the annual flow and timing of cash in and out of a business. It has four major sections: sources of cash, uses of cash, a flow of funds summary, and the loan balances. As its name implies, the cash flow statement contains any and only cash transactions. It does not include noncash items such as depreciation or farm-produced feed, for example. Although some items not found on an income statement, such as loan payments, are found on a cash flow statement because they are part of the flow of cash, they are not part of the expenses and incomes for an income statement. Examples of cash flow statements are figure 8.1 in this chapter and figures 7.6 and 7.7 in chapter 7.

Projected Versus Actual Cash Flow

A projected cash flow statement or budget for a future period is used by a manager for two main purposes: planning for the future and adjusting plans according to actual results.

- Projecting future cash flows allows a farmer to adjust the timing of sales, expenses, loan receipts, and loan payments before production starts.
- Comparing projected to actual cash flows allows a farmer to adjust the timing of sales, expenses, loan receipts, and loan payments as production is taking place and to change plans and expectations in the midst of the current production period. This process of comparing is explained later in this chapter under "Financial Control."

A projected cash flow is called a cash flow budget in common usage. This change in the name is probably due to the projection showing our expectations of the amounts we have budgeted for spending and receiving, that is, the cash going out and the cash coming in.

Cash flow budgets for planning are prepared on a monthly or quarterly basis. The projected quarterly cash flow statement for the example farm in this chapter is shown in figure 8.1. Statements of actual cash flow often use the annual format shown in figure 7.7.

Projecting a Cash Flow Budget

Projecting a cash flow budget requires a great deal of information and effort but, in return, it provides very crucial information to the farmer. To ensure an accurate statement, an orderly process, such as the steps listed below, needs to be followed. As an example, let us follow the projection made by the example Iowa farmers, Dave and Sue Sanderson, for their dairy farm. The result of their work is shown in figure 8.1.

CASH RECEIPTS	Jan.-Mar.	Apr.-Jun.	Jul.-Sept.	Oct.-Dec.	ANNUAL
Milk sales	248,000	244,000	239,000	235,000	966,000
Calf sales	5,500	5,800	5,500	5,110	21,910
Corn sales	0	1,400	0	0	1,400
Soybean sales	19,000	0	0	22,350	41,350
Alfalfa sales	1,250	0	0	0	1,250
Government payments	4,239	556	0	4,830	9,625
Other farm income	0	0	0	0	0
Miscellaneous	0	0	0	0	0
Capital Sales:					
Cull cows	20,000	19,000	18,500	18,940	76,440
Machinery	0	0	0	0	0
New Term Debt	0	95,000	0	0	95,000
Total cash receipts	297,989	365,756	263,000	286,230	1,212,975

CASH OUTFLOW	Jan.-Mar.	Apr.-Jun.	Jul.-Sept.	Oct.-Dec.	ANNUAL
Operating Expenses					
Seed	11,950	0	0	7,160	19,110
Fertilizer	20,000	3,250	0	14,800	38,050
Crop chemicals	0	12,500	0	4,250	16,750
Crop insurance	2,860	0	0	0	2,860
Drying fuel	0	0	0	5,700	5,700
Crop miscellaneous	600	2,200	700	800	4,300
Machinery expenses	2,200	8,700	2,000	4,700	17,600
Purchased feed	82,600	65,200	72,300	35,800	255,900
Veterinary	10,325	10,325	10,325	10,325	41,300
Livestock supplies	14,875	14,875	14,875	14,875	59,500
Breeding fees	4,375	4,375	4,375	4,375	17,500
Fuel, utilities, repairs	14,000	14,000	14,000	14,000	56,000
DHIA & accounting	2,625	2,625	2,625	2,625	10,500
Hired Labor	43,000	52,000	47,000	52,000	194,000
Land rent	33,350	0	0	33,350	66,700
Real estate taxes	14,319	0	0	14,319	28,638
Farm insurance	5,900	3,300	1,100	5,300	15,600
Trucking and Marketing	6,000	2,000	5,000	10,000	23,000
Miscellaneous	4,500	4,500	4,500	4,500	18,000

Other Outflows	Jan.-Mar.	Apr.-Jun.	Jul.-Sept.	Oct.-Dec.	ANNUAL
Capital purchases					
Breeding Livestock	0	0	0	0	0
Machinery	0	127,500	0	0	127,500
Buildings & improvements	0	0	0	0	0
Other assets	0	0	0	0	0
Proprietor withdrawals	15,000	15,000	15,000	15,000	60,000
Term Loan Payments	0	0	0	0	0
Principal	44,023	14,532	0	14,023	72,578
Interest	39,061	5,850	0	13,563	58,473
Total cash outflow	371,563	362,732	193,800	281,465	1,209,560

Figure 8.1. Projected cash flow budget for Sandersons' dairy farm. (*Continued on next page.*)

FLOW-OF-FUNDS SUMMARY	Jan.-Mar.	Apr.-Jun.	Jul.-Sept.	Oct.-Dec.	ANNUAL
Beginning cash balance	36,500	5,000	5,000	33,666	36,500
Total cash receipts	297,989	365,756	263,000	286,230	1,212,975
Total cash outflow	371,563	362,732	193,800	281,465	1,209,560
Cash difference	-37,074	8,024	74,200	38,431	39,915
Borrowing this period	42,074	0	0	0	42,074
Payment on operating loans					
Principal	0	2,261	39,813	0	42,074
Interest	0	763	722	0	1,484
Ending cash balance	5,000	5,000	33,666	38,431	38,431
LOAN BALANCES AT END OF PERIOD					
Term debt	791,310	871,778	871,778	857,755	857,755
Operating debt	42,074	39,813	0	0	0

Beginning loan balances:			
Term debt	835,334	Operating loan interest rate:	7.25%
Operating debt	0	Desired minimum balance:	5,000

Figure 8.1. (*Continued.*)

1. *Develop the crop and livestock production plan for the whole farm.*

   The Sandersons have a total of 770 crop acres on their farm. They own 310 crop acres and rent an additional 460. They also have a 350-cow dairy and facilities about 12 years old. They have the equipment for corn, corn silage, soybeans, and alfalfa. Based on their estimates of feed requirements and of costs and returns, they decide to grow 250 acres of corn for grain, 135 acres of corn silage, 175 acres of soybeans, and 210 acres of alfalfa. With a three-year rotation, they plan to have 70 of those 210 alfalfa acres in an establishment year using a herbicide treatment, not an oat cover crop.

2. *Estimate the crop production levels and livestock feed requirements.*

   This involves both the quantities produced and needed and the timing of those quantities. End-of-year inventories may need to be maintained for feed until the next year's crop production is harvested.

   For their dairy herd, the Sandersons estimate they will need 39,550 bushels of corn, 2,800 tons of corn silage, and 2,100 tons of alfalfa hay. They estimate they will have enough production of corn and corn silage to meet all their feed needs, but they estimate they will need to buy 1,100 tons of alfalfa hay in the winter besides their own production. They also estimate how much of last year's corn and hay they will have in storage and be able to sell.

3. *Estimate cash receipts from livestock and crops.*

   Based on these production and feed estimates and projected market prices, they estimate the amount of each product they will sell next year. For example, they think they will sell $966,000 in milk and $41,350 in soybeans next year. Since they plan to feed most, but not all, of the corn they grow, their estimated corn sales for next year are only $1,400.

4. *Estimate other income such as interest, government payments, and insurance payments.*

   The Sandersons estimate the total government program payments will be $9,625. This total includes part of the payments based on this year but paid next year and the part of next year's payments that will be paid next year. This category would also include nonfarm income if that is to be included in the farm cash flow.

5. *Estimate cash operating expenses.*

   This includes both variable expenses, such as purchased feed, fertilizer, and labor, and fixed expenses, such as real estate taxes, insurance payments, and loan payments. From their records, published budgets, county offices, and insurance agents, the Sandersons estimate their expenses for the crops and dairy. They estimate they will spend $19,110 for seed next year, $38,550 for fertilizer, $225,900 for purchased feed, $23,000 for trucking, and so on.

6. *Estimate personal and, if included with the farm, nonfarm cash expenses.*

   This includes either family living expenses directly or withdrawals for family living. Together they estimate they will work about three thousand hours in labor time plus more time in management. Based on their living expense records, they want to pay themselves $60,000 for their own labor and management.

7. *Estimate the purchases and sales of capital assets such as machinery, breeding livestock, buildings, land, and ownership in other businesses.*

   The value of the cull cows expected to be sold is estimated as part of the enterprise budget for dairy. They plan to buy a new combine in June using $32,500 in cash, a $95,000 loan, and their two-year-old combine as a trade-in. The loan is entered in the cash receipts section. The $32,500 cannot be seen directly but is part of the $127,500 capital purchase in the cash outflow section. They do not plan to make any other capital purchases or sales next year.

8. *Calculate the scheduled principal and interest payments on current loans and capital leases.*

   The Sandersons have four term loans. To be sure, they check with their creditors to verify their calculations of next year's principal and interest payments. For example, their real estate payment is due in December with $14,023 in principal and $13,563 in interest. Their loan for their dairy facilities is due in February. One of their machinery loans is due in March; the other one is due in June. These four term loans will have a combined balance due at the beginning of next year of $835,334. They will not carry any operating loan into next year. They can obtain an operating loan at an annual rate of 7.25% with a credit line of $100,000.

9. *Estimate the beginning balances of all cash accounts such as checking, saving, and marketing accounts. Determine what desired minimum cash balance is needed to carry from one period to the next.*

After checking their current balances and estimating what they will receive and spend for the rest of this year, they estimate that they will start next year with a cash balance of $36,500. The Sandersons decide they would like a minimum cash balance of $5,000 at the end of each quarter to have a cushion for unexpected fluctuations in the cash flow.

10. *Enter this information into the cash flow format and complete the flow of funds summary.*

As part of the flow of funds summary, project the timing of new debt and potential payments on new debt. Estimate the ending loan balances for each period in the cash flow statement.

All the information on cash receipts, cash outflows, their beginning cash balance, and their beginning loan balances are entered into the quarterly cash flow statement as shown in figure 8.1. Since this is a projection, they do not know the precise amount of each receipt and expenditure, so the totals are divided among the four quarters based on when the Sandersons think they may occur. Fertilizer expenses, for example, usually occur in the late winter and early spring except fall applied fertilizers and any prepayments made in late fall for next year. Seasonal patterns in prices are reflected in some differences. The timing of some amounts is known quite well, real estate taxes, principal and interest payments, for example, so these can be entered in the specific quarter. The timing of other amounts cannot be determined very well and they are simply divided evenly between the quarters—veterinary expenses, for example.

As a first step in completing the flow of funds summary, the Sandersons estimate the cash difference for each quarter. For example, they start with a cash balance of $36,500, add total projected cash receipts of $297,989, subtract total projected cash outflows of $371,563, and find a negative cash difference of $37,074 in the first quarter.

Since the cash difference is negative, they know they need to borrow to maintain their desired minimum balance of $5,000 at the end of the quarter. They do not need to borrow to pay any operating interest since they are not bringing an operating debt into the year. So they estimate they will need to borrow $42,074 to cover the negative cash difference and to have $5,000 left at the end of the quarter.

In the second quarter, the cash difference is $8,024. Since this is greater than the desired minimum balance ($5,000) plus the interest on the operating loan from the first quarter, they will not need to borrow any more. The operating interest of $763 is estimated by multiplying the operating loan balance at the end of the first quarter ($42,074) by the annual interest rate of 7.25% and dividing by four since the cash flow is quarterly. To avoid paying interest when they do not need to, the Sandersons estimate they will be able to pay $2,261 in principal on the operating loan they took out at the end of the first quarter. This $2,261 is estimated by subtracting the interest payment of $763 and the minimum balance of $5,000 from the cash difference of $8,024.

In the third quarter, the cash difference is large enough to pay off the remaining operating debt and have an ending balance greater than the minimum.

Loan balances are calculated by adding new borrowings to the beginning balances and subtracting any principal payments. For example, the beginning total term debt of $835,334 is decreased by a principal payment of $44,023 in the first quarter so the balance at the end of the first quarter is $791,310. In the second quarter, the term debt is increased by a loan of $95,000 listed in cash receipts and decreased by a principal payment of $14,532. So the term debt balance at the end of the second quarter is $871,778. The operating balance starts at $0, increases to $42,074 when that amount is borrowed in the first quarter, and decreases to $39,813 and then to $0 as it is paid off.

With this projected budget, the Sandersons can see when their cash is expected to flow in and to flow out. They can also see that their ending cash balance will have a slight increase by the end of the year: from $36,500 to $38,431. The direction is right, but the quantity is not large, especially when the only additional capital purchase is for a new combine. However, they know they can probably meet their desired level of pay ($60,000).

Other uses of the projection are outlined in the next section. These other uses include trying to see if the loan payments due in the first part of the year could be rescheduled to the end of the year and avoid the $1,484 in operating interest costs shown in this projection.

Uses of a Cash Flow Budget

As just noted, the main use of a projected cash flow is to project the amount and timing of both new borrowing and loan repayments though the year. With the projected cash flow, partners, family members, and financial advisors can offer better advice to the farmer. The projected cash flow budget can also be used in other areas, such as those listed below.

- Evaluate changes in the timing of product sales to coincide with cash needs and, thus, reduce borrowing needs
- Consider changes in the timing of expenditures and scheduled debt repayments to reduce borrowing needs

- Develop a borrowing and debt repayment schedule that fits the project cash flow. This could reduce interest expenses and allow for a more orderly process of arranging for financing needs.
- Assess the value of potential discounts for cash payments for inputs obtained by rearranging the timing of cash inflows and outflows instead of borrowing
- Calculate potential tax impacts of alternative decisions and decide to take advantage of or avoid those impacts
- Obtain a better balance of short-term and long-term debt by observing the impact of alternative debt structure plans on the projected cash flow
- Observe the interdependence of farm, nonfarm, and personal cash flows; take advantage of synergies; and avoid potential conflicts
- Estimate the sensitivity of the initial plan to unexpected changes in prices, crop yields, animal productivities, and so on, to determine if adjustments are needed to develop a more robust plan
- Monitor and control the financial side of the farm, as described in the next section.

FINANCIAL CONTROL

Control is one of the four main management functions: planning, organizing, directing, and controlling. Control is the process of determining and implementing the necessary actions to make certain that plans are transferred into desired results. Through effective control, we are better able to achieve the goals and objectives that have been established.

Financial control takes place in three parts: establishing standards, measuring performance, and taking corrective actions, if needed. Standards are established during planning (especially during enterprise and whole-farm planning, as discussed in chapters 4 and 5). Examples of financial standards include expected interest rates, needed borrowings, and timing of borrowing needs. Measuring financial performance takes place in two ways: monitoring financial conditions and recording actual performance. The three kinds of corrective actions are: (1) rules to change the plan if conditions change (repair instead of replace a combine if interest rates increase, for example); (2) rules to change the implementation of the financial plan (change how much equity capital is used if new sources are found, for example); and (3) change the goals and standards (change the expected interest rate as the Federal Reserve changes its policies, for example).

The *three types of control* (preliminary, concurrent, and feedback) are used in finance also. For example, a projected cash flow statement can show the expected size and timing of needed operating loans, so preliminary control can take place by negotiating for interest rates before the need is immediate.

Monitoring of the actual cash flow through the year is one example of con-current control that allows early detection of deviations from expectations and implementation of needed corrective actions. Year-end financial analysis of actual performance provides feedback information for planning for the future.

Financial control revolves around the *three financial statements*: the cash flow statement, the income statement, and the net worth statement. For con-current control, the cash flow statement is the most useful, as shown in the next section. For feedback control, all three financial statements are useful.

Cash Flow Deviation

The projected cash flow budget is compared with the actual cash flow to devel-op the *cash flow deviation report*. This report shows the source, direction, and size of deviations from the budget or plan. It is used to monitor and control the cash flow of the business and as an early warning system for changes in the plans for profitability, liquidity, and solvency. The cash flow deviation report may not show the problem to be corrected, but it can show where to look. While a cash flow statement contains all the cash transactions for a firm dur-ing a period, a cash flow deviation report shows only the deviations between planned and actual cash flow that violate previously set minimums for devia-tions in absolute amounts and in percentage changes. This is the "management by exception" method where only those items that deviate by a certain amount or percentage, say $2,000 or 10%, are listed in the report because they are *exceptions*. They deviate enough that they need some management attention.

As an example, let us consider again the Sandersons' dairy. For their cash flow deviations report, they set deviation minimums of $1,000 and 10%. An abbreviated example for the second quarter and the six months from January through June is shown in table 8.1.

Table 8.1.   Cash Flow Deviations Report for the Sandersons' Dairy

	Second quarter				Year to date			
	Budget	Actual	Deviation		Budget	Actual	Deviation	
			$	%			$	%
Milk sales	244,000	271,583	27,583	11.3	492,000	521,948	29,948	6.1%
Crop chemicals	12,500	1,633	−10,687	−86.9	12,500	10,763	−1,737	−13.9%
Crop misc.	2,200	2,483	283	12.9	2,800	3,024	224	8.0%
Purchased feed	65,200	38,760	−26,440	−40.6	147,800	216,865	69,065	46.7%

The first item they see is good news: milk sales are going up. As of July 1, sales are up 11.3% for the second quarter and 6.1% for the year to date. Their records confirm what they had been noticing, milk production is close to the planned amounts and the actual milk price has been higher than the price used to project the cash flow. While the deviation is less than 10% year to date, the amount ($29,948) is large enough to check on its effect on cash flow, credit needs, and the ability to alter plans in a favorable way.

At first glance, the second item in the deviation report, crop chemicals, is startling. The actual amount spent in the second quarter is $10,867 or 86.9% lower than planned. Then, they note the year-to-date actual expenditure is closer to the budget. Here, the deviation may be due to a sale on herbicides occurring in January that had not been anticipated. So the physical amount of herbicide used is close to the plan but due to the timing of the sale, the expenditure is both lower and earlier.

The third item, miscellaneous crop expenses, is an example of the deviation violating one minimum, 10%, but not violating the absolute minimum ($1,000). Although even small differences can be important, deviations such as these could be ignored (in favor of larger problems to correct and opportunities to explore) after considering the sources of the deviations.

The fourth item, purchased feed, is much lower in the second quarter than planned: down 40.6%. However, the Sandersons were reading crop reports and saw lower hay production estimates pointing to increases in hay prices. So they decided to buy more hay than planned and to buy it earlier in the first quarter rather than in the second quarter. The year-to-date deviation shows an increase compared with the plan, which supports the story of buying more and buying earlier. The deviation may mean that other parts of the cash flow need attention. For instance, a larger operating loan may have been needed in the first quarter to pay for the increased hay purchases in that quarter. However, the milk price increases causing the increases in milk sales may have offset that need. Further evaluation of the full cash flow statement would answer these overall questions.

Deviations should be evaluated for their sources, for needed adjustments in production and marketing plans, and for impacts on the financial plan (borrowing more or less, for example). Rather than pouring over two full cash flow statements (projected and actual), a manager can use a cash flow deviation report to survey the deviations quickly, identify problems or opportunities, and move on to other management issues. The year-to-date section is necessary so that a month or quarter is not evaluated in isolation from other months. Timing of sales and purchase may be altered from the plan to take advantage of better prices. This can cause major deviations in one month but they are not a deviation in a longer view. It may be good to be able to look ahead, too, in case a sale has been made earlier or a purchase has been delayed.

In each overall category (or set of rows) in a cash flow statement, the management question varies slightly but is still in general terms: what caused the deviation? As an example, let us look at the following categories.

Receipts and Expenditures. If there is a deviation, a manager needs to evaluate how large the difference is and whether the deviation merits any more management time to learn what may have caused the difference.

Borrowings and Loan Payments. Evaluate these with an eye on the ending cash balance for the whole farm. If the ending cash balance is too low, the farm may need to borrow more or make some adjustments elsewhere in the business. These adjustments may include selling earlier than planned, delaying expenditures until the cash is available, or decreasing expenditures. If the ending cash balance is too high, the farmer could evaluate the options of borrowing less, paying more on current loan balances, or building cash reserves for poor cash flow years.

Cash Difference. If this deviates too far from expected levels, the deviations may be from the receipts and expenditures for that period or a carryover from another period.

Beginning and Ending Cash Balances. If other deviations do not "net out," we may need to check with our banker about borrowing more money or prepaying some loans. Deviations in the ending cash balance may need special action, whether the causes are differences in receipts or expenditures.

In all cases, the manager needs to decide if the deviations are due to timing, mistakes, price, unplanned purchases, quantity, unplanned sales, or other events. The appropriate actions will vary with individual farms.

Summary Points

- Financial management is the process of obtaining, using, and controlling capital.
- Capital comes in two forms: equity capital and debt capital.
- Equity capital is the money of the owners, partners, and other investors.
- Debt capital is a liability or other financial obligation on which interest and other fees have to be paid.
- Owner-operator equity is the largest source of capital for U.S. farmers.
- Taking on debt increases risk, but can also improve farm income, wealth, and progress toward other goals.
- Leasing and contracting allow a farmer to obtain the use of capital assets without using equity or debt capital but they are financial obligations.

- Principal and interest payments are calculated by multiplying the loan principal by an amortization factor. The amortization factor is based on the loan's interest rate, the number of payments per year, and the number of years of the loan.
- Since different loans can have different interest rates, fees, and other terms, the effective actual percentage rate (APR) can be calculated and used to compare the cost of alternative sources of credit.
- Cash flow management is critical for both financially strapped farms and profitable farms.
- The four major sections of a cash flow statement are sources of cash, uses of cash, flow of funds summary, and loan balances.
- A projected cash flow statement is used for planning for the future and adjusting plans according to actual results.
- Cash flow budgets are projected in a series of steps starting with estimating the physical quantities and timing of production and inputs and selecting appropriate prices and costs.
- Financial control involves establishing standards, measuring performance, and taking corrective actions, as needed, to assure that plans are transferred into desired results.
- Preliminary, concurrent, and feedback controls are used in financial control.
- Actual cash flows can be compared to a projected cash flow budget to develop a cash flow deviation report. This report can be used to identify those categories deviating from the plan by more preset levels and, thus, needing management attention.

REVIEW QUESTIONS

1. Why are farmers interested in increasing their risk by borrowing capital?
2. How can the alternative sources of credit be compared for their true or actual cost of borrowing?
3. What is the annual principal and interest payment for a loan of $250,000 with these terms: annual payments, 20 years, 7.5% interest, a 2% loan origination fee, and a $200 appraisal fee?
4. What is the effective actual percentage rate, or APR, for the loan in the previous question?
5. What are the four major sections of a cash flow statement?
6. How can both projected and actual cash flows be used by farmers?
7. What are the steps for projecting a cash flow budget?
8. How can a projected cash flow be used by farmers to improve their cash flow management and their profitability?
9. How can the projected and actual cash flow statements be used in financial control?
10. Describe "management by exception" and how it can be used by farmers.

9

Investment Analysis

Machinery, buildings, land, and other capital assets are just as critical to farming as are seed, feed, and labor, but they are acquired, controlled, and used in different ways. Inputs, such as seed, fertilizer, and feed, are purchased and used up in the production process. They cannot be recovered. Labor is hired or "rented" for a certain period only (per hour, per job, or per year, for instance). While labor may be "lost" if it is not used in that certain period, labor is not used up in production. Labor is not owned; it cannot be bought and sold as other inputs or assets. In contrast, capital assets (machinery, buildings, and land) can be purchased, leased, or rented. In the case of machinery, it can also be obtained by hiring custom operators for specific jobs. Capital assets are not used up completely during production, and they can be acquired in several ways. Thus, the decision to use and how to use capital assets is more complicated than the decision to use other inputs.

In this chapter, the decision to use or acquire capital assets is analyzed first as an ownership question, that is, as an investment or capital budgeting problem. Then the investment or ownership option is compared with leasing and renting. The chapter starts with a review of the time value of money, which is critical to understanding how to analyze the benefits and costs of assets that last more than one year. In the rest of the chapter, the process of investment analysis or capital budgeting is explained and shown through examples. Three machinery investment examples are discussed: investing in a combine for custom work, leasing versus owning a tractor, and purchasing a grain drill versus custom hiring the drilling operation.

THE TIME VALUE OF MONEY

One basic concept used in investment analysis is the time value of money. Some people want to spend or invest more money than they currently have, and they are willing to pay others a price to get that money. These "other people" apparently have sufficient money to meet immediate concerns and are willing to be paid not to spend all the money they currently have. In this view, money is a commodity and the market involves time. The time value of money is based on how much people are willing to pay to use someone else's money for a specified period. The time value of money is expressed as an interest rate.

The time value of money is used in one of four ways depending on the investment question being asked. The four methods are compounding, discounting, the present value of an annuity, and amortization (or capital recovery). These are described in the following sections. Understanding discounting and amortizing is critical to understanding and performing investment analysis.

Compounding

Compounding is the simple process of adding interest to a beginning amount. Compounding is what occurs in a savings account; after each period (perhaps daily), the interest to be paid is calculated based on the beginning balance and the interest rate and added to the saving account to calculate the value at the end of the period. This is called compounding because in the second and later periods, interest will accrue to both the initial amount and the interest earned in the early periods. We know the present value and the interest rate, but we do not know the future value.

For example, what is $500 worth after three years at 7% per year? To answer this, we calculate the interest earned in the first year on the initial $500, then the interest earned in the second year on the initial $500 plus the interest earned in the first year, and then the interest earned during the third year on the initial $500 plus the interest earned in both the first and second years. This is compounding because we earn interest in subsequent years on the interest earned in previous years.

In this example, we deposit $500 in the account today (that is, year 0). During the first year, we earn 7% interest on the initial $500. The first year's interest is $35, which is calculated by multiplying $500 by 0.07. Thus, the total value in the bank at the end of the first year would be $535 as shown in table 9.1.

During the second year, we earn 7% interest on $535, the value at the beginning of the second year. The interest for the second year is calculated to be $37.45, and the total value is $572.45 at the end of the second year. During the third year, we earn 7% interest on the beginning value of $572.45. The interest earned during the third year is calculated to be $40.07, and the total value is $612.52 at the end of the third year.

Table 9.1.   Compounding $500 for 3 Years at 7% Interest

Year	Value at the beginning of year	Interest earned during the year	Value at the end of the year
0	—	—	$500.00
1	$500.00	$500.00 × .07 = 35.00	$500.00 + 35.00 = 535.00
2	$535.00	$535.00 × .07 = 37.45	$535.00 + 37.45 = 572.45
3	$572.45	$572.45 × .07 = 40.07	$572.45 + 40.07 = 612.52

Instead of having to prepare a table like that shown in table 9.1 for every compounding problem, the future value can be calculated using the same ideas expressed mathematically. By looking at the process in table 9.1, we can see that the future value of $500 at 7% can also be calculated by multiplying $500 by 1.07 three times, as in the following equation. Note that 1.07 is equal to 1 plus the interest rate written in decimal form.

$500 \times 1.07 \times 1.07 \times 1.07 = \$612.52$

We could also write this mathematical relationship using 1.07 raised to the power of 3 since $500 is multiplied by 1.07 three times.

$500 \times [(1.07)^3] = \$612.52$

Written in a general form for any compounding problem, the future value (FV) of a present value (PV) after n periods at a certain interest rate, i, can be calculated by substituting the actual values for PV, n, and i into the following equation:

$$FV_{(n,i)} = PV \times [(1 + i)^n]$$

In most applications, instead of calculating the expression $[(1 + i)^n]$ for each problem, the calculation of the future value is simplified by using a table of compounding factors (cf) calculated for many pairs of interest rates and number of periods. The compounding factors (cf) are substituted into the equation for the future value (FV) as shown below.

$$cf = [(1 + i)^n]$$

$$FV_{(n,i)} = PV \times cf$$

For the example used above, the compound factor for 7% and three periods is calculated as one plus the interest rate raised to the third power: $(1 + .07)^3$

$= 1.07 \times 1.07 \times 1.07 = 1.2250$. The future value is the present value multiplied by the calculated compound factor:

$\quad FV_{(3,.07)} = 500 \times 1.2250 = 612.50$

Thus, the future value of \$500 that earns 7% for three years is \$612.50 using the compounding factor of 1.2250. The difference of two cents (between \$612.52 in the table above and \$612.50) is due to rounding when the compounding factor for 7% and three years (i.e., 1.2250) is used rather than multiplying \$500 by 1.07 three times.

The compounding factors [i.e., $(1 + i)^n$] for several interest rates and periods are listed in the appendix in table D1. For the example, the compounding factor for 7% and three years (i.e., 1.2250) is found at the intersection of the column for 7% and the row for three periods.

Discounting

When we know a future value but do not know its present value, we work backward by subtracting the value (i.e., interest) that could accrue over time from the future value to estimate the present value. We discount the future value to estimate a present value. Since we know we can earn interest over time but we do not know the future perfectly, we purposely use the verb "discounting." Due to this uncertainty, the term "discount rate" is used instead of interest rate when we are discounting future values.

Discounting can be thought of as the opposite of compounding, the opposite of accruing money in a savings account. For instance, we may know how much we have to pay in the future, and by discounting that future amount, we can estimate how much we need to invest today to have that future amount. For example, we may know what the future balloon payment on our loan will be and want to know what we need to put into a savings account now to pay that balloon.

Discounting is used in investment analysis to compare today's investment with the future income from that investment. We estimate the future income and then convert that into a present value. Those future incomes are not compared directly to the initial investment because they are in the future. So we discount them in both the financial and English meanings of "discount" because they are expected in future years and they are less certain than the price of the initial investment.

In discounting, the present value (PV) of a known future value (FV) received in n periods at a given interest rate, i, is calculated by multiplying the future value by the present value factor (pvf) for the appropriate interest rate and period. We can use the following equation to calculate present values.

$$PV_{(n,i)} = FV \times [1/(1 + i)^n] = FV \times pvf$$

For example, how much should we pay for an investment that will pay us $800 in four years and we could invest the money in another investment that earns 12%. The present value factor for 12% and four years is calculated as one divided by the quantity one plus the interest rate raised to the fourth power: $[1/(1 + 0.12)^4] = 1/(1.12 \times 1.12 \times 1.12 \times 1.12) = 1/1.5735 = 0.6355$. Thus, the present value of an investment that would pay $800 at the end of four years is $508.40 at 12% interest. If we could earn 12% on our money in another investment, we should not pay more that $508.40 for this investment that will pay us $800 in four years.

$$PV = \$800 \times 0.6355 = 508.40$$

The present value factors (pvf) for several discount rates and periods are listed in appendix table D2. For example, the discount rate for 12% and four years (i.e., 0.6355) is found at the intersection of the column for 12% and the row for four periods.

As a check on this discounting process, suppose we invested $508.40 in an account that earned 12%. Let us use compounding to find what we would have at the end of four years. At the end of the first year, we would have earned 12% on our original $508.40 for a total of $569.41, as shown in table 9.2. During the second year, the amount increases to $637.74; the third year, to $714.27. During the fourth year, the amount increases to $799.98. (The difference of $.02 between $799.98 and $800 is due to rounding of the present value factor instead of using the full equation and raising 1.12 to the power of 4.)

Present Value of an Annuity

In some investment problems, the investment needs to be compared with an annuity to be received in the future, that is, the same amount per period for

Table 9.2. Compounding Check of the Discounting Example: $508.40 for 4 Years at 12% Interest

Year	Value at the beginning of year	Interest earned during the year	Value at the end of the year
0	—	—	$508.40
1	$508.40	$508.40 × .12 = 61.01	$508.40 + 61.01 = 569.41
2	$569.41	$569.41 × .12 = 68.33	$569.41 + 68.33 = 637.74
3	$637.74	$637.74 × .12 = 76.53	$637.74 + 76.53 = 714.27
4	$714.27	$714.27 × .12 = 85.71	$714.27 + 85.71 = 799.98

several periods. For example, a land lease with a fixed cash payment for several years is an annuity to the owner. In this type of problem, we are estimating the present value (PV_A) of an n period annuity (A) at a given interest rate, i. To calculate this present value, the annuity is multiplied by the present value annuity factor (pvf_A).

$$PV_{A(n,i)} = A \times \{[1 - (1 + i)^{-n}]/i\}$$

$$= A \times pvf_A$$

For example, suppose a person expected to receive $150 at the end of each of the next three years and could receive 12% in an alternative investment. The present value annuity factor (pvf_A) is:

$$pvf_A = [1 - (1 + 0.12)^{-3}]/0.12 = 2.4018$$

and the present value of the annuity in this example is:

$$PV_A = 150 \times 2.4018 = \$360.27$$

If this person could receive an interest rate of 12%, this person would be indifferent between receiving $360.27 now and $150 in each of the next three years.

The present value factors for annuities (pvf_A) for several discount rates and periods are listed in appendix table D3. For example, the present value factor for an annuity for 12% and three years (i.e., 2.4018) is found at the intersection of the column for 12% and the row for three periods.

Income Capitalization for Land Value Estimation

Estimating the value of land held in perpetuity is a special case of the present value of an annuity. If we own the land in perpetuity, the number of periods is infinite and the formula shown above for the present value of an annuity simplifies to the income capitalization formula shown below.

$$PV = R / i$$

where PV is the cash income value of the land, R is the estimated average annual return to the land, and i is the chosen discount or income capitalization rate. For example, if cash rent were $85 per acre and taxes, insurance, depreciation, and other expenses amounted to $23 per acre, the returns to land could be expressed as an annuity of $62 per acre. (Note that interest and principal are not included in those figures.) With a discount rate of 5%, the land would have a value of $1,240 per acre using the income capitalization method: 1,240 = 62 / 0.05. This method of valuing land is discussed more in chapter 10, "Land Purchase and Rental," along with several points of concern and caution surrounding this simple formula and the complex process of valuing and buying land.

Amortization or Capital Recovery

As discussed in chapter 8, loans that last longer than one year and have annual payments, or loans that have multiple payments per year, have their loan principal amortized over the length of the loan. In this situation, we know the loan principal, the quoted interest rate, the length of the loan, and the number of payments per year. We want to calculate the periodic payment necessary to pay back the loan and the interest due. This periodic payment is also known as the principal and interest (PI) payment. The PI is calculated by the following formula:

$$PI_{(n,i)} = \text{loan principal} \times \text{amortization factor}_{(n,i)}$$

where PI is the periodic principal and interest payment, the loan principal is the face value of the loan, n is the number of periods (that is, number of years times the number of payments per year), and i is the interest rate specified in the terms of the loan. The amortization factors can be found in appendix table D4 for the specified loan terms, or by using the following formula:

$$\text{Amortization factor}_{(n,i)} = \{i/[1 - (1 + i)^{-n}]\}$$

In one of the examples, discussed in chapter 8, a farmer needs a loan of $250,000 to expand her operation and could obtain financing from an insurance company with these terms: annual payments, 20 years, 7% interest, 2.5% loan origination fee, and no other costs. Under these terms, each equal annual principal and interest (PI) payment is:

$$PI_{(20,.07)} = \$250,000 \times 0.0944 = \$23,600$$

where the amortization factor$_{(20,.07)}$ = $0.07/[1 - (1 + 0.07)^{-20}]$ = 0.0944. This amortization factor can be found at the intersection of the column for 7% and the row for the 20th period in appendix table D4.

For some questions, we need to estimate the future annuity that is equivalent to a present total amount. For example, instead of calculating depreciation and interest payments for a depreciable asset (such as a tractor or building), the initial cost of that capital asset can be amortized over its useful life; this is called *capital recovery*. (The IRS has specific rules on how and when this method can be used for tax purposes. What is described here is how to use it for management planning, not tax preparation.) With capital recovery, we estimate the annual capital costs of an asset as the n period annuity [$A_{(n,i)}$] that is equivalent to its present value (i.e., the construction cost or purchase price) at a given interest rate.

$$A_{(n,i)} = PV \times crf$$

$$crf = PV \times \{i/[1 - (1 + i)^{-n}]\}$$

For example, the annual capital recovery costs (ACRC) for a tractor that costs $80,000 are estimated to be $15,896 using a 9% discount rate and a useful life of seven years.

$$crf = 0.09/[1 - (1 + 0.09)^{-7}] = 0.1987$$

$$ACRC = \$80,000 \times 0.1987 = \$15,896$$

The amortization or capital recovery factors (crf) can be found in appendix table D4.

INVESTMENT ANALYSIS (A.K.A. CAPITAL BUDGETING)

The decision to invest in machinery, buildings, land, and other capital assets requires an analysis process that is different from that used to buy seed, fertilizer, feed, and other operating inputs. If we buy seed or feed, we usually expect to receive the results of that purchase within a few months; we may do some budgeting to project if the purchase will be worthwhile but the budgeting is a simple, straightforward process. A capital asset, however, is bought once, and the income from that asset occurs for several years (in perpetuity for land). To decide if we want to buy the capital asset, we make an *investment analysis*, or, in other words, we do *capital budgeting*. Both terms refer to the same process of determining whether an asset is a profitable or wise investment: multiyear budgeting; discounting of future benefits and costs to estimate a present value of those future benefits and costs; and comparing that present value with the initial investment in the asset.

Investment analysis can be described as a series of five steps.

1. *Identify investment alternatives.*

At this initial point, all reasonable or seemingly reasonable alternatives are selected for an initial consideration. A first check is a determination of whether the potential investments fit with the strategy identified and described in an earlier part of the business plan. (See chapter 2, "Strategic Management," for a fuller description of this process.) Some alternatives are dropped from consideration immediately for strategic, personal, and a host of other reasons. Others are kept in consideration for further analysis. Depending on the manager and the situation, only one alternative at a time may be analyzed in depth. The other alternatives that survived the first scrutiny are kept as a backup in case the first choice is ultimately rejected.

2. *Estimate streams of receipts and costs through time.*

   The initial investment needs to be calculated plus the receipts and costs in each year of the planning horizon. Occasionally, as when buying bare land, the expected receipts and costs may be the same in each year. (The effects of risk and uncertainty are discussed in chapter 11, "Risk Management.") With other investments, such as buildings, machines, or fruit trees, for example, the annual receipts and costs (and the potential tax impacts) will change over time.

3. *Evaluate the economic profitability and financial feasibility of investments.*

   An investment is economically profitable if it returns a reasonable profit on the initial investment. It is evaluated in one of three ways: the payback period, net present value, and internal rate of return. Financial feasibility is the investment's own ability to generate sufficient cash flow to cover any debt incurred to make the investment. An investment may be economically profitable but still be financially infeasible because of cash flow problems. For example, planting young apple trees may require outside capital during their development years before they start bearing fruit. Similarly, building a new livestock facility may require outside financing until the animals, equipment, and people bring the operation into the desired ranges of efficiency and productivity.

4. *Sensitivity analysis through scenario analysis.*

   After the initial determination of economic profitability and financial feasibility, the evaluation should be repeated after changing the initial set of assumptions about productivity, prices, costs, interest rates, and other variables. This sensitivity or scenario analysis should determine the importance of different variables that affect the receipts and costs, for example. Both profitability and feasibility should be reevaluated with these new assumptions.

5. *Select investment(s).*

   As with any managerial decision, the information from investment analysis and capital budgeting needs to be evaluated, along with any other pertinent information, and a choice made. If investment alternatives such as the stock market were not considered earlier, they should be evaluated at this point. Although this was done when they were initially selected for analysis, the alternatives again need to be compared with the business strategy. As the alternatives were examined and evaluated more closely, new information may have been found that may change the evaluation of how well one alternative fits the chosen strategy compared with alternatives.

Implementation of the investment choice will vary with the magnitude or importance of the investment. The investment in a new, improved tractor may be viewed as an important but relatively minor investment left to the crop and livestock production aspects of the business to acquire and place into service. In contrast, the decision to build a larger dairy facility will likely be such a major investment that the same ideas for strategy implementation and control (described in chapter 2) are useful and needed to assure proper implementation of such an investment.

Measuring Economic Profitability

Economic profitability can be measured in three ways. **Payback period** is the number of years required to recover the initial cost of the investment. It can be estimated with either discounted or undiscounted future after-tax net returns. **Net present value (NPV)** is the sum of the present values of future after-tax net cash flows minus the initial investment. **Internal rate of return (IRR)** is the discount rate that sets the net present value of the investment to zero.

For an example of calculating these three methods, let us consider an investment with an initial outlay of $20,000. This simple asset has a life of five years, a salvage value of $1,000 at the end of five years, and these estimates of annual net after-tax revenue: Year 1—$2,000, Year 2—$4,000, Year 3—$6,000, Year 4—$9,000, and Year 5—$9,000.

Payback Period

The payback period is the number of years required to recover the initial cost of the investment. It is simple to calculate and has a high emphasis on liquidity. Investments are chosen if the payback period is less than or equal to some established maximum. If investment funds are limited, investments are chosen according to the shortest payback periods.

The payback period is calculated in the following steps:

1. Sum the annual after-tax net cash flows until the cumulative sum will exceed the initial outlay if the next year's net cash flow is added to the sum. The after-tax net cash flows can be either discounted or undiscounted; both are used in business.
2. Estimate the portion of the next period's cash flow necessary to recover the initial investment exactly.

3. The payback period is the number of years of after-tax net cash flow needed to pay back the initial investment, that is, the number of years found in
   step 1 and the fraction of a year estimated in step 2.

For the simple example described above, the sum of the undiscounted net
revenue in the first three years totals $12,000 (table 9.3). To recoup the initial
investment of $20,000, only $8,000 of the next year's undiscounted net revenue of $9,000 is needed. Thus, the

Undiscounted payback period = 3 + 8,000 / 9,000 = 3.89 years

The undiscounted payback period has two deficiencies. First, it ignores any
revenue after the initial investment is recovered. For example, while one
investment might take longer to pay back the initial amount when compared
with a second investment, the second investment would seem to be preferred.
However, the first investment may continue to produce significant revenue
after the initial investment is paid back while the revenue from the second
investment declines. Thus, the first investment may be better, but evaluating by
the payback period alone would miss that extra revenue. The second deficiency of the undiscounted payback period is that it ignores the time value of
money. Two investments with different patterns of revenue over time may be
incorrectly ranked using only the undiscounted payback period.
The second deficiency of the payback period can be overcome by using the
discounted net returns. In this alternative estimation of the payback period, the
future net returns are discounted back to a present value and the length of time

Table 9.3. Estimation of Discounted and Undiscounted Payback Periods

Year	Future value (FV) of the undiscounted net after-tax revenues	Accumulated sum of undis-counted FV	Present value (PV) of FV (at i = .08)	Accumulated sum of PV
1	2,000	2,000	1,852	1,852
2	4,000	6,000	3,429	5,281
3	6,000	12,000	4,763	10,044
4	9,000	$8,000 of the $9,000 is needed to pay back the initial $20,000	6,615	16,659
5	10,000		6,810	$3,341 of the $6,810 is needed to pay back the initial $20,000

required to pay back the initial investment is estimated using the present values of the future revenues. The discounted payback period will always be longer than the undiscounted payback period because the discounted future returns are used and they are smaller because they are discounted. In the example, the

Discounted payback period = 4 + 3,341 / 6,810 = 4.49 years

Net Present Value

The net present value (NPV) is the sum of the present values of future net cash flows minus the initial investment. The net present value measure accurately considers the timing of cash flows. It is the preferred measure of economic profitability for investment decisions. With the net present value measure of investment profitability, the criterion is to accept any investment if NPV > 0 when it is evaluated at the weighted cost of capital. If capital is limited, the choice is made by comparing different combinations of potential investments for the highest total NPV. Mathematically,

$$NPV = \sum_{t=o}^{n} \left[\frac{1}{(1+r)^t} \right] Y_t = Y_o + Y_1 \left[\frac{1}{(1+r)^1} \right] + \dots Y_n \left[\frac{1}{(1+r)^n} \right]$$

where NPV = net present value, r = interest rate or discount rate, and Y_t = estimated net revenue in period t. The initial investment can be either included as Y_o (as a negative amount) in the formula above or, as shown in the table below, subtracted explicitly from the sum of present values.

In the example, with a discount rate of 8%, the present value of the net revenues is $23,465 (table 9.4). After the initial investment of $20,000 is paid, the net present value is $3,465.

Internal Rate of Return

The internal rate of return (IRR) is the discount rate that sets the net present value of the investment to zero.

Table 9.4.   Estimation of Net Present Value

Year	Net revenue	Discount factor*	Present value
1	2,000	0.9259	1,852
2	4,000	0.8573	3,429
3	6,000	0.7938	4,763
4	9,000	0.7350	6,615
5	10,000	0.6806	6,806
		Sum of the present values of the net revenues	23,465
		Minus the initial investment of	−20,000
		Net present value (NPV) at 8%	3,465

*The discount factors are taken from the 8% column in appendix D, table D2.

$$IRR = r^*; \text{ where } NPV = \sum_{t=o}^{n} Y_t \left[\frac{1}{(1 + r^*)^t} \right] = 0$$

With IRR, the decision criterion is to make any investment with an IRR greater than the cost of capital for the farm. For example, suppose a farm's weighted cost of borrowed and equity capital is 12%, the decision criterion is that all investments with an IRR greater than 12% should be made. If capital is limited, the IRR provides a quick measure of which investments provide the best relative returns. The NPV and IRR measures should give the same results.

IRR can be estimated in three ways. First, by trial and error, that is, calculating and recalculating the NPV at different discount rates until the one is found for which NPV = 0. The second method is graphical. Two estimates of NPV at different discount rates are placed on a graph and a line drawn through them and across the discount rate axis. The point at which the line crosses the discount rate axis is where the NPV = 0 and is, thus, the IRR. The third method is to use a function on a computer spreadsheet or financial calculator.

In the example, the NPV are calculated for a few discount rates between 6% and 15% (table 9.5). From the results of this trial and error approach, we estimate that the IRR is approximately 13% since the NPV at 13% is only 8.

For IRR to be estimated graphically, NPV is placed on the vertical axis and r on the horizontal axis as shown in figure 9.1. A pair of r and NPV are plotted and a line is drawn through the horizontal axis connecting the pair. For better estimations, the two discount rates should not be close. In the example investment, 12% and 15% provide a positive NPV and a negative NPV and, thus, a more accurate slope for the line. The IRR is the point where the line crosses the horizontal axis since NPV is zero at that point. In the example graph, the line crosses at approximately r = 13% so the IRR is approximately 13%.

Investment Choice Example

Suppose a farmer had $20,000 to invest in two alternatives: A and B. Investment A is estimated to provide a net cash return of $6,000 per year for 5 years but to have no salvage or terminal value. Investment B is estimated to provide $5,000 per year for 5 years and to have a terminal value of $8,000 at the end of the fifth year. (All values are received at the end of the year.)

Using the methods just described, the net present value of investment A is estimated to be $3,956, the internal rate of return to be 15.2%, the undiscount-

Table 9.5. Estimated NPV of the Example Investmen at Various Discount Rates

Discount rate (r):	.06	.08	.09	.12	.13	.15
Estimated NPV:	5,086	3,465	2,710	639	8	−1,174

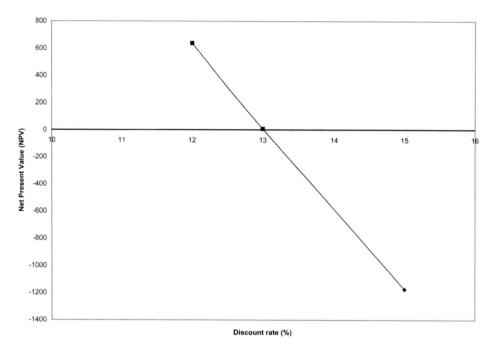

Figure 9.1. Graphical estimation of IRR for the example investment.

ed payback period to be 3.3 years, and the discounted payback period to be 4.0 years (table 9.6). For investment B, the net present value is estimated to be $5,409, the internal rate of return to be 16.3%, the undiscounted payback period to be 4.0 years, and the discounted payback period to be 5.0 years (table 9.7).

Table 9.6.   Analysis of Example Investment A

Year	Net cash revenues	pvf at 8%	Present value of net cash revenues	Accumulated undiscounted net cash revenues	Accumulated present value of net cash revenues
1	6,000	0.9259	5,555	6,000	5,555
2	6,000	0.8573	5,144	12,000	10,699
3	6,000	0.7938	4,763	18,000	15,462
4	6,000	0.7350	4,410	24,000	19,872
5 + SV*	6,000	0.6806	4,084	30,000	23,956
	Sum of present value of net cash revenue		23,956		
	Minus the initial cost (year 0)		−20,000		
	Net present value		3,956		
	Internal rate of return		15.2%		
	Undiscounted payback period		3.3 years		
	Discounted payback period		4.0 years		

*SV = salvage value of 0 for investment A.

Table 9.7.   Analysis of Example Investment B

Year	Net cash revenues	pvf at 8%	Present value of net cash revenues	Accumulated undiscounted net cash revenues	Accumulated present value of net cash revenues
1	5,000	0.9259	4,630	5,000	4,630
2	5,000	0.8573	4,287	10,000	8,917
3	5,000	0.7938	3,969	15,000	12,886
4	5,000	0.7350	3,675	20,000	16,561
5 + SV*	13,000	0.6806	8,848	33,000	25,409
Sum of present value of net cash revenue			25,409		
Minus the initial cost (year 0)			−20,000		
Net present value			5,409		
Internal rate of return			16.3%		
Undiscounted payback period			4.0 years		
Discounted payback period			5 years**		

*SV = salvage value of $8,000 for investment B.
**The discounted payback period is set at an even 5 years since the discounted salvage value is needed to reach $20,000 and that is not received until the end of year 5.

This investment analysis shows that investment B is preferred when using the measures of NPV and IRR since they are higher for B compared with A. However, the higher annual net cash revenues for A do provide a faster payback period for A. Using the discounted payback measure, B has to wait until the asset is sold after the fifth year to receive the last $36.45 for the full payback. Thus, if a person would prefer to pay back the investment quickly or considers the salvage value of B in five years as too uncertain, investment A would be chosen. If another person is not as worried about liquidity and places a premium on rates of return, investment B would be chosen.

Combine Investment Example

Suppose a farmer is considering the purchase of a combine to use for custom work as one way to increase farm income. The initial purchase price of the combine is $157,000: $116,400 for the basic combine unit, $21,700 for the 8-row-30" corn head, and $18,900 for the soybean head (Lazarus). The combine is expected to last eight years with a salvage value of $37,837 after eight years. Based on expected usage of 1,069 acres of soybeans and 1,018 acres of corn, the direct operating costs per acre are estimated to be $5.95 for soybeans and $6.41 for corn. Labor expenses per acre are estimated to be $2.39 for soybeans and $2.51 for corn. By estimating the housing and insurance as 3% of the initial purchase price, total costs per acre (excluding depreciation and interest costs) are $10.70 for soybeans and $11.28 for corn. These costs are expected to increase at the rate of 5% per year.

This farmer thinks he could sign enough contracts to harvest 2,000 acres (1,000 each of corn and soybean) starting in the first year. The local custom harvest market also provides $23 per acre for custom harvesting soybeans and $25 for corn. These custom rates are expected to increase at the rate of 5% per year.

The marginal tax rate for this farmer is 28%. For taxes, he will take the full Section 179 expensing option with this combine. He has chosen a discount rate of 8%. By including a charge of $9 per hour for his labor, the annual net cash flow can be viewed as the return to the investment in the combine and not as a mixture of return to labor and investment.

Using this information and these numbers, the annual net cash flow and its present value are estimated (table 9.8). Cash income is the number of acres harvested multiplied by the appropriate custom rate—increased by 5% each year. Cash expense is the number of acres harvested multiplied by the appropriate total operating cost—increased by 5% each year. Taxes are calculated by subtracting cash expenses and depreciation from cash income and then multiplying by the marginal tax rate, 28%. The salvage value is estimated from engineering data. The annual net cash flow (ANCF) is the cash income minus the cash expense and taxes plus the salvage value. The present value factor (pvf) is based on the discount rate of 8%. The present value of ANCF is the multiple of ANCF and pvf.

Based on these calculations, the investment in a combine for custom work has a net present value (NPV) of $22,849 and an internal rate of return (IRR) of 11.3% (table 9.8). The payback periods are 5.7 years (undiscounted) and 7.4 years when the cash flow is discounted.

This investment seems profitable, but is it financially feasible? Suppose the farmer could obtain a loan for 75% of the purchase price (that is, a loan of $117,750) and the bank offered loan terms of 9% for 5 years. Using the amortization formula, the annual payment is calculated to be $30,273 which is greater than the ANCF in any of the first 5 years. Before declaring the investment not financially feasible, however, the tax savings due to the deductibility of the interest payments needs to be accounted for.

The tax savings is calculated from the marginal tax rate for this farmer and the interest portion of each year's loan payment. The interest portion is calculated from the loan balance at the beginning of the year. After adjusting the loan payment to the after-tax payment, we can see that these loan terms (9% and 5 years) result in a deficit in four of the first 5 years (table 9.9). The combine cannot pay for itself. If the farmer could negotiate a loan length of just one more year, to 6 years, the combine purchase would be financially feasible even if the interest rate stayed at 9%.

Table 9.8. Present Value Analysis of a Combine Purchase

Year	Cash income	Cash expense	Taxes	Salvage value	Annual net cash flow (ANCF)	pvf at 8%	PV of ANCF
1	48,000	21,970	–1,795		27,825	0.9259	25,763
2	50,400	23,069	181		27,150	0.8573	23,276
3	52,920	24,222	2,165		26,533	0.7938	21,062
4	55,566	25,433	3,652		26,481	0.7350	19,464
5	58,344	26,705	4,074		27,565	0.6806	18,761
6	61,262	28,040	4,517		28,705	0.6302	18,090
7	64,325	29,442	4,982		29,901	0.5835	17,447
8	67,541	30,914	7,861	37,837	66,603	0.5403	35,986
				Present value of the net cash flows			179,849
				Minus the initial cost of investment			–157,000
				Net present value of investment			22,849
				Internal rate of return			11.3%
				Payback period (undiscounted)			5.7 years
				Discounted payback period			7.4 years

267

Table 9.9.   Financial Feasibility of the Combine Purchase

Year	ANCF	Loan payment	Interest payment	Principal payment	Tax savings	After tax payment	Surplus or deficit
1	27,825	30,273	10,598	19,675	2,967	27,306	519
2	27,150	30,273	8,827	21,446	2,471	27,802	−651
3	26,533	30,273	6,897	23,376	1,931	28,342	−1,809
4	26,481	30,273	4,793	25,480	1,342	28,931	−2,450
5	27,565	30,273	2,500	27,773	700	29,573	−2,008
6	28,705						28,705
7	29,901						29,901
8	66,603						66,603

Machinery Purchase Versus Lease

Suppose a farmer has a choice of either purchasing or leasing a tractor. If purchased, the tractor would cost $70,000, have annual cash operating expenses of $6,000, and a salvage value of $10,000 after eight years. The loan terms were 20% down, 10% interest, six years, and one payment per year. Leasing would require an initial payment of $10,000, lease payments of $12,500 at the end of each year (including the first year), and the same operating expenses of $6,000 per year. The leased tractor has no terminal value for the farmer. For this farmer, let us say the marginal tax rate is 28%.

For this problem, the question is which alternative has the lowest cost. The farmer has already determined that a tractor is needed and that the same level of tractor services can be found by either purchasing or leasing; neither alternative will provide more profit than the other. Thus, the question is simplified to which alternative provides the lowest cost for the same tractor services. Including production and sales would be redundant and add complexity in this problem. (In the previous example, the combine purchase was being considered to create income; thus, custom work income had to be included in the analysis.)

The information above is used to calculate the present value of the after-tax costs over the life of the tractor for purchasing and for leasing. The cash outlay is the initial payment and the costs associated with either owning or leasing the tractor. Tax benefits come from being able to deduct those expenses and, thus, lower taxable income. For this problem, a 10% discount rate is used to change the future values in years one through eight into a comparable present value of the cash outlays.

In this example, purchasing is the cheaper option. The total, after-tax costs of purchasing the tractor are estimated to be $70,285 over the eight years and the total, after-tax costs of leasing are estimated to be $78,261 (tables 9.10 and 9.11).

Perennial Crop Establishment

Perennial crops such as fruit and nut trees usually involve a period of years before production starts and especially before the annual value of production exceeds the annual costs. Compared to annual crops, a producer may endure perhaps several years of negative cash flow, not just one year, before positive

Table 9.10. Present Value of the Costs of Purchasing a Tractor

Year	Cash outlay	Salvage value	Tax savings	After-tax cash outlay	pvf at 10%	PV of after tax cash outlay
0	14,000		4,200	9,800	1.0000	9,800
1	18,858		4,897	13,961	0.9091	12,692
2	18,858		5,991	12,867	0.8264	10,633
3	18,858		5,136	13,722	0.7513	10,309
4	18,858		4,462	14,396	0.6830	9,832
5	18,858		4,192	14,667	0.6209	9,107
6	18,858		3,894	14,964	0.5645	8,447
7	6,000		3,567	2,433	0.5132	1,249
8	6,000	10,000	−176	−3,824	0.4665	−1,784
Present value of the after-tax cash outlay for purchasing						70,285

Table 9.11. Present Value of the Costs of Leasing a Tractor

Year	Cash outlay	Salvage value	Tax savings	After tax cash outlay	pvf at 10%	PV of after tax cash outlay
0	10,000		2,800	7,200	1.0000	7,200
1	18,500		5,180	13,320	0.9091	12,109
2	18,500		5,180	13,320	0.8264	11,008
3	18,500		5,180	13,320	0.7513	10,007
4	18,500		5,180	13,320	0.6830	9,098
5	18,500		5,180	13,320	0.6209	8,270
6	18,500		5,180	13,320	0.5645	7,519
7	18,500		5,180	13,320	0.5132	6,836
8	18,500	0	5,180	13,320	0.4665	6,214
Present value of the after-tax cash outlay for leasing						78,261

cash flows are obtained. Other than the longer period of negative cash flow, investment analysis of the decision to invest in establishing a perennial crop follows the same procedures as the investment examples shown earlier in this chapter.

As an example, let us consider removing the old orchard and establishing a new almond orchard on 40 acres near Modesto, California, in the northern part of the San Joaquin Valley. We'll assume this farm already exists, so initial investment in investment year 0 is minimal, but significant costs are incurred during year 1 for land preparation, trees, planting, and so on. Almond nut meat production is expected to start with a small yield of three hundred pounds per acre harvested in the third year (table 9.12). Nut meat production increases up to the established yield of two thousand pounds per acre in the seventh year. The nut meat is projected to be worth $1.25 per pound. In the eighth year, both production and costs stabilize and are projected to remain the same for the remainder of the orchard's life. This farm has noncash expenses associated with establishing and producing almonds including shop building, land, irrigation system, and other equipment. Establishment costs start to be recovered in the eighth year. Tax impacts are estimated assuming the farmer has a marginal tax rate of 28% on net income after both cash and noncash expenses. The after-tax net cash flow is calculated as the cash income minus the cash expense plus the tax impact. Using a discount rate of 10%, the net present value (NPV) of establishing an orchard is $3,170 per acre. The IRR of the after-tax net cash flows is 21.6%. These estimated returns to almond establishment can be compared to estimates of establishing other orchards and to estimates of growing annual crops for 25 years.

Calculating the Effective Actual Percentage Rate (APR)

As mentioned in chapter 8, one method for calculating the effective actual percentage rate (APR) is to use the procedures for estimating the internal rate of return (IRR). To do this the stream of payments (akin to the after-tax net cash flows used above) is set with the net amount borrowed as the initial amount and the periodic payments as negative amounts for each of the periods under the terms of the loan. This stream of payments is entered into the IRR calculation procedures in a spreadsheet or calculator. The resulting IRR is the APR for the stream of payments and, thus, the terms of the loan.

As an example, let us consider again the two loans the farmer was considering in the last chapter. In the first loan, the local bank was willing to provide a loan of $250,000 at 8.5% annual interest for 20 years with annual payments. The annual payment would be $26,425. To obtain these terms, the farmer must pay a 1% loan origination fee ($2,500) plus a $250 appraisal fee. Thus, the net amount borrowed is $247,250. To calculate the APR using the IRR procedure,

Table 9.12. Investment Analysis of Establishing an Almond Orchard

Year	Almond meat production (pounds)	Cash income	Cash expense	Noncash expenses including cost recovery	Tax impact	After tax annual net cash flow (ANCF)	Present value factor (10%)	Present value of ANCF
				($/acre)				
1	0	0	2,702	756	−968	−1,734	0.9091	−1,576
2	0	0	601	750	−378	−223	0.8264	−184
3	300	375	927	764	−368	−184	0.7513	−138
4	800	1,000	1,384	765	−322	−62	0.6830	−42
5	1600	2,000	1,528	765	−82	554	0.6209	344
6	1800	2,250	1,567	765	−23	706	0.5645	399
7	2000	2,500	1,589	765	41	870	0.5132	446
8	2000	2,500	1,633	1,098	−65	932	0.4665	435
9	2000	2,500	1,633	1,098	−65	932	0.4241	395
10	2000	2,500	1,633	1,098	−65	932	0.3855	359
11	2000	2,500	1,633	1,098	−65	932	0.3505	327
12	2000	2,500	1,633	1,098	−65	932	0.3186	297
13	2000	2,500	1,633	1,098	−65	932	0.2897	270
14	2000	2,500	1,633	1,098	−65	932	0.2633	245
15	2000	2,500	1,633	1,098	−65	932	0.2394	223
16	2000	2,500	1,633	1,098	−65	932	0.2176	203
17	2000	2,500	1,633	1,098	−65	932	0.1978	184
18	2000	2,500	1,633	1,098	−65	932	0.1799	168
19	2000	2,500	1,633	1,098	−65	932	0.1635	152
20	2000	2,500	1,633	1,098	−65	932	0.1486	138
21	2000	2,500	1,633	1,098	−65	932	0.1351	126
22	2000	2,500	1,633	1,098	−65	932	0.1228	114
23	2000	2,500	1,633	1,098	−65	932	0.1117	104
24	2000	2,500	1,633	1,098	−65	932	0.1015	95
25	2000	2,500	1,633	1,098	−65	932	0.0923	86
Net present value of investment (NPV) at 10%								3,170

Source: Based on Duncan et al. 2002.

the stream of payments starts with receiving $247,250 in year 0 (as in an investment problem) and continues with 20 years of paying $26,425. Plugging this stream into a spreadsheet procedure with 247,250 as a positive number and 26,425 as a negative, the IRR procedure calculates the APR to be 8.66%. This is slightly different from the 8.64% calculated in chapter 8 due to rounding.

As you may recall, this farmer could also obtain financing from an insurance company at these terms: annual payments, 20 years, 7% interest, 2.5% loan origination fee, and no other costs. Under these terms, the farmer pays a 2.5% loan origination fee ($6,250) but has no other fees. So, the net amount borrowed is $243,750 and the annual payments are $23,600. To calculate the APR using the IRR procedure, the stream of payments starts with receiving $243,750 in year 0 and continues with 20 years of paying $23,600. Plugging this stream into a spreadsheet procedure with 243,750 as a positive number and 23,600 as a negative, the IRR procedure calculates the APR to be 7.33%. This is slightly different from the 7.32% calculated in chapter 8 due to rounding.

Since the loan terms are unchanged, calculating the effective APRs using the IRR procedure yields the same estimates and the some conclusion. That is, the loan from the insurance company is still the lower cost loan even though it has a higher fee associated with it. However, as before, the final choice between these two loans also has to consider the availability of funds to pay the higher fee when the loan is received versus the benefits of lower annual payments. And this part of the decision still needs the preparation of a cash flow of the potential expansion and the impact on the balance sheet of the farmer.

Summary Points

- Investment analysis (or capital budgeting) is needed since capital assets such as machinery, buildings, and land last longer than one year and the benefits and costs occur in more than one year.
- The time value of money is calculated in four ways: compounding, discounting, the present value of an annuity, and amortization (or cost recovery).
- Investments need to be analyzed for both economic profitability and financial feasibility.
- Economic profitability of an investment is measured by the payback period, net present value (NPV), and internal rate of return (IRR).
- Payback period is the number of years required to recover the initial cost of the investment. It can be estimated with either discounted or undiscounted future after-tax net returns.

- Net present value (NPV) is the sum of the present values of future after-tax net cash flows minus the initial investment.
- Internal rate of return (IRR) is the discount rate that sets the net present value of the investment to zero.
- Investment analysis can be used to compare ownership with leasing of capital assets.

REVIEW QUESTIONS

1. Why isn't a dollar spent today worth the same as a dollar to be received in three years?
2. What is the difference between the interest rate earned on a savings account and the discount rate used in investment decisions?
3. Compare and contrast economic profitability and financial feasibility in investments.
4. Why is the timing of receipts and expenses critical in the analysis of an investment?
5. Define the payback period used in investment analysis.
6. Why is the discounted payback period preferred to the (undiscounted) payback period?
7. What is the main disadvantage or problem with the payback period, whether it is discounted or not?
8. Define the net present value (NPV) of an investment. How is NPV calculated?
9. Define the internal rate of return (IRR) of an investment.
10. Given the investment example with an after-tax initial investment of $74,500 and the after-tax net cash flows listed below, calculate net present value (NPV) using a discount rate of 10%, estimate the internal rate of return (IRR), and estimate the (undiscounted) payback period.

Table 9.13. Estimated After-Tax Net Cash Flows for Example Investment

Year	After-tax annual net cash flow
1	−43,300
2	12,680
3	34,000
4	49,200
5	65,600
6	62,400

10

Land Purchase and Rental

Land is a critical asset for farming. Unlike machinery, buildings, and other depreciable assets, land is not expected to wear out (excluding the problems caused by erosion or contamination). So income from land is expected to accrue to the owner in perpetuity, that is, forever. This requires a different method to evaluate potential land investment and control. The first part of this chapter shows how to estimate the value of land to be purchased and the maximum rational bid for land.

The use of land also can be obtained by leasing or renting. The second section of this chapter explains the rental options commonly used by farmers, how to compare a tenant's costs with a landowner's costs, and how this information can be used for negotiating a rental agreement.

PURCHASING LAND

The value of land comes from several sources. The income the land can produce and the appreciation in land market values are the two sources of value that we consider in this text. Land value also comes from factors such as natural resources, recreation, hunting, open-space interest, water quality impacts, and historical considerations.

Land value can be estimated by either the market value or income capitalization method. With the market value approach, the value of land is determined by comparing it with similar pieces of land that have sold recently. With the income capitalization method, the estimated income per year is assumed to last into perpetuity and a simple formula is used to capitalize that income into a present value of the land. These methods are described below.

These two methods, market value and income capitalization, provide potential buyers and current owners with estimates of what they think a specific piece of land may be worth to them. They need to compare their estimated values with what the land market is asking and decide whether to buy or sell.

Market Valuation

Market valuation involves comparing the land of interest with the sales of comparable pieces of land. The best comparable sales are close to the land of interest and similar in land quality, buildings, proximity to markets, access to roads, and other factors that affect economic returns. However, even the best comparable sales will not be identical so adjustments in the value need to be made. A first step in estimating market value would be to estimate the value of the bare land by subtracting the value of any buildings from the total value of the sale. Then the quality of the land (soil type, topography, drainage, crop yield potential, etc.) can be assessed and used to adjust the price of the bare land. The value of any buildings on the land needs to be appraised and added to the total value. Adjustments are also made for differences in the size of the parcels, distance to the buyer's current operation, and other factors that affect economic returns.

Income Capitalization

Using the income capitalization method, land value is calculated from the expected annual return to land and the discount rate (or income capitalization rate). As discussed briefly in chapter 9, the income capitalization formula is:

$$PV = R / i$$

where PV = the present value of the future stream of income, R = the annual net returns to land, and i = the income capitalization rate. This is usually done on a per acre basis but could also be used on a whole-farm basis.

Annual net returns are estimated using enterprise budgets and projecting yields, prices, and costs into the future. Since these obviously will vary in the future, a sensitivity analysis is necessary to evaluate the variability of the land price and the importance of different factors. Future conditions for price, yield, and costs also need to be forecast and incorporated into the estimate of the annual net returns.

The discount rate for land valuation is called the *capitalization rate* (or *cap rate*, in shorthand) in recognition that it is chosen with different factors in mind than a discount rate chosen for depreciable assets. Some obvious factors affecting the choice of the capitalization rate are other interest rates (such as on U.S. Treasury bonds), projections of inflation, the riskiness of production on the land, and the real growth of the economy in this area. If the net returns do not include taxes, the cap rate should be chosen on a pretax basis; if taxes are included, the cap rate should be on an after-tax basis.

Another way to choose a desired cap rate is to consider the sources of increasing land value and decide how much each source should contribute to the total return. Land value usually comes from two main sources: income production and land market appreciation. If a landowner wants to earn a total of, say, 15% on the market value of the land and land values have been increasing an average of 11% per year in the area, the income from the land should provide the other 4%. Thus, in this example, a cap rate of 4% would show the income potential of the land. And if the market continues to increase at the average rate, the owner will receive the other 11% when the land is sold.

The example of income capitalization in chapter 9 was of a landowner who received $85 in cash rent per acre, paid $23 in taxes, insurance, depreciation, and other expenses (but not including any interest and principal for the land), and received a net return from land of $62 per acre. With an income capitalization rate of 5%, the land would be valued at $1,240 per acre:

$$PV = 62/.05 = 1,240$$

For another example, consider a farm couple, the Robinsons, who are considering buying the farm they have been renting since they started farming in 1996. The current owners have offered to sell the farm to them privately for $2,500 per acre. They have become very familiar with its yield history and are confident their future net returns would be similar to those they have had in the past. They have had the 300 acres in a 50-50 corn-soybean rotation and expect to continue that rotation in the future. The Robinsons want to average $18,000 or $60 per acre from their crops for living. (They also have some other farm-related business activities for additional income.) Their yields, prices, production costs, and net returns since 1996 are listed in table 10.1.

For the six years the Robinsons have rented this farm, their average net cash returns have been $127 per corn acre and $138 for soybeans. With an average government payment of $41 per acre, their average net cash return has been $174 per acre for the 50-50 corn-soybean rotation. To estimate the resulting land value, they must first subtract their estimated real estate tax ($11 per acre), general insurance ($5 per acre), and a charge for their labor and management ($60 per acre). This results in an estimated net return after all expenses of $98 per acre (i.e., 174 − 11 − 5 − 60 = 98).

Using a cap (or capitalization) rate of 5%, the value of the land using the income capitalization method is estimated to be $1,960 per acre (i.e., 98 / .05 = 1,960). Based on this income capitalization value of $1,960 per acre, the

Table 10.1. Yields, Prices, Costs, and Net Returns for the Robinsons' Farm

	1996	1997	1998	1999	2000	2001	Ave.
Corn yield, bu./acre	135	147	178	163	151	150	154
Corn price, $/bu	3.25	2.78	2.12	1.86	1.75	1.69	2.24
Corn production costs, $/acre*	203	215	234	228	200	201	214
Corn net returns by year, $/acre	236	194	143	75	64	53	127
Soybean yield, bu/acre	43	48	57	47	46	45	48
Soybean price, $/bu	7.37	7.12	5.68	5.38	5.16	4.55	5.88
Soybean production costs, $/acre*	137	149	157	152	126	126	141
Soybean net returns by year, $/acre	180	193	167	101	111	79	138
Average government payment, $/acre	13	19	43	62	69	39	41
Average rotation net return, $/acre	221	212	198	150	157	105	174

*These are total cash production costs and do not include cash rent or their own labor.

offer price of $2,500 is too high. The Robinsons do not think they could afford to pay that price and still cover all the costs plus their desired living expense.

If the current owners were to say that $2,500 per acre was a firm asking price (say due to local land market conditions) and the Robinsons still wanted to buy the farm, they have several possibilities to consider that might allow them to raise the estimated value under the income capitalization method. They could change their production techniques to increase yields or decrease production costs (while still maintaining yield), change their marketing methods to increase prices received, decrease their income expectations, or use some combination of these ideas. If they dropped their income expectations to $30 per acre and left the other figures unchanged, the income capitalization value becomes $2,560 per acre.

Perhaps their estimate for the capitalization rate is too high. Choosing the cap rate is definitely not as easy as calling to ask the price of diesel. Decreasing the capitalization rate to 4% would increase the income capitalization value to $2,450. The question now becomes whether they have chosen the right cap rate.

So, should the Robinsons buy the farm for the asking price? Well, that depends on what else is happening in the land market. If land values are forecast to appreciate (i.e., increase), the increase in value can be factored into the price that could be paid. Other factors (besides net returns from production) that affect land values include the demand for land for farm enlargement; the growth of towns, cities, and industrial areas; and government fiscal, monetary, and tax policies. These factors are not easily incorporated into the income capitalization formula (and should not be included in that formula in the way it is presented here).

One way to incorporate these other factors is to modify the income capitalization formula to evaluate a shorter time period (say ten years) and account for the present value of a forecasted price (say ten years in the future). In this modification, the land produces income for a specified number of years (not in perpetuity) and has an estimated market value at the end of this period. The modified formula has the value of farming for a set number of years (which is the value of the land held in perpetuity minus the value of the income flow after the end of the specified period) and adds the present value of the anticipated price (AP) at the end of the period:

$$PV = \sum_{t=1}^{n} [R_t / (1 + i)^t] + AP / (1 + i)^n$$

where R_t is the estimated net return in each year from 1 to n, AP is the anticipated price of land, n is the year of the anticipated price, and i is the discount rate of future income and not the income capitalization rate used earlier.

As an example of using this modified formula, suppose a land auction is scheduled in the near future. A neighboring farmer wants to know the maximum bid price that is reasonable for this land. She determines that, for the standard corn-soybean crop mix, the expected annual net income per acre is $112.50. She decides the appropriate discount rate is 8%. If land prices continued to increase at the historical rate of 4% per year, she calculates that today's typical price of $1,150 per acre in her area will be $1,700 in ten years.

Using the modified formula, the present value of farming for ten years is $755 per acre and the present value of the anticipated land price in ten years is $787. The maximum bid price is thus, $1,542 per acre.

$$PV = \sum_{t=1}^{10} [112.5 / (1 + .08)^t] + 1,700 / (1 + .08)^{10}$$
$$= 755 + 787$$
$$= 1,542$$

Financial Feasibility

This discussion of estimating the value of land is analogous to the discussion in the last chapter on evaluating the economic profitability of investing in depreciable assets such as machinery and buildings. As also described in the last chapter, the next step in deciding to buy land would be to evaluate the financial feasibility of the purchase using the estimated returns and the available loan terms. A sensitivity analysis also needs to be done to determine the impact of the variability in the land price and the importance of various factors affecting the land value and financial feasibility.

As an example of evaluating the financial feasibility of buying a farm, let us consider the Robinsons' potential purchase of the farm discussed earlier. Suppose their banker offers them a loan at 7.5% for 30 years if they pay 25% down. If the owners will not reduce the price below $2,500 per acre, the total price for the 300 acres is $750,000. The down payment would be $187,500, the loan would be $562,500, and the annual loan payment would be $47,628. The after-tax impact of the loan payment is reduced by their ability to deduct the interest portion as a cost and thus reduce their taxes due. They are in the 28% marginal bracket. Even with the 25% down payment, the Robinsons realize they would not be able to obtain their income goal of $60 per acre. Some sensitivity analysis shows they will have to reduce their desired income goal to $35 in order to make the loan payment and pay all other costs. This is done by subtracting the loan payment from the annual net cash flow and adjusting for the tax savings due to the additional savings (table 10.2). Notice that the surplus becomes a deficit in year nine even after the Robinsons' reduced their income goal to $35 per acre. Also note that the annual net cash flow does not change; this is one assumption of the income capitalization method. Only 15 years are shown in table 10.2 for space considerations.

LAND RENTAL

Renting land is an alternative to owning. Renting allows a farmer to increase the size of his or her operation without having to commit (or even have) the money needed to buy the land. For starting farmers, renting is a very good way to enjoy the benefits of a larger operation without taking on large amounts of debt. Renting also provides a farmer with more flexibility to move his or her operation to different geographical areas. For example, a young farmer could initially expand by renting land over a relatively large area and then, as land becomes available closer to the home farm, stop renting land at the farthest points in favor of land closer.

Over half of the cropland in the United States is rented. The landowners may be the tenant's parents or siblings, retired farmers, surviving spouses, or off-farm investors. Landowners either manage the rental arrangements themselves or hire a professional farm manager to handle part or all aspects of land management.

The two most common rental agreements for farmland are a cash lease and a crop share lease. Flexible cash rent agreements are also available but not as widely used as a straight cash lease.

Table 10.2.   Example of Financial Feasibility Using Robinsons' Farm

Year	Annual net cash flow	Loan payment	Interest payment	Principal payment	Tax savings	After-tax payment schedule	Surplus (+) or deficit (−)
1	36,900	47,628	42,188	5,440	11,813	35,815	1,085
2	36,900	47,628	41,779	5,848	11,698	35,929	971
3	36,900	47,628	41,341	6,287	11,575	36,052	848
4	36,900	47,628	40,869	6,758	11,443	36,184	716
5	36,900	47,628	40,363	7,265	11,302	36,326	574
6	36,900	47,628	39,818	7,810	11,149	36,479	421
7	36,900	47,628	39,232	8,396	10,985	36,643	257
8	36,900	47,628	38,602	9,025	10,809	36,819	81
9	36,900	47,628	37,925	9,702	10,619	37,008	−108
10	36,900	47,628	37,198	10,430	10,415	37,212	−312
11	36,900	47,628	36,415	11,212	10,196	37,431	−531
12	36,900	47,628	35,575	12,053	9,961	37,667	−767
13	36,900	47,628	34,671	12,957	9,708	37,920	−1,020
14	36,900	47,628	33,699	13,929	9,436	38,192	−1,292
15	36,900	47,628	32,654	14,973	9,143	38,484	−1,584

Share Rental Agreements

With a *share lease*, the owner agrees to share in some of the direct growing costs. Typically, the owner shares in the seed, fertilizer, crop insurance, drying, and transportation costs. With a typical share lease, the tenant pays for fuel, oil, repairs, hired labor, and machinery depreciation, and the owner pays the real estate taxes, general insurance, and land loan interest. Some share owners also share in the costs of weed control; others say that the tenant pays for all weed control and decides whether to use chemical or mechanical weed control. With the improved efficacy of herbicides, the common approach today is to share weed control costs. Usually, the owner takes ownership and control of his or her share of the crop at harvest. In a traditional crop share lease, the tenant and owner are responsible for the storage and marketing of their own shares in the crop. Since the owner is receiving the physical crop as the rental payment under a share lease, the owner is assuming some of the risk of the resulting value of the crop and the tenant is relieved of some of the risk. Thus, the owner can benefit from good weather and good prices more with a share lease but is open to the problems caused by poor weather and poor prices. The tenant loses some of the potential benefit of good weather and good prices but, with a share lease, is able to give some of the risk to the owner.

In a share rent agreement, the tenant farms the land with the owner paying a share of the direct production costs and receiving a share of the physical product. Once the physical yield is divided between tenant and owner after harvest, they are responsible for their own marketing decisions. Since he or she will benefit from good yields, a landowner usually takes a more active management role with a share rent agreement, that is, the landowner may want to help decide which varieties to plant, fertility levels, planting and harvesting schedules, and so on. Compared to a cash rent, the landowner takes on more risk of what the yields and prices will be, so in an average or typical year, the landowner should receive a higher return to land than a typical cash rent. Landowners like share rent if they like to be more involved in the farming operation; they do take on more risk of bad years, but they also have the chance to enjoy good years. Tenants, especially young farmers, may appreciate the decreased risk they face with share rent agreements since some of the price and yield risk is shifted to the landowner. Common shares or percentages in share rent agreements are 50-50 and 60-40 for the tenant and landowner respectively but these shares do vary by locality.

For a share rental agreement to be fair, the tenant's and landowner's shares of production should be equal (or nearly equal) to their shares of all expenses. If production is not shared in the same way as expenses, some inputs may not be applied at economically correct levels. For instance, if a tenant receives 50% of production but is expected to pay for 60% of the fertilizer expenses, the tenant will not realize the full benefit of the crop's response to fertilizer and, thus, decide not to apply as much as the landowner may want. In reality, shares usually match well due to gradual adjustments over time. Also, in tight land markets, tenants strive to keep renting the land they currently rent. Thus, they will make sure fertilizers and other production inputs are applied at levels to achieve good production and to keep landowners pleased.

Cash Rental Agreements

With a cash lease, the tenant pays the owner a fixed amount per year and then owns all of the produced crop to use or sell as the tenant determines. Under a cash lease, the tenant pays all the direct growing costs for that crop (seed, fertilizer, pest control, fuel, crop insurance, transportation, and so on); the landowner pays all the costs associated with owning the land (real estate taxes, land loan interest, general insurance, building depreciation, and so on). With a cash lease, the tenant assumes all the risks of producing and marketing the crop; the owner assumes only the risk of the tenant not paying the specified, fixed rent.

In a cash rent agreement or cash lease, the tenant farms the land and pays the landowner a fixed cash rent. The landowner does not pay any production costs and does not receive any physical crop to market. Compared to the share rent, the tenant with a cash rent agreement has more freedom to make production decisions. The tenant makes all crop marketing decisions. With a cash lease, the landowner does not participate in management decisions except perhaps in setting guidelines for crop rotations, fertility levels, erosion control, and other concerns related to maintaining soil and environmental quality; these guidelines are usually written into the rental agreement. Landowners have less risk with cash rent; they do not have an immediate or direct worry about what the yield and price will be in the rental year. The only risk a landowner may have with a cash lease is the risk of the tenant not being able to pay the rent. Because there is less risk, the landowner should expect to receive less net return to land with a cash rent agreement than with a share rent agreement. Landowners who do not want the worry of making marketing decisions or the risk of a bad year may want a cash lease. Tenants who want more freedom in production decisions and who can take on the additional risk may appreciate a cash lease over a share rent agreement. Typical cash rents vary geographically and by soil quality.

Historically, the timing of cash rent payments has been half in the spring and half in the fall. However, some landowners, and perhaps an increasing number of them, are requiring payment of the entire rent in the spring to avoid the need to file a landlord's lien with the county in order to have protection in case the tenant defaults. Having to make the full payment in the spring increases the cash flow needs and financing requirements of the tenant since the tenant cannot delay half of the cash rent until after harvest time and the chance to sell some of the crop to pay rent.

Flexible Cash Rental Agreements

Flexible or variable cash rents are also available for landowners and tenants. With these leases, the final cash rent is determined after harvest when the current year's yield and/or price are known. Landowners may enjoy this type of lease because they do not have to market their own crop but they can enjoy the higher returns in good years. Tenants may enjoy this type of lease because they can shift some risk to the landowner and maintain control over more of the production decisions. A tenant who wants all the grain for livestock but also wants to decrease his or her risk may want to consider a flexible cash lease. Some common variations of the flexible cash lease are described below.

Base rent multiplied by the ratio of current year's price to a stated base price.
The tenant and landowner specify the base rent and base price in the lease using typical or expected prices, yields, and costs. The lease also specifies how the current year's price is to be determined. This could be done by

choosing a certain period (September 15 to November 15, for example) and calculating the average price at a specific location, the local elevator, for example. Say the tenant and landowner agree that the base rent will be $90 per acre and the base price will be $2.00 for corn. If the current year price turns out to be $2.15, the rent is increased to $96.75 per acre ($96.75 = 90 \times 2.15/2.00$). The procedure could be modified slightly to account for a typical crop rotation by specifying the base rent and base prices for both corn and soybeans and determining the annual rent based on the average of the corn and soybean price ratios. In a multiple year lease, the rules of the base rent and base price would remain the same, but the annual cash rent would vary due to the current year's price varying.

Base rent with stated adjustments for price changes. Rather than changing the annual rent for any change in the current year's price, this form of the flexible cash lease describes how the base rent will be adjusted if the prices moves out of a specified range. Perhaps the rent changes are made only when the price moves above a specified price. This form of flexibility results in less change in the annual rent than the first alternative. Some tenants and landowners may prefer to have more stability in their cash flows.

Fixed amount of commodity. The lease defines the rent in terms of physical yield, say number of bushels or tons, and also defines a procedure for determining the current year's price. The rent for each year is the set physical yield multiplied by the current year's price. In a multiple year lease, the number of bushels would remain the same, but the annual cash rent could vary as the current year's price varies.

Base rent multiplied by the ratio of current year's price to a base price and by the current year's yield to a base yield. This variation in flexibility is similar to the first option described except the yield is also included. As in the earlier example, say the tenant and landowner agree that the base rent will be $90 per acre and the base price will be $2.00 for corn plus they also agree that the base yield for corn will be 150 bushels per acre. If the current year price turns out to be $2.15 but the actual yield is 145, the rent is increased to $93.52 per acre ($96.75 = 90 \times 2.15/2.00 \times 145/150$). This annual rent is lower than the earlier example because the actual current year's yield is lower than the base yield.

Stated percentage of the current crop's value. This variation is very similar to a share rent agreement except a pricing procedure is specified rather than the landowner taking physical possession of the crop.

Minimum base rent plus a percentage of increased value. With this variation, the tenant agrees to pay a fixed, minimum base rent plus a specified percentage of an increase in value based on the current year's yield and price. Two similar but slightly different methods for determining the flexible portion in this lease are (1) as a percentage of the yield above a base yield with the price chosen on a certain day or period and (2) as a percentage of the current year's crop value over a specified base. In the first option, the bonus or variable portion is based on the actual yield exceeding a base and then valued using the current year's price. With this option, the variable portion will be positive if the yield is high and the price is low. If the yield is below the base and current prices are quite high, the landowner will not receive a higher payment even though farm income may be higher. With the second option, the bonus or variable portion of the rent is based on each year's combination of yields and prices. Thus, if low yields and high prices happen in the same year, the total value could still exceed the specified base value and the landowner still receives a positive variable rent payment.

Rent Negotiation

When negotiating to rent or lease land, astute farmers and landowners need the same two pieces of information. First, they need to know how competitive the rental market is. By knowing how competitive the market is, each side can decide how hard it can push the other party in the negotiations. In most areas, the land rental market is very competitive. When the market is competitive, landowners have more power to negotiate better rental terms for themselves as owners. They have this power because they know other renters are looking for more land to rent and could be willing to pay higher rents. If the land market is less competitive, the landowner does not have as much power as in a very competitive market so the tenant has more ability to negotiate rental terms.

The second piece of information both parties need to know is their own and the other's estimated revenues and costs. Although they may not know the other's revenue and costs perfectly, having an estimate allows each side to know which costs are hard, which costs are softer and could be changed, and which are desired returns to land, labor, management, and risk. This knowledge of revenues and costs allows each side to decide how hard to negotiate without breaking the negotiating process and losing a relationship. For example, how high can a landowner raise rent before no renter is willing to work for no pay? Or, how much of a rent decrease can a tenant ask for and still keep a landowner's return to land at a reasonable rate? Also, by knowing their own costs, tenants know when the potential rent becomes too high for their desired profit levels and, thus, when they need to consider other options.

As an example, let us look at a recent estimate of the costs of producing corn in southwestern Minnesota (table 10.3). The landowner wants to receive $100 per acre as an opportunity cost of owning the land, which is about a 4.0% return on the land value from production. The landowner also is expecting to

Table 10.3.   Comparison of Tenant's and Landowner's Costs for Corn for Grain in Southwest Minnesota for 2002

	Total value	CASH RENT		SHARE RENT	
		Tenant's share	Owner's share	Tenant's share	Owner's share
REVENUE					
Yield (bu/ac)	160	160			
Price ($/bu.)	2.15	2.15			
Gov't payment ($/ac)	26	26			
Total ($/ac)	370	370	0		
DIRECT EXPENSES	($/ac)	($/ac)	($/ac)	($/ac)	($/ac)
Seed	38	38		19	19
Fertilizer	50	50		25	25
Chemicals	24	24		12	12
Crop Insurance	10	10		5	5
Drying fuel	6	6		3	3
Fuel & oil	11	11		11	0
Repairs	22	22		22	0
Miscellaneous	6	6		6	0
Operating interest	11	11		7	4
Total direct expense	178	178	0	110	68
OVERHEAD EXPENSES					
Hired labor	8	8		8	
Real estate taxes	13		13		13
Farm insurance	5		5		5
Utilities	3		3		3
Interest (opportunity cost)	100		100		100
Depreciation	30	30		30	
Miscellaneous	7				7
Total overhead expenses	166	38	128	38	128
Labor & management	30	30	0	30	0
Total listed expenses	374	246	128	178	196
Net Return before rent	− 4	124	−128		

SO, for cash rent negotiation (given these returns and costs):
    The maximum cash rent for tenant is: 124
    The minimum cash rent for owner is: 128
SO, for share rent negotiation (given these costs):
    The share of these costs and, thus, the fair share of revenues is: 48%   52%

have an additional return from land appreciation, but will realize that at the time of selling the land. For their unpaid labor and management, the tenants want to receive $30 per acre.

By splitting the revenues and costs between the tenant and the landowner, the potential net returns to each can be estimated. If the tenant received the estimated yield and price and paid the estimated costs and desired labor and management returns, the tenant's maximum cash rent is the difference between estimated revenue and estimated costs or $124 per acre in this example. In order for the landowner to pay the estimated costs and receive the desired returns to the land investment, labor, and management, the minimum cash rent the owner could receive is the total of these costs, or $128 per acre. The tenant and landowner can use these estimates (1) to compare with what the land rental market is saying and (2) as starting points for rent negotiation.

For share rent negotiation, the fair share of the production is the share each party has in the total costs. For this example, the tenant and the landowner are sharing the costs of seed, fertilizer, crop chemicals, and crop insurance equally. Given the costs in this example, the fair share for the tenant is 48% and for the landowner is 52%. These percentages are quite close to the market in this area where a 50-50 share rent agreement is common.

For both types of leases, these estimates would be the beginning positions. The dynamics of the local land market, the unique conditions of each party, and the negotiating power of each person would determine the final rental rates. If either or both parties wants higher returns or the expected prices and yields are low, the tenant's maximum could be lower than the owner's minimum. In this case, a tenant could change production techniques to increase yields or decrease production costs (while still maintaining yield), change marketing methods to increase prices received, decrease income expectations, or use some combination of these ideas. Or the owner may have to adjust his or her desired returns.

Typical Lease Terms

In addition to the names of the owners and tenants and the legal description of the land, written lease agreements contain many more terms and clauses describing how the land will be farmed and who has what rights, obligations, and duties. The typical lease terms are described in the following section, but each lease agreement and pair of landowners and tenants may change the general lease agreement by adjusting standard terms and by adding new terms and clauses.

General terms include the time period covered, how long the lease is in effect, how it will or can be reviewed and terminated, and how it can be amended or altered. The landowner usually retains the right of entry and access at reasonable times and without interfering with the tenant's regular operations to consult with the tenant, to make repairs, and to perform farming operations after the agreement is terminated. Clauses are usually included that say the agreement does not create a partnership relationship, that the tenant does not have the right to sublease any part of the farm, that the owner can transfer the title to the farm subject to the provisions of the lease, and that the lease is binding upon the heirs, executors, administrators, and successors of both the landowner and the tenant. Additional clauses, such as a landlord's lien, may be added to cover specific situations and conditions.

Clauses are usually included to stipulate how the land can be used. The landowner may want to specify what crops can be grown and perhaps the acres of each, any restrictions on a tenant's crop choice, participation in government programs, and how government program payments will be shared. Permission and restrictions on livestock and pastures may be included as needed.

Several clauses cover the expectations of the tenant regarding the operation and maintenance of the farm. The tenant usually agrees to provide the labor to maintain the farm in good condition, control noxious weeds, control erosion, and repair erosion control structures. The tenant also agrees to pay all costs of operation except those specified to be shared in a share rental arrangement and those costs specified elsewhere in the agreement. The tenant also agrees not to do several actions without written permission from the landowner: not to spend more than a specified amount for maintenance and repairs, not to plow pastures, not to cut live trees, not to pasture new seedlings in the year they are first seeded, not to cause violation of terms in the landowner's insurance policies, not to add improvements or incur expenses for the landowner, not to add plumbing or wiring, and not to damage erosion control structures. The tenant agrees to pay when leaving reasonable compensation for damages to the farm if responsible.

The agreement also usually includes several expectations of the landowner regarding the operation and maintenance of the farm. The landowner is expected to replace or repair as promptly as possible the dwelling or any other building or equipment regularly used by the tenant that is damaged or destroyed by causes beyond the control of the tenant or to make rental arrangements in lieu of replacements. The landowner is also expected to provide materials to the tenant for normal maintenance and repairs, to provide skilled labor that the tenant is unable to perform satisfactorily, and to reimburse the tenant for materials up to the limit specified in the agreement. The landowner is usually required to allow the tenant to make improvements (at the tenant's expense) that are removable and that do not mar the condition or appearance of the farm and to allow the tenant to remove the improvements upon the termination of the lease. In a share agreement, the landowner is expected to pay directly or to reimburse the tenant for the landowner's specified share of production expenses and for any field work done for crops to be harvested in the year following termination of the lease.

Both the tenant and the landowner agree not to obligate the other party for extension of credit or for payment of debts. How the costs of establishing permanent pastures and of lime and other long-lived fertilizers are to be shared by the tenant and landowner is specified in a typical agreement. Procedures for arbitration of differences are also specified in most agreements.

Other terms depend on the specific conditions of the farm and needs of the parties, the geographical area, and other unique characteristics. The rent for and use of dwellings and livestock facilities are specified as needed. The ownership of mineral rights may be specified in a rental agreement in areas where mineral rights are an important part of total land value.

SUMMARY POINTS

- The value of land comes from several sources.
- Land value can be estimated by either the market value or income capitalization method.
- With the market value approach, the value of land is determined by comparing it with similar pieces of land that have sold recently.
- With the income capitalization method, the estimated income per year is assumed to last into perpetuity and the present value of the land is estimated by dividing the estimated income by the capitalization rate.
- Land investment involves questions of both economic profitability and financial feasibility.
- Renting land is an alternative to owning.
- The two most common forms of rental agreements are the cash lease and share rent.
- With a share lease, the landowner agrees to share in some of the direct growing costs and receives a share of the production.
- If costs are not shared in the same proportion as production, optimal levels of fertilizer and other inputs may not be applied.
- With a cash lease, the landowner receives a fixed cash payment, does not pay for any direct production costs, and does not receive any production.
- Various forms of flexible cash rents are available but not widely used.
- Estimating and comparing the costs and returns of both the landowner and the tenant, and knowing current market rental conditions, provide both parties a good base from which to negotiate land rental rates.
- The typical lease specifies the payment level and terms and also includes several clauses describing expectations of the landowner and the tenant concerning treatment of the land, ending the agreement, and other conditions.

REVIEW QUESTIONS

1. Jeremy and Kathy Hansen are considering buying a farm they have been renting for several years. The farm has been offered to them privately for

$2,050 per acre. Since they are very familiar with its yield history and pro-
duction costs, they are confident their future net returns would be similar to
those they have had in the past. They have had the 150 acres in a 50-50
corn-soybean rotation and expect to continue that rotation in the future.
Their yield and financial records for the past six years are in table 10.4.

In addition to this land, the Hansens own and farm another 300 tillable
acres. They have been striving to average $30,000 a year for living and debt
repayment from their current cropping enterprises and want to continue
that expectation. They are both 56 years old and are thinking of retiring
within ten years. Thus, they are considering the purchase as a potential part
of their retirement plan.

a. What is the average net return per acre for corn, for soybeans, and for
    the 50-50 corn-soybean rotation?
b. What cost should they use for their own labor and management?
c. Based on this information, what is your estimate of the annual return to
    land?
d. Using a cap (or capitalization) rate of 5%, what is the value of the land
    using the income capitalization method?
e. If the Hansens were considering selling the 150 acres in ten years for
    their retirement, what is the value of the land? Assume the asking price
    of $2,050 is an accurate reflection of the land market and that land prices
    increase by an average of 6% per year for the next ten years. Use a dis-
    count rate of 7% and the modified income capitalization method.
f. Do you think the Hansens should buy the farm? Why or why not?

Table 10.4. Yields, Prices, and Costs for the Hansens

Corn yield, bu./acre	108	172	140	140	156	186
Corn price, $/bu	2.06	2.21	2.32	3.31	2.67	2.08
Soybean yield, bu./acre	45	53	52	50	53	51
Soybean price, $/bu	6.11	6.07	5.72	7.28	7.02	5.82
Production costs:*						
Corn, $/acre	193	182	198	196	203	211
Soybean, $/acre	109	114	118	127	133	136

*These are their total production costs on their rented land. These costs do not include the rent they paid. Howev-
er, the costs do include the potential real estate taxes they would have had to pay as owners. Principal and interest payments are not included. Their own labor expenses are not included.

2. Carl Krill is debating whether to rent more land. He thinks some will be coming available close to his farm. Using the budget information below and the procedures described in the text, fill in table 10.5 and decide whether Carl should rent this land or not.

Table 10.5.   Carl Krill's Estimated Corn Production Costs and Rental Evaluation

	Owner operator	Cash rent evaluation		Share rent evaluation	
		Tenant	Owner	Tenant	Owner
RECEIPTS					
Yield (bushels per acre)	145				
Price ($ per bushel)	2.15				
Gross sales ($ per acre)	312				
Gov't transition payment	30				
Total income ($ per acre)	342				
DIRECT COSTS	($/acre)				
Corn seed	27				
Fertilizer	38				
Chemical	35				
Crop insurance	5				
Fuel & oil	9				
Repairs	21				
Drying fuel	7				
Miscellaneous	3				
Operating interest	7				
TOTAL DIRECT COSTS	152				
OVERHEAD COSTS					
Hired labor	8				
Interest on land	80				
Machinery depreciation	38				
Real estate taxes	11				
Farm insurance	5				
Miscellaneous	8				
TOTAL OVERHEAD COSTS	150				
Operator labor and mgt.	30				
TOTAL LISTED COSTS	332				
NET RETURN ($/acre)	10				

Estimated beginning negotiating positions (using the figures listed above):

Maximum cash rent feasible for tenant:			xxx	xxx	xxx
Minimum cash rent needed by landowner:		xxx		xxx	xxx
Estimated share in total expenses:		xxx	xxx		

With a cash lease, Mr. Krill will receive all production and the full government payment and pay for all direct costs plus these overhead costs: hired labor, machinery depreciation, and miscellaneous expenses. Under a cash lease, the landowner will pay the real estate taxes and farm insurance, plus she wants $80 as the opportunity interest payment for the value of the land.

Since a 50-50 share rent agreement is very common in his area, Mr. Krill uses those terms to evaluate a share lease. Under these terms, the landowner will receive 50% of the production and 50% of the government payment; the owner also will pay 50% of the seed, fertilizer, chemical, and drying costs as well as the overhead of real estate taxes and farm insurance; the owner wants the full $80 in opportunity interest on the land.

a. What is the maximum cash rent feasible for the tenant?
b. What is the minimum cash rent needed by the landowner?
c. What are the tenant's and landowner's estimated shares in total expenses?

11

Risk Management

Every decision in farm management involves some risk. We cannot avoid risk, but by taking risks, we have the chance to accomplish our strategic and financial objectives—our goals. The farm manager needs to incorporate risk into his or her management process so risks can be considered explicitly.

The goal of risk management is to balance a farm's risk exposure and tolerance with the farm's strategic and financial objectives, such as income, wealth, environmental quality, and personal goals. This balancing is done after considering the sources of risk, the methods of reducing risk, the ability and the willingness to take risks, and the income potential of alternative strategies. The goal of risk management is not to reduce risk only; other objectives would not be met then. Risk management involves choosing how we use our resources (time, land, money, etc.) best to achieve our personal and business objectives.

This chapter has several parts. After presenting the main sources of risk faced by a farmer, many management options are identified and discussed briefly. Then crop insurance options are discussed in more depth. In the rest of the chapter, the methods for making risky decisions are presented along with the process of developing scenarios for assessing the future, analyzing different potential paths, and considering the impact of different scenarios on objectives.

For most of us, risk and uncertainty refer to the same thing: variation and change that cannot be completely controlled. Sometimes distinctions are made between risk and uncertainty. **Risk** is used when the decision maker knows all the possible outcomes of an action and the objective probability of each outcome. **Uncertainty** is used when the decision maker knows part or all of the possible outcomes, but cannot quantify the probabilities. In this book, risk and uncertainty are used interchangeably. For a working definition of risk, let us say a manager faces risk when decisions must be made but the outcome of one or more alternative actions is unknown.

SOURCES OF RISK

Any business or person faces risk coming from many sources. These can be grouped into five main sources: production, marketing, financial, legal, and human resources. Let us look at each of these briefly.

Production risk. Production risk is not knowing what your crop yield, animal productivity, or other production will be. The major sources of production risk are weather, pests, diseases, technology, genetics, machinery efficiency and reliability, and the quality of inputs.

Marketing risk. We do not know with certainty what prices will be. Unanticipated forces, such as weather or government action, can lead to dramatic changes in crop and livestock prices.

Financial risk. Financial risk has four basic components: (1) the cost and availability of debt capital, (2) the ability to meet cash flow needs, (3) the ability to maintain and grow equity, and (4) the increasing chance of losing equity by larger levels of borrowing against the same net worth. While the first three components are influenced by internal and external forces, the fourth is very much affected by farmers' decisions on how much debt they take on compared with their equity.

Legal risk. Legal issues are involved with almost every aspect of farm management and day-to-day activities of farm operation. The legal issues most commonly associated with agriculture fall into four main areas: business structure and tax and estate planning, contractual arrangements, tort liability, and statutory compliance, including environmental issues. All these contribute to the potential for extensive risk exposure for a farm business.

   Political risk is the risk of changing policies—both governmental and institutional policies. Changes in federal farm policy or in a bank's lending policy, for example, can increase the risk of not achieving a farm's objectives and goals. Policies are usually expressed through rules and regulations that farmers may follow and enjoy the benefits or violate and face penalties and lack of potential benefits. Risk can be present in the inability to follow the rules and in not knowing certain rules.

Human resource risk. People can bring both increased risk exposure and increased ability to deal with risk. Death, divorce, injury, illness of a principal owner, manager, or employee of a farm can disrupt how the farm performs or even survives. If people are not managed well, risk will increase due to improper operation and application of production and marketing procedures, for example. However, the proper hiring of the right people for the right jobs and the use of training are examples of using people to reduce risk exposure. Moral risk is due to the behavior of people inside and outside the business. Moral risk is caused not just by corrupt and criminal behavior, but also by devious and less-than-truthful behavior by individuals and other companies.

Managing Risk

With all these possible sources, how does a farmer decide which sources need attention first? One way is to rank the sources based on two criteria: (1) potential impact on the business and (2) the probability of the risk happening. The potential impact can range from a small impact to a catastrophic impact. The probability of happening ranges from low to high. By evaluating each possible source of risk on each criterion and putting the two criteria together, a farmer can group the sources into those that need immediate attention and action down to those that need no attention or action. The graph in figure 11.1 can help a manager decide which sources of risk have the highest priorities and thus need attention before other sources.

An extreme example of using these criteria and prioritizing risk involves a lightning strike. If a person is struck by lightning, catastrophic results may occur. If the sky is clear while planting, for example, the chance of a lightning strike is extremely low. The combination of a potential catastrophic impact but extremely low probability of happening puts this occurrence in the extreme lower-right corner of figure 11.1. Thus, very little action is needed to protect the farmer from a lightning strike. A farmer could continue to work in the field with little concern. However, suppose a thunderstorm is forecast. This raises the probability of a lightning strike out of the lower-right corner but only enough, say, to require a farmer to be more attentive to the sky and potential development of a storm. Suppose a severe thunderstorm has developed and is approaching directly toward the farmer in the field, the probability of a light-

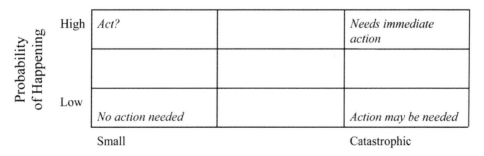

Figure 11.1. Prioritizing risks.

ning strike is rather high. Catastrophic results may occur if a person is struck, so the situation is now in the upper-right corner of figure 11.1. Immediate action is needed. A farmer planting in such a situation would most likely move to a protected area or at least into an area that reduces the chance of being struck by lightning.

The problems of price and yield variability are not as dramatic as a lightning strike, but if a farmer suffers a poor yield in a low-price year, the results may be catastrophic for the finances of that farmer. Based on current market information and past yield history, a farmer can decide how much protection is needed. That is, how much crop insurance to buy, how much hedging to do, whether to buy an insurance product such as Crop Revenue Protection (CRP), which includes both yield and price protection. One set of market and weather conditions may suggest a high probability of a catastrophe happening and thus a higher need for action. Another set of market and weather conditions may not create such a heightened need for protection. The cost of the crop insurance and hedging options will also have to be balanced with the potential losses.

Other examples show how farmers can be in different sections of figure 11.1 and thus respond to similar risks in different ways. The destruction of a livestock confinement building due to fire would be catastrophic for most farmers but fire usually has a low probability of occurring. Thus, the purchase of insurance to cover losses due to fire (and, within the same policy, several other potential causes) is probably sufficient. A farmer with buildings on a flood plain has a higher probability of being flooded and a higher need to buy flood insurance compared with a farmer on high ground. In the Midwest, crops on sandy soils have a greater chance of crop failure due to lack of moisture, so a farmer may consider the need for irrigation on those soils before irrigating loam and clay soils, for example. Farms with more employees have a higher probability of problems due to human resource risk and thus a higher need for good hiring and management procedures.

Ways to Handle Risk

Once we have decided which sources of risk need our attention as managers, we can manage or handle this risk one of five ways, or in combinations of the five ways.

1. Retain risk with no protection from downside risk, as in holding an unpriced commodity.
2. Shift risk by using a contractual arrangement, such as insurance, where someone else takes on some of the chance of a negative occurrence in exchange for a premium. The more risk shifted, the higher the cost.

3. Reduce risk. Keep fences in good repair to keep livestock off the highway and have a marketing plan that locks in some level of guaranteed return, for example.
4. Self-insure. Maintain emergency reserves funded from previous years' profits.
5. Avoid risk by not selecting a particular enterprise, not pushing either end of planting windows, and not increasing your debt-to-asset ratio beyond your comfort level.

These are the main methods to deal with and manage risk and uncertainty in decision making. Let us look at a variety of, but not an exhaustive list of, more specific methods for dealing with risk and uncertainty.

Buy Insurance

Buying insurance is paying someone else to take your risk. Several types of insurance are due to different types of risk: crop, health, fire, life, and liability. Government programs to support farm income is one form of insurance. Also, purchasing options on the futures market to set a minimum price could be viewed as insurance. Current options for crop insurance are discussed in more detail in the next section of this chapter.

The decision to buy insurance involves estimating potential losses, the probability of losses, the cost of the insurance, and both the *ability* to bear the loss and the *impact* on the business. The ability to bear the loss considers whether a loss would cause business failure. This is different from the impact on the business that looks at cash flow deviations and other changes that could cause deviations from the plans that are undesirable, but not complete failure.

Hedging

Hedging is trading cash price risk for basis risk. Besides knowing the market conditions, we also need to be aware of the transaction costs and potential margin calls before making the final decision to hedge or not.

Forward Contracts

Forward contracts usually have both a fixed quantity and price in the contract. A farmer usually does not contract his or her whole crop due to the potential of low production and, thus, the need to buy replacement quantities at a price higher than the contract price to fulfill the contract.

Options

In a very simplistic sense, options allow a farmer to buy a price minimum without potential margin calls. Options allow a farmer to protect his or her ability to cover direct costs.

Other Choices

A farmer's *choice of activities* can have a large impact on how much risk is taken on by the farmer.

Diversification decreases risk if activities are chosen that tend to have good and poor years in different years. The success of diversification depends upon the correlations between the choices and the variations of the potential returns. Diversification can be done by choice of commodity, variety, geographical location, time (e.g., planting date), nonfarm activities and investments, and any combination of these choices.

Crop and variety selection can provide drought and disease resistance and other tolerances.

By renting versus owning land, machinery and livestock, a farmer decreases the commitment of term debt and capital and decreases risk by increasing flexibility. This is especially true when renting land and buildings instead of buying due to the large debt requirement for buying.

Sharecropping instead cash renting transfers some production and price risk from tenant to landowner.

Retaining experts to monitor business should increase management information and decrease surprises. These experts include veterinarians, crop scouts, lawyers, and management consultants.

Obtaining more and better information by subscribing to market news services, attending meetings, calling those who (should) know, and other methods of obtaining information rather than relying on current sources can reduce the risk of surprises and future unknowns.

Shortening lead times in production allows a farmer to recapture costs sooner. Examples of shorter lead times are annual crops versus permanent crops, hogs versus cattle, and buying heavier feeders versus lighter feeders.

Learning new skills and knowledge can cause earlier awareness of important internal and external changes and conditions. For example, proper machinery operation and repair knowledge may decrease downtime during critical work periods.

Environmental control can decrease risk by controlling the microclimate of the enterprise. Examples include irrigation, frost control, hot caps, plastic mulch, greenhouses, strip farming, and contour plowing.

Redundancy of resources. For example, a farmer can decrease the chance of being without a working tractor by having two smaller tractors instead of one big tractor or without electricity by having a backup generator for livestock confinement buildings.

Financial reserves saved in good years to reduce the financial impact of poor years—"saving for a rainy day." Farm and Ranch Risk Management

(FARRM) accounts have been proposed and debated in Congress. As proposed, FARRM accounts would provide additional incentives and benefits to farmers by allowing them to place financial reserves into these accounts and reduce their taxable income by that amount in "good" years. Then, during a "bad" year, the money could be withdrawn and would be subject to taxation in that year—possibly at a lower tax rate. Currently, the proposal also includes the restriction of not being able to keep the money in the FARRM account for more than five years.

Reversibility of decisions and commitments to allow later changes, such as the ability to buy back a contract.

Flexibility in plans means a farmer is willing and able to change to meet new conditions and opportunities. For example, buildings designed for flexibility, not specialty, can allow a farm to change enterprises. A basic pole barn customized for hog finishing could be easier to change to beef finishing compared with a specialized hog confinement building. Other examples of flexibility include leasing/renting versus owning and annual crops versus permanent crops. Flexibility also means knowing where your decision points are, such as: do I sell my calves now as planned or feed them to yearlings?

Risk and uncertainty cannot be reduced or avoided completely. However, we can learn how to make decisions in the face of risk and uncertainty. One way to deal with uncertainty is to *forecast* potential events, especially if this involves estimating a range of possibilities, not just a point. Also, as discussed at the end of this chapter, **scenarios** of the future can be developed and used to deal with a lack of knowledge of the future and analyze the success or failure of potential plans under each scenario. After forecasting and scenario development, the manager also needs to have a set of tools that allows him or her to make decisions in a risky environment.

In actuality, *combinations* of the above methods are used. For examples, one farmer may specialize in one product and use "fancy" or detailed forecasting to reduce the risk of unforeseen price changes. Another farmer may choose to be more diversified and not spend so much time and resources on forecasting the future.

CROP INSURANCE

Federal legislation in 1994 and 1996 eliminated much of the risk protection the federal government provided, including target prices and the potential of deficiency payments based on those target payments. Formal federal disaster programs were also eliminated. This has dramatically changed the government's role and shifted the responsibility of risk management to producers. Current discussions in Congress are aimed at modifying and improving the present insurance options available to farmers.[1]

Currently, the Federal Crop Insurance Corporation (FCIC) provides several crop insurance programs through local crop insurance agents: Multi-Peril Crop Insurance (MPCI), Catastrophic Risk Protection (CAT), Crop Revenue Coverage (CRC), Revenue Assurance (RA), Income Protection (IP), and Group Risk Protection (GRP). These programs are described briefly below after some general information.

Yield levels for all FCIC crop insurance coverage are based *upon actual production history* (APH) *or* a percentage of an established county yield *or* a combination of both. APH requires a minimum of four years of production records and accumulates to a maximum of ten years of production records. Farmers who have less than four years of production records will be required to use a County T Yield for those years (up to four) without actual production records.

The County T Yield *is not* based upon Farm Service Agency (FSA) program yields. The County T Yield *is* based upon an established county yield for a given insured crop in a given county. County T Yields vary by county and crop but not within counties. Once the farmers have four or more years of actual production records, they will no longer be required to use the County T Yield. A reduction factor is applied to County T Yields to arrive at an adjusted APH Yield. The reduction factor increases from 0.65 to 1.00 as the number of years of actual production records increases.

Each of the FCIC crop insurance coverage programs has an administrative fee. These fees differ from program to program. All but one of the FCIC crop insurance coverage programs have a premium attached to them. CAT does not include a premium. For those programs with a premium, they vary by county, crop, and level of coverage. Premium costs are partially subsidized by the federal government. Premium discounts vary over time.

Unit Structure

Unit structure is an insurance coverage selection that enables the farmer to combine crops for coverage purposes. Unit structure choices differ by insurance coverage program. Four types of unit structures are available: basic, optional, enterprise, and whole-farm.

In *Basic Unit Structure*, all 100% share land is grouped together in one unit. Land shared with each different landowner is another basic unit of its own.

In *Optional Unit Structure*, acreage of a given insurable crop can be divided into separate units such as by section, rented acres grouped by landowner, owned land divided by section, and fields with different planting patterns as long as the pattern does not cross section line. Optional units must have separate production records.

In *Enterprise Unit Structure*, all insurable acres of the same insurable crop are lumped together into one unit regardless of site location and rental arrangements.

In *Whole-Farm Unit Structure*, all insurable acres of all insured crops are lumped together into one unit, regardless of land location and rental arrangements.

Replant and prevented planting coverage is available for MPCI, CRC, RA, and IP, but not CAT or GRP. Late planting insurance coverage is reduced 1% per day for the first 25 days that planting is delayed beyond the established final planting deadlines, up to a maximum reduction of 25%. Previously, the maximum reduction was 40%.

Market price (MP) guarantees are per bushel payment levels used to calculate CAT and MPCI coverages only. These change each year. When selecting MPCI coverage levels, farmers can decide upon different percentages of the market price guarantee for their coverage calculation. MP guarantees have nothing to do with actual crop market prices. They are used in coverage calculation only for CAT and MPCI.

Catastrophic Risk Protection (CAT)

CAT is the minimum level of crop insurance coverage. It covers losses to crop yield only. The APH and/or County T Yield is used in the coverage calculation. There is no premium for CAT coverage, but there is an administrative fee of $60 per crop per county. CAT insurance will cover yields reductions that are 50 percent or more below established yields. The payment rate will be 55% of market price guarantees. Only basic units apply under CAT coverage.

Multi-Peril Crop Insurance (MPCI)

MPCI provides comprehensive protection against weather-related causes of loss and certain other unavoidable perils. It protects against losses to crop yield only. Coverage is available on more than 60 crops in primary production areas through the United States. The coverage is available at 50% to 75% of the APH for the farm. An indemnity price election from 60% to 100% of the FCIC expected market price is selected at the time of purchase. Premium amounts per acre vary by crop and county and are partially subsidized by the federal government. MPCI includes the choice of either basic or optional unit structure.

Crop Revenue Coverage (CRC)

CRC protects against losses from *both* yield and price fluctuation by converting the bushel guarantee per acre to a dollar guarantee per acre. A loss results when the calculated revenue is less than the final guarantee. The difference between these two figures times the insured's share results in a payable indemnity. Losses are based on the minimum or harvest guarantee (whichever is higher) and the calculated revenue. CRC includes the choice of basic, optional, or enterprise unit structures. CRC yield coverage choices are the same as for standard MPCI from 50% to 75% (and 85% in some counties). APH and/or County T Yields are used in the coverage calculation for CRC.

All price levels used in CRC calculations are average Chicago Board of Trade (CBOT) futures prices on specific contract months depending on the crop. Local elevator prices have no impact on CRC guarantees or payments, nor do the farmers' actual sales. Sale of the crop is not even required.

The minimum revenue guarantee with CRC is established by the March 15 closing date. That guarantee cannot be reduced, but it can be increased if harvest CBOT futures prices are higher than the base CBOT price. If the harvest CBOT price is higher than the initial base price, the higher level becomes the new revenue guarantee with CRC. If the harvest CBOT price is lower than the initial base price, the bushel threshold where losses are paid increases because losses are determined by a revenue guarantee. The calculated production revenue is the actual production times the final calculated harvest price based on CBOT futures. It the calculated production revenue is lower than the final guaranteed revenue, an indemnity payment is made on the CRC policy.

CRC premiums are partially subsidized by the federal government. They are based upon the base price and remain the same even if the harvest price is higher than the base price. Administrative fees for CRC are the same as for MPCI coverage.

Revenue Assurance (RA)

RA is similar to CRC and provides coverage to protect against loss of revenue caused by either low yields, low prices, or a combination of both. Currently, RA coverage applies to corn and soybeans only, and it may not be available in all areas. If it is available, it is available in any percentage from 65% to 75% for basic, optional, and enterprise unit structures, and from 65% to 80% for the whole-farm unit structure. The calculations for APH and the use of County T yields with factors are the same for RA as they are for MPCI.

The RA price reference for its insurance coverage guarantee is based upon 100% of the CBOT options contract projected price. The insurance coverage guarantee may increase during the insurance coverage period. RA, unlike CRC, has no maximum upward price movement for the insurance coverage guarantee. The basis for the RA insurance coverage guarantee is:

APH yield × coverage level % × projected harvest price (from CBOT options contracts)

If the fall harvest price is greater than the projected harvest price, an indemnity payment is made. Therefore yield variability is based on APH rules, and price variability is based on CBOT options contracts. Local elevator prices, basis levels, and individual farmer's grain sale prices have no impact on guaranteed price levels for RA coverage.

RA premiums are partially subsidized by the federal government. Administrative fees for RA coverage are the same as for MPCI options.

Income Protection (IP)

IP insurance coverage protects against reductions in gross income when either a crop's price (projected spring commodity price) falls or yield declines from early season expectations. It is similar to CRC and RA.

Group Risk Protection (GRP)

GRP insurance covers yield and price loss. Coverage is based on an index of the expected county yield for a given crop. When the county yield for the insured crop falls below the yield level chosen by the farmer, the farmer receives a loss payment. This coverage is similar to MPCI except that it targets farmers without production records and, therefore, unable to verify APH.

MAKING RISKY DECISIONS

To choose between options in a risky environment, a manager has to have some framework or process to make these risky decisions. This section presents several methods and rules for making risky decisions.

A risky decision has five parts:

1. Actions: What are the choices available to us?
2. Events (states, conditions): What can't be controlled, but will affect the results? What is random or unpredictable?
3. Payoffs: What are the measured consequences of our actions under each event?

4. Probabilities: What are the chances that the events will occur?
5. Criteria on how to choose the best course of action.

For simple situations, three simple ways can be used to organize the first four pieces of data: payoff matrices, regret matrices, and decision trees. These come from what we call game theory, which is why we use the words "actions," "events," and "payoffs."

Payoff Matrices

When a decision involves a choice of action and a result (or payoff) depending upon an event (rainfall or price level, as examples), a payoff matrix may be useful for some people. This method usually involves only one decision and one set of events. The potential returns are called "payoffs" in games and in game theory; so, the name "payoff matrix" is used.

A payoff matrix could be used to answer the question of selling cattle now or selling them next week. Selling now means we will get a certain price and weight. Selling next week means the price may change, and we will pay more in feed costs. Are the chances that the price will rise worth the additional feed costs?

A payoff matrix such as the one in table 11.1 can help order the information and help the manager make a decision. On the left side are the possible events: price down, steady, and up. On the top are the manager's two choices: sell now or sell next week. Within the table (or matrix) are the expected returns for each choice under each event (i.e., price level). In the example, if the manager decides to sell now, he will receive a net return of $31 per head regard-

Table 11.1. Example Payoff Matrix and Regret Matrix

Price change	(prob.)	Payoff matrix		Regret matrix	
		Sell now	Sell next week	Sell now	Sell next week
		($)	($)	($)	($)
5¢ down	(.2)	31	−20	0	51
Steady	(.5)	31	30	0	1
5¢ up	(.3)	31	80	49	0
Minimum return		31	−20	xxx	xxx
Maximum regret		xxx	xxx	49	51
Expected return or regret		31	35	14.7	10.7
Simple average		31	30	16.3	17.3

less of what the price does during the next week. If he decides to wait and sell next week and the price is steady, he will receive a net return of $30 after paying for the extra feed and other costs and benefiting from a heavier animal. If price would go down five cents, he would lose $20 per head. If the price goes up, he would receive $80.

The data in a payoff matrix are evaluated by judging which action will provide the highest payoff or returns. The various methods of evaluating or deciding are presented in the later section on decision criteria.

Regret Matrices

A regret matrix uses the same information as a payoff matrix but looks at the decision in a different way. A payoff matrix shows the potential returns (or payoffs) for having chosen a particular action for each event. A regret matrix shows the potential regret a manager might have after choosing a specific alternative. That is, a regret matrix shows the potential lost return (or payoff) of not having chosen the best action for each event.

Using the data in the payoff matrix, a regret matrix can be formed (table 11.1). First, consider the action of selling now. If the price were to go down, the regret would be 0 because selling now is the best action if the price were to go down. Selling now is also the best option if the price were to remain steady; so the regret is also 0. However, if the price were to go up, the manager would regret having sold now rather than waiting and selling next week. The size of the regret is the difference in potential returns: $80 minus $31 (or $49). In other words, the manager would regret selling now if the price were to go up because the return from selling next week at the higher price is $49 higher than the return from selling now.

The potential regret due to deciding to sell next week rather than now can be calculated in a similar way. If the manager would decide to sell next week and the price goes down, the manager would regret the decision because the return is $51 lower than selling now. If the price were to be steady, there is a regret of $1 if the manager waited to sell because the return is $1 lower. If the price were to go up, there would be no regret for having waited to sell because the return is higher.

The regret matrix is evaluated in a similar but opposite way from the payoff matrix. We want higher payoffs and lower regrets.

Decision Trees

When dealing with both production and marketing risks, a decision tree may be the easiest way to organize the data. A decision tree consists of branches and nodes; it grows on its side. Square nodes are used to denote decisions or alternative actions, and circular nodes denote the things that depend on chance, the events. The decision tree framework is illustrated by the cattle sale example (fig. 11.2).

The decision tree is drawn in chronological sequence from left to right with the alternative actions (e.g., selling now or next week) branching from the decision node denoted by a square and the events (price level) branching from events or chance nodes denoted by circles. At the decision node, it is the manager's choice whether to go down the "sell now" branch or the "sell next week" branch. The "sell next week" action branch has three event branches: price movements of "down," "steady," or "up." The estimated net returns are shown at the end of each branch.

The choice between using a payoff matrix, regret matrix, decision tree, or other information organizing method depends on the question being studied and the personal preference of the manager. It could very well be that just the process of organizing the information will give the manager enough information to decide. Also having an idea of how to organize this information may allow a manager to sort things out in his or her mind without having to draw a decision tree such as is the one in figure 11.2.

Decision Criteria

Although we can organize our data with payoff matrices, regret matrices, and decision trees, we still need some process to sort it all out and some criteria by which we can decide. Several different rules or criteria are available for our use. We will look at six of them.

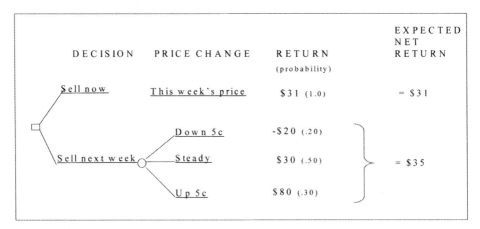

Figure 11.2. A decision tree.

The first three decision rules do not require probabilities: maximin, minimax, and simple average. Two decision rules do require probabilities: maximum expected returns and safety-first rules. The sixth rule compares the probability of success between alternatives.

1. *Maximin* (for returns or payoffs). The manager chooses the action that has the largest of the minimum returns. When using the maximin rule and the data in the example payoff matrix (table 11.1), the manager would choose to sell now because selling now has the largest minimum return: $31 versus −$20.

   The advantage of this decision rule is that it protects the lower end of the income. The disadvantage of this decision rule is that it ignores the highest potential returns that may occur and the average returns that may occur.

2. *Minimax* (for regrets). The manager chooses the action that has the smallest of the maximum regrets. When using the minimax rule and the data in the example payoff matrix (Table 11.1), the manager would choose to sell now because selling now has the smallest maximum regret: $49 versus $51.

   The advantage of this rule is that it avoids those actions that would cause large regrets. The disadvantage of this decision rule is that it ignores high-income potential.

3. *Maximum Simple Average*. The manager cannot decide which events are most likely to happen; no probabilities are estimated. Thus, the manager chooses the action that has the largest simple average of all possible returns.

   Using this decision rule, the feedlot manager chooses to sell now because it has a simple average return of $31 compared with the simple average of $30 for selling next week (table 11.1).

4. *Maximum Expected Returns (or Minimum Expected Regrets)*. With this decision rule, the manager weights each potential return by the probability or chance of the event or combination of events occurring. The best action is chosen based on the largest total of these weighted potential returns. This method accounts for variability in returns and the potential that some events will be more likely to occur than other events.

   In the cattle selling example (table 11.1), probabilities were chosen for each of the potential price events. These probabilities are .2, .5, and .3 for down, steady and up, respectively. With these probabilities, the expected net return from selling next week is $35 ($0.2 \times -20 + 0.5 \times 30 + 0.3 \times 80$). The expected return from selling now is $31. Using the maximum expected return rule, the feedlot manager would choose to sell next week.

   In the example decision tree (fig. 11.2), the same potential returns and probabilities are used. Thus, the expected net return from selling next week is $35 and the expected return from selling now is $31. Again, using the maximum expected return rule, the feedlot manager will choose to sell next week.

The same estimated probabilities can also be used with the regret matrix (table 11.1). With these probabilities, the expected regret from selling now is $14.7 ($0.2 \times 0 + 0.5 \times 0 + 0.3 \times 49$). The expected regret from selling next week is $9.7 ($0.2 \times 51 + 0.5 \times 1 + 0.3 \times 0$). Using the minimum expected regret rule, the feedlot manager would choose to sell next week.

Several methods for estimating probabilities are described in appendix C at the end of the book.

5. *Safety-First Rule.* The safety-first decision rule contains multiple objectives. The first objective is to obtain a minimum return with a certain probability. Any action that does not meet this minimum return with a certain probability is eliminated from further consideration. The second part of the safety-first rule is to choose the best action from those remaining. Usually the maximum expected returns rule is applied on the remaining actions. A manager may use a "Safety-First" rule to say that minimum levels of cash flow and security are needed before risks can be taken. Using the safety-first rule in the cattle feeding example (table 11.1), the manager may choose to sell now to avoid the chance of the negative return if the cattle are sold next week and the price declines.

6. *Probability of Success.* Another way to look at a risky decision and choose among alternatives is to evaluate the chances of successfully accomplishing the goal of, say, paying all expenses or meeting income goals. The alternative with the highest probability of success is the alternative chosen. In table 11.1, for example, the probability of having a positive return, that is, successfully covering all expenses, is 100% for selling now. However, due to the 20% chance of a loss, selling next week has an 80% chance of having success.

SCENARIOS FOR MANAGEMENT PLANNING

Using only one view of the future to plan major directions and investments could be a mistake with devastating results. Instead, alternative scenarios or descriptions of the future can be developed to help understand the potential impacts of different paths that future events may take. These scenarios can be used to help estimate results and subjective scores for proposed strategies under different views of the future. Scenarios are useful for forecasting the future because they help us (1) identify which factors and forces are and will be important, (2) focus on the forces in the marketplace and other environments, (3) see the future even with imperfect information, and (4) not blindly accept one view of the future.

Developing and Utilizing Scenarios

The development of good scenarios of the future requires a good understanding of the forces involved in the current situation and how those forces are affecting trends and movements into the future. The procedures described below begin by identifying the uncertain elements in the situation and the causal factors that will affect those uncertainties. Thus, rather than just choosing a set of variables in a seemingly random process, these procedures will help a manager develop a set of more rational, internally logical, and consistent scenarios.

1. Identify the uncertainties
   a. Examine and classify each element in the situation as: *constant*: unlikely to change; *predetermined*: predictable, trends are stable; *uncertain*: depending on unresolvable uncertainty.
   b. All three elements are part of each scenario, but the uncertain elements are used to create and differentiate the scenarios.
   c. Evaluate both trends and discontinuities in the uncertainties. This involves using the knowledge and expertise from many disciplines and fields.
2. Classify the uncertainties
   a *Independent uncertainties* are independent of other elements affecting the decision or they are essentially independent even if one could find some weak but plausible connection. For example, interest rates for farmers are essentially independent of other elements, such as crop prices and labor availability.
   b. *Dependent uncertainties* are determined by independent uncertainties. For example, the corn price is a dependent uncertainty because it is largely determined by the production levels in the main parts of the corn belt and by the demand for corn.
   c. Only independent uncertainties are used to develop scenarios. Dependent uncertainties are known once assumptions about the independent uncertainties are made.
3. Identify causal factors for independent uncertainties
   a. Decide what affects the independent uncertainties. For example, what determines how much corn is produced in Illinois and Iowa? What affects the performance of the general economy, which can affect the Federal Reserve's interest rate decisions? A manager needs to make assumptions about how these causal factors will behave.
   b. For reasons of practicality, we limit how "far back" we go to identify causal factors. For example, we may choose different levels of corn production in the main part of the corn belt rather than split that into the dif-

ferent factors of costs, weather, and other factors. Practicality says we can comprehend only so much detail; we deal with the lack of detail by the number of scenarios we analyze.

   c. These casual factors become the "scenario variables" from which we define the potential scenarios of future events.

4. Develop internally logical and consistent scenarios
   a. To do this effectively, we need a knowledge of how the farm and its environment works.
   b. Estimate the "second-order" effects of assumptions. Variables may be interrelated. For example, if we assume bad weather in the corn belt and thus poor corn yields, we should not assume high soybean yields in the same area.
   c. A good scenario will not have conflicting assumptions and second-order effects.

5. Analyze scenarios for
   a. Returns for each alternative action
   b. Future results on the industry or firm
   c. Implications for structural attractiveness (i.e., profit)
   d. Competitive advantages for firm
   e. Impacts on the factors identified in the strategy tests listed earlier

Scenarios can be used in at least two ways. First, they can be used to help develop scores for proposed strategies as described earlier. Second, they can also be used within the decision framework of payoff and regret matrices: scenarios can be thought of as events and the different strategies as actions.

As an example of developing scenarios, let us consider a couple producing hogs in the Midwest. Their farm has the capacity to farrow and finish about four thousand hogs. They have and are willing to sell feeder pigs if the market suggests that is the best option. They also farm about 480 acres (160 owned and 320 rented) with the help of both full- and part-time employees.

They had done well earlier but recent years have been rough on their operation. They have endured extremely low prices and have had disease problems that required them to depopulate and rebuild the herd. So both productivity and profitability have been low and their financial position has eroded tremendously. The bank is threatening not to renew their operating loan unless they can show a high likelihood that the bank will be repaid and their financial position will improve.

Right now they are considering three main options. First, they could continue raising both hogs and crops. Second, since the hogs are the biggest financial drain, they would consider renting out their buildings, raising only crops, and taking an off-farm job. Third, since the hogs are the biggest potential source of income, they also would consider the possibility of not renting the land and focusing on the hogs. These three options are broad definitions of what they could choose to do. They are also open to other options and to fine-tuning these three options if needed.

However, they are unsure of what the future will look like. They do not know what the prices may be, whether disease may hit them again, and so on. Using the steps outlined above, they have developed a set of scenarios to help evaluate their situation and the potential of alternative strategies for the future (table 11.2). This is not a complete list of elements and scenarios, but they do show the process of scenario development. This example also will need to be reviewed and likely changed for different producers and different years.

Table 11.2.   Identification of Causal Factors for Scenario Development for a Hog Farm

Step 1. Identify the uncertainties AND

Step 2. Divide the uncertainties into independent and dependent

Elements	Constant, Predetermined, or Uncertain?	Are Uncertainties Independent or Dependent?
Hog prices in market	Uncertain	Dependent on hog supply in market and pork demand
Hog supply in market	Uncertain	Dependent on other producers' actions
Other producers' actions	Uncertain	Independent
Market access	Uncertain	Dependent on other producers' actions and processors' offerings
Pork demand	Predetermined	
This farm's hog supply	Uncertain	Dependent upon farrowing decisions and disease level
This farm's farrowing decisions	Uncertain	Dependent on hog prices in market and farrowing capacity
Farrowing capacity	Constant	
Disease level	Uncertain	Independent (even after following) sanitation procedures
Feed prices in market	Uncertain	Dependent upon market crop production and demand
Crop production	Uncertain	Dependent upon weather and expected crop prices
Weather	Uncertain	Independent
Labor supply and wages (by local market)	Predetermined	
Government programs	Predetermined (with some uncertainty)	

Step 3. Identify causal factors for uncertainties

Independent uncertainty	Causal factors
Other producers' actions	Long-run price expectations, expected packer behavior, changes in pork demand
Disease level	Infected visitors, wind, failed safety procedures
Weather	Position of jet streams

Several scenarios could be developed from the causal factors just identified. I have described five of them below. Others would be variations of these or perhaps different intensities of each factor.

Scenario A ("most likely"): Other producers continue to expand faster than producers exit, thus production has a net increase in production that is in balance with increases in pork demand so hog prices remain at current levels. Management can control disease so farrowing productivity and production efficiencies are good. Weather is normal so yields are normal so crop and feed prices are at normal levels.

Scenario B ("lower prices"): Also very likely, perhaps equally likely as scenario A. The net increase in hog production is greater than the increase in pork demand so hog prices feel a downward pressure from current levels. Management can control disease so farrowing productivity and production efficiencies are good. Weather is normal so yields are normal so crop and feed prices are at normal levels.

Scenario C ("disease hits"): Despite management's efforts, disease breaks out pushing productivity and efficiency down. Other producers continue to expand faster than producers exit, thus production has a net increase in production that is in balance with increases in pork demand so hog prices remain at current levels. Weather is normal so yields are normal so crop and feed prices are at normal levels.

Scenario D ("widespread drought"): The potential for widespread drought is realized in the next year. Crop production is down, crop and feed prices are up considerably. Other producers continue to expand faster than producers exit, thus production has a net increase in production that is in balance with increases in pork demand so hog prices remain at current levels. Management can control disease so farrowing productivity and production efficiencies are good.

Scenario E ("good times"): The net increase in hog production is not as great as the increase in pork demand so hog prices rise from current levels. Management can control disease so farrowing productivity and production efficiencies are good. Weather is normal so yields are normal so crop and feed prices are at normal levels.

Other variations could include drought over most of the corn belt but not on this farm. Another variation is the opposite: drought and poor yields on this farm but the drought is not widespread. The number and need for more scenarios will vary with the need for details and the time available for analysis and interpretation. The number of scenarios needs to be balanced with the potential for information overload, rejection of alternative scenarios, and focusing on only one scenario thus resulting in an incomplete picture of the future.

Choosing the Number of Scenarios

The number of scenarios needs to be limited due to the confusion and possible rejection of the whole process if too many options and alternatives are analyzed. One way to reduce the number of scenarios is to reduce the number of scenario variables to only those crucial variables with large impacts on the results being watched (say, net farm income or cash flow).

Another way to reduce the number of scenarios is to reduce the number of assumptions about each variable. The choice of assumptions is affected by four factors.

1. The need to encompass the uncertainty to give credibility to the analysis. If we do not cover all the possible alternatives to what may happen, the analysis may be faulty due to incompleteness.
2. The regularity of the impact of the variable. If the analysis is very sensitive to small changes in the scenario variable, more scenarios are needed. If the results are not very sensitive, fewer scenarios will be needed to analyze the impact of the variable.
3. The owner's or manager's beliefs about the future or about the impact of variables will determine the number of scenarios needed. Some beliefs need to be tested and evaluated to see what impact they may have and whether that impact is important. As a result, some beliefs need to be changed.
4. The practicality of analysis will restrict the number of scenarios. More scenarios means more work needs to be done. Time is needed for both doing the analysis and interpreting the results. The comprehension ability and endurance of the intended audience also need to be considered.

The number of scenarios can be kept to a minimum by choosing an analysis sequence that yields management insight for a selection of a strategy. For instance, the first scenario to be analyzed should be the one considered to be the most likely. Then a scenario polar to the most likely one should be analyzed. After that, analyze more scenarios until the impacts of the important scenario variables are understood. Limit the number of scenarios, but be sure to include major discontinuities in the variables (such as major shifts in demand or supply). Include enough scenarios to illuminate the range of possible futures that will affect strategy formulation and to communicate, educate, and stretch managers' thinking about the future. Willis (1987) suggests that we look for themes in the trends, variables, and other elements and then develop scenarios around those themes.

Another way to choose the number of scenarios is to evaluate the advantages and disadvantages of each number of scenarios. These are pointed out below.

One: The "most likely" scenario. This ignores other possibilities and may omit important events.

Two: Two scenarios are better especially if they are polar scenarios. Two can be very useful if they are equally likely to happen, that is, they are "deadly enemies." Trying to capture all possible events in two scenarios can be a problem.

Three: Three scenarios are also better but they may degenerate to the most likely, optimistic, and pessimistic scenarios. Then the optimistic and pessimistic scenarios are often "ignored" and the problems of only one scenario return. Also, what is optimistic? What is pessimistic?

Four: Four scenarios can be useful if they are developed to have similar probabilities. This many scenarios can start to show and encompass uncertainty. Four also allows the evaluation of the opportunity costs of strategies (i.e., the regrets of choosing this strategy over another).

At what point does confusion start? What is too many? What can management comprehend? These questions need to be answered for each situation and each manager or management team.

Once strategies are planned, what might happen to change the plan? At this point in planning, scenarios can be developed and used in an iterative process. A manager or management team can play the "devil's advocate" on what may happen and evaluate the consequences for the business under a potentially final strategy. This iterative process can test specific questions about the impact of future events and what, if any, contingency plans are needed within the final chosen strategy.

SUMMARY POINTS

- Every decision involves some risk, so risk needs to be incorporated into the decision process.
- The goal of risk management is to balance a farm's risk exposure and tolerance with the farm's strategic and financial objectives.
- Risk can be described as coming from five main sources: production, marketing, financial, legal, and human resources.
- In order to decide which risks need attention first, sources of risk can be prioritized by considering their probability of happening along with the potential impact. Those risks that have a large potential impact and a high probability of happening need immediate attention.
- Risk can be managed or handled in one of five ways: retain risk, shift risk, reduce risk, self-insure, or avoid risk.
- A manager has many specific options for managing risks including, insurance, hedging, contracting, diversification, and financial reserves.

- Crop insurance has become an important risk management tool, especially in combination with the futures market so both yield and price are protected—Crop Revenue Coverage (CRC), for example.
- Payoff matrices, regret matrices, and decision trees can be used to organize information and, thus, help a manager make a risky decision.
- Decision criteria for risky decisions include maximin, minimax, maximum simple average, maximum expected returns, minimum expected regrets, safety-first rule, and probability of success.
- Managers can develop scenarios to better understand the impact of alternative views of the future and, thus, to make more informed decisions.

REVIEW QUESTIONS

1. List at least two potential sources of risk that a farmer could encounter in each major risk category. If you are familiar with a specific farm, list the sources of risk for that farm.

Production risk:	a.	b.
Marketing risk:	c.	d.
Financial risk:	e.	f.
Legal risk:	g.	h.
Human resource risk:	i.	j.

2. For each of the specific risks you listed in question 1, consider their probability of happening and potential impact and place the letter of each in the graph (fig. 11.3).

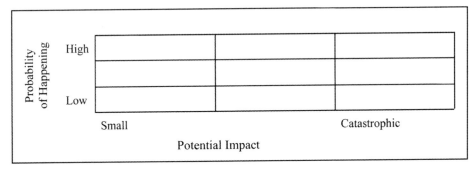

Figure 11.3. Prioritizing risks.

3. For two sources of risk you identified as high priority (that is, the upper-right corner of the graph) in question 2, choose two methods that could be used to reduce the farm's exposure to risk due to this source.

4. Compare and contrast a payoff matrix and a regret matrix. What information does a regret matrix provide that a payoff matrix does not?

5. For many years Glenda and Paul Christianson, like most of their neighbors, have been growing corn and soybeans in a 50-50 rotation using a chisel plow as their primary tillage. Now they are considering switching from this conventional tillage system to either an alternative system that still uses the chisel plow but adds alfalfa to their rotation or a ridge till system. The ridge till system does not include alfalfa. After talking with their neighbors who are using the other systems and researchers at the local university research center, they estimated their potential returns under the three tillage systems for each of the past six years. They put these estimates into a table with one row for each year and a column for each of their tillage choices. This table became their payoff matrix and is shown here (table 11.3). Develop the regret matrix for their decision.

Table 11.3.   Christianson's Estimated Payoff Matrix for Tillage Systems

	PAYOFF MATRIX			REGRET MATRIX		
Year	Conven-tional	Altern-ative	Ridge Till	Conven-tional	Altern-ative	Ridge Till
	($/year)			($/year)		
1	1,563	2,765	13,503			
2	19,411	18,436	11,588			
3	20,025	15,774	15,749			
4	−25,274	4,894	−23,103			
5	16,783	12,174	14,762			
6	21,818	19,261	20,667			
Minimum return				xxx	xxx	xxx
Maximum regret	xxx	xxx	xxx			
Average return or regret						

6. For the tillage decision in question 5, choose which alternative should be chosen using each of these decision criteria:
   a. Maximin
   b. Minimax
   c. Maximum expected returns
   d. Safety-first(use the safety rule of no more than a 10% chance of a negative payoff)
   e. Based on these results, which alternative would you choose and why?
7. How can the probability of success be used in making risky decisions?
8. Gene Meyer has estimated his variable costs of finishing feeder pigs to be $51 per head. He thinks the next group of feeder pigs will cost $45 per head (for a 45 lb. animal). He sells his hogs at 230 lb. and expects a 1% death loss. After listening to market analysts and looking at recent trading at the

Chicago Mercantile, Gene has developed his conviction weights for finished hog prices in four months. These are shown in table 11.4.

a. Using the conviction weight method, estimate the probability for each price interval. (This method is described in appendix C.)
b. What is his estimated total cost per cwt for the finished market hog in four months? Include the cost of the feeder pig, variable costs, and death loss.
c. What is the probability of success for covering the total cost of the feeder pig plus his variable costs?
d. Do you think Gene should buy these pigs? Why or why not?

Table 11.4.   Estimating Price Probabilities Using Conviction Weights

Market hog price intervals ($/cwt)	Midpoint	Conviction weights	Estimated interval probabilities	Estimated cumulative probability
30–34.99	32.50	50		
35–39.99	37.50	90		
40–44.99	42.50	100		
45–49.99	47.50	70		
50–55.00	52.50	20		
TOTALS				

9. How can scenarios be used in making risky decisions?

NOTES

1. This section is based on the most current information from USDA and Minnesota Extension Service. However, policy available, rules, and other information are subject to change. Crop insurance agents should be consulted for the latest information and consultation for individual situations.

12

Production Contract Evaluation

Contracts have been used in farming for a long time and in many ways. There are land rental agreements, marketing agreements, forward contracts, futures contracts, security agreements, loans, contracts for deed, labor contracts, contracts to produce vegetables and seed crops, and contracts to deliver sugar beets as part of a processing cooperative. In recent years, the use of production contracts has become increasingly common and important for livestock, grains to meet specific characteristics, new venture cooperatives for grain processing, and new products. This chapter concentrates on production contracts, since they are a relatively new phenomenon for many farmers.[1]

Production contracts can be made between two farmers or between farmers and agribusinesses. Contracts usually specify certain quality requirements, price, quantity, and services provided. Through combining both market and production functions, contracting generally reduces all participants' exposure to risk. Farmers enter into contracts for various reasons, including income stability, improved efficiency, market security, and access to capital. Processors enter into contracts to control input supplies, improve responses to consumer demand, and expand and diversify operations.

A contract is an agreement between two or more persons to do something or to not do something. Four basic requirements must be met for an agreement to be a legally enforceable contract:

1. *The parties to a contract must be identifiable and legally competent.* If both parties are not known, there is no method of determining who is obligated to meet the terms of the agreement. Also, a party must be of legal age and mentally competent.
2. *The subject matter of the agreement must be legal.* For example, a loan agreement that has a rate exceeding the legal maximum is not enforceable.

319

3. *There must be mutual agreement.* One party must make an offer that is accepted by another party.
4. *There must be consideration given.* Something is given for something received. This consideration may consist of money, goods, or services, or merely the promise of future consideration.

A contract may be either verbal or written, but there are a few important types that must be written. To be legally enforceable, the following contracts must be in writing:(1) contracts that cannot be completed within one year, (2) contracts for transfers of interests in real estate, (3) contracts to sell goods valued at $500 or more, (4) credit agreements, and (5) loan guarantees.

Even though not all contracts have to be written, putting an agreement in writing decreases the likelihood that the parties will misunderstand the agreement, their responsibilities, and what each party expects to receive. A written contract also provides tangible evidence of an agreement and the terms of that agreement. All terms and agreements should be made part of a written agreement. Verbal agreements made but not included in the written terms are rarely enforceable. The language of a contract must be very specific. Ambiguous language, such as "reasonable efforts," "best of one's ability," or "a timely manner," can easily mean different things to different people. A contract should either avoid terms such as these or specify the criteria to define what these terms mean in the context of the contract.

TYPES OF PRODUCTION CONTRACTS

Production contracts specify the quality and quantity of the commodity to be produced, the quality and quantity of production inputs to be supplied by the contracting firm (e.g., processor, feed mill, or other farm operation), the quality and quantity of services to be provided by the grower (i.e., contractee), and the type, magnitude, and schedule of compensation that the grower will receive. The manner in which contracts are constructed affect the legal relationship between the producer and the contractor.

Production contracts may be structured as **sales contracts** or **market-specific production contracts**. These types of contracts are very similar to forward contracts. With sales contracts, the farmer produces the crop or livestock and agrees to sell the product at harvest to the contractor. The farmer retains ownership until harvest and consequent sale. According to legal counsel, these contracts are usually subject to the provisions of the Uniform Commercial Code (UCC) relating to sales contracts. A buyer in a sales contract benefits mainly because delivery schedules are specified and thus more stable allowing the buyer to stabilize further processing or transportation and take advantage of efficiencies not available when deliveries are not stable. The seller benefits from risk reduction of finding a market and usually receives premiums above a spot-market price. The market-specific or sales contract usually transfers minimal control across stages (that is, between the producer and processor).

A **production-management contract** increases the control of the buyer (say a vegetable processor or seed company) over the production process. Production-management contracts are often used when the seller (i.e., producer), through production decisions, can affect the value of the product to the buyer or when the seller, through marketing decisions, can affect the value of the product to the buyer.

In these contracts, buyers usually gain control over decisions once made in open production, such as the timing of planting decisions, variety of seed planted, weed control methods, and harvest timing. By taking more control than in a market-specific contract, the contractor or buyer usually takes on more of the producer's price risk. The buyer benefits from the control of the quality and timing of the product. For instance, a vegetable processor can gain control over the timing of the planting of specific seed varieties and the timing of harvest and, thus, presumably, benefits from a more stable flow of a product with a more consistent quality.

Production-management contracts may be considered **bailments.** A bailment is the legal relationship that exists when someone else is entrusted with the possession of property, but has no ownership interest in it. A common bailment is a grain storage contract in which the elevator stores but does not own the farmer's grain. Crop production contracts structured as bailments provide the contractor with additional protection against the unauthorized distribution of seeds, crops, livestock, and genetics that remain the property of the contractor.

Production contracts may also be structured as **personal service contracts** or **resource-providing contracts**. These contracts specify that the producer is to provide services, not commodities, to the contractor. For example, the producer will provide the services of a hog finishing building and management knowledge to the owner of the hogs. The hogs will reside in the producer's building but the contractor will retain ownership of the hogs. According to legal counsel, the UCC provisions relating to sales of commodities will not be applicable to a personal sales contract. Some production contracts may be *leases* of facilities especially if the contracts relate to production of livestock.

EVALUATING PRODUCTION CONTRACTS

Evaluating a contract involves answering three basic questions. First, does the contract improve profitability? Second, how does the contract change the risks that the farmer faces? Third, how does the contract change legal relationships and obligations? If the economic and legal implications of the contract are not fully understood, a farmer should seek expert advice to help evaluate the contract. That expert advice can be obtained from lawyers, business and financial consultants, and other farmers experienced with the same type of contract.

Before a farmer signs a contract, he or she should know and understand the following points. If there are questions on any of the points, the contract should not be signed.

1. All of the terms and features and how they will affect the business
2. How profit will be affected
3. How risk is reduced, how risk is increased
4. How the contract works under extreme price and production conditions
5. The other party to the contract and, if possible, their financial condition and ability to meet their obligations within the contract
6. How prices and quantities will be determined. How any equations behave in extreme market conditions
7. Who will determine prices and quantities
8. The implications and obligations if production falls below contracted quantities
9. Whether and how the contract can be terminated by any party involved
10. Who, in the other parties, to communicate with before signing and throughout the life of the contract

Profitability

The best, and perhaps only, way to measure profitability is to estimate two sets of budgets: one for operating under the contract and another for operating independently. The procedures for estimating budgets are described in chapter 4. Except for how prices, sales quantities, and costs are chosen, the same budgeting procedures should be used for evaluating contract profitability. The terms of the contract will specify product prices and quantity as well as the timing of the payments. Costs will be determined by either the product specifications described in the contract or by the production and capital investment specifications in the contract. If the contract is for several years and, especially, if it requires capital investments, either a budget using long-term prices and costs or a budget for each year should be developed. The budgets developed using the terms of the contract should be compared to budgets developed as if the farmer(s) were operating independently.

Risk

Avoiding risk is a major reason for farmers to sign contracts, but contracts can introduce new risks too. The contract usually specifies a product or service price, a quantity that can be sold, or both. Signing a contract also usually means that the farmer will have a market for his or her product. However, the contract needs to be evaluated to see if the farmer is exposed to other sources of risk that weren't present or were less important without the contract. These risks can be potential liabilities, greater moral risk, increased risk of not being able to meet product specifications, and so on.

As an example, consider recent hog contracts that guarantee a certain cash price or a range of prices for the hogs sold under the contract. This sales price guarantees stable cash flow and income for the length of the contract. These contracts also contain a clause about the packer maintaining an account and tracking the times the contract price was higher than the cash market price and when the contract price was lower. Under this contract, the producer decreased the risk of price and income variation but the producers were also exposed to a new risk of large liability at the end of the contract. They became painfully aware of this when the cash hog market reached very low levels in 1998–99 and these packer accounts became very large against the producer.

Risk also increases when a building is built as part of the contract. Livestock and poultry contracts often involve a building purchase. This is a fixed asset and a fixed financial obligation for many years. Before signing, a manager needs to evaluate how that fixed obligation may affect them if the other party does not perform as the contract says or encounters financial difficulties. This risk is sometimes called hold up and asset fixity. The problem is having a single-purpose building and only one buyer of the product. If the other party does not deliver, the building sits empty. The asset is fixed or held up. Contracts usually have a clause saying that the building expenses will be paid for during the life of the contract, but the farmer's expected return from that investment is nonexistent. The problem is exacerbated if the other party is not able to even pay the building costs. The farmer can be left with a financial obligation, a building, and not many options for its use. This type of risk can be evaluated by considering the financial status and reputation of the other party. If this is questionable or not available, this hold-up risk is higher.

Recent crop contracts for non-GMO grains may seem very good in terms of a better, more stable price and quantity. However, some contracts could expose producers to other forms of risk due to signing a guarantee of being, for example, 95% free of GMO grain (or 99%, 99.5%, or even 100%). The risk is changed from market price fluctuations to one of potential contamination of the grain in storage and shipment and thus not meeting the product specifications in the contract.

Legal

Contracts have been used for centuries and their terms have developed very specific legal meanings. These specific meanings can have large impacts on the potential profitability of a contract to a producer, the risks to which they are exposed, and the legal rights and obligations under the contract. Certain differences in wording, for example, can change a producer from a contractor to an employee, which can have major impacts on the legal obligations and responsibilities. The different types of production contract also have different legal impacts on both the producer and contractor. Several of these legal concerns and specific wordings are noted in the checklist in the next section. As noted earlier, if the terms in the contract are not fully understood, outside expert advice is needed before signing.

A PRODUCTION CONTRACT CHECKLIST

The following checklist is developed to help farmers evaluate potential production contracts. The list is adapted and expanded from the lists developed by Gerhardson (1999) and by a task force established by the Iowa attorney general (Production Contracts Task Force 1996a, 1996b). However, no checklist can raise every question and not every question will be relevant for every producer since times, conditions, and points of law change. The decision to sign a contract is each party's own responsibility.

In the following list, the term "producer" is used to identify the party who will be obligated to produce a product, provide a service, or buy an input. In this discussion of production contracts in farm management, the "producer" is usually thought of as the farmer. The term "contractor" is used to identify the party other than the producer or farmer. The contractor may be asking the producer to produce a certain product according to specifications in the contract, to provide a specified service, or to buy a certain input. The contractor may be a processor, a product buyer, an input supplier, or another farmer. If both parties are farmers, each farmer would evaluate the contract using this same checklist but from his or her own viewpoint as a producer or contractor.

The list consists of both items to check when done and questions to be answered. The first item, "I have read ALL the terms of the contract," for example, is an item to be checked as done; it isn't a question of how much has been read, it is a question of whether it all has been read. Questions to be answered involve questions with simple answers, "Is the contract in writing?" for example, and questions requiring considerable analysis, "Does the contract improve the prospects for producer profitability?" for example. By considering the items and answering the questions in this comprehensive list, a producer will gather considerable information from which he or she can make a much more informed decision on whether to sign a contract, that is, enter into a contractual relationship.

First of All:
    I have read ALL the terms of the contract.
    I understand ALL the terms of the contract.

Get It in Writing:
    Is the contract in writing?
    Is the written agreement different than the verbal promises made?
    Is the ENTIRE agreement in writing?
    Is the agreement clearly written?
    When does the agreement become an enforceable contract?
    How do you enforce the contract?

In General:
    What is the legal form of the contract?
    Can I negotiate the terms of the contract?
    Can I join a marketing association that will negotiate contract terms with
        the contractor?
    Why is the contractor considering contracting?
    Why is the producer considering contracting?
    Does the contract fit into the producer's long-term goals and business
        strategy?
    Have neighbors been informed and consulted if the contract involves live-
        stock production, new genetically modified organisms, or other inputs
        and products that may affect them?

Parties to the Contract:
    Who are the parties to the contract?
    Is the contractor a subsidiary of a larger company?
    What are the contractor's credentials?
        If the producer has concerns about being paid, will the contractor provide
            a financial statement?
        Will the contractor provide a list of producers with whom the contractor
            has contracted in the past?
    Does it appear that the contractor is committed to contracting in this area in
        the future?
        Has the contractor made investments in fixed assets or relocated man-
            agement to this region?
        Is contracting part of the contractor's core business?
    What are the producer's credentials?

How can they be verified?

If the contractor has questions about the producer's ability to perform the contract, is the producer required (and willing and able) to release a financial statement and names of individuals who will verify the producer's financial stability and management abilities?

Can the contract be assigned or transferred by the producer to other producers?

Can the contract be assigned or transferred by the contractor to others such as a lender?

Do other parties have to approve the contract, such as landlords, lenders, spouses?

Duration, Termination, and Renewal of the Contract:

What is the duration of the contract?

Is the duration of the contract adequate to recover any investments in facilities and equipment?

Can the contracted be terminated early?

Under what conditions can the contractor terminate the contract?

Who determines whether those conditions are met?

Are there objective standards or is termination at the discretion of the contractor?

How much notice must the contractor give to the producer before terminating the contract?

Is the producer given an opportunity to cure any problem before the contractor terminates? How much time is the producer given to cure a problem? What options are specified for the producer to cure a problem?

What are the producer's rights after termination of the contract?

Will the producer be paid for work done up to the date of termination?

In the event of termination, does the producer have an option to buy the product (livestock or crops) if it is owned by the contractor?

Can the contract be terminated before the producer's investment is fully recovered?

If so, what is specified to compensate the producer for any remaining, unrecovered investment?

Under what conditions can the producer terminate the contract?

If the producer gets sick, is disabled, or dies, can the contract be terminated?

Can the producer terminate the contract if the contractor fails to deliver inputs on time or make payments on time?

What happens if the contractor files for bankruptcy?

What happens if the producer files for bankruptcy?

Does the contract excuse nonperformance on the part of the contractor caused by "Acts of God," meaning occurrences out of human control?

Does the contract excuse nonperformance on the part of the producer caused by "Acts of God," meaning occurrences out of human control?

Legal Relationship Created by the Contract:

Does the contract describe a sale of goods or a sale of services?

Who owns the product during production?

If the contract describes a sale of goods, does the contract specifically address whether the Uniform Commercial Code applies to the contract?

Does the contract describe the producer as a "merchant who deals in goods of the kind"?

    Does the contract make the claim that the producer holds himself out as having knowledge or skill particular to the practices of goods involved in the transaction?

If the contract describes a sale of services, does the contract address whether the producer is considered an independent contractor or an employee of the contractor?

    Does it address whether the producer is a partner, joint venturer, or agent of the contractor? Does the contract establish some other relationship, such as landlord/tenant?

Price, Payment, and Compensation:

On what basis is the producer being paid?

Are the terms describing pricing/compensation clear?

Is the method of calculating payment clearly defined?

Which payment criteria are out of the producer's control?

    Is the schedule of payments firmly set?

    Will this schedule satisfy the producer's need for cash flow?

May the contractor change the calculation of payments or the schedule of payments?

Are there penalties for late payments by the contractor?

    If so, are they clearly defined with a firm schedule of when they would be paid?

Can the payments be assigned to a lender or other third party?

Will the full payment be made before the product leaves the producer's facilities?

Will the name of the producer's lender be on the payment checks?

Are the weighing and grading procedures outlined in the contract clearly defined?

Who has control over weighing and grading?

How may the producer verify weighing and grading if it is performed by the contractor?

Does the contract include incentive payments?

What does the producer have to do to receive the incentive payments?

How are the payments calculated and when are the payments made?

Can the producer examine the computations used to determine the incentive payments and verify any methods used?

What are the producer's costs of production?

When the costs of production are deducted from the payments, is the contract profitable?

If the producer does not have cost of production records, have the Extension Service or other professionals provided assistance in arriving at estimated production costs?

Does the contract improve the prospects for producer profitability?

Does the contract attempt to modify the producer's right to an agriculture production lien?

Production Issues:

What are the written obligations of the producer under the contract?

What specific production inputs and practices are required of the producer?

What are the written obligations of the contractor under the contract?

What specific production inputs and operations will be provided by the contractor?

What normal and necessary production inputs and operations are not specifically listed? And who supplies them?

Who is responsible for the quality of inputs (feed, animals, seed, equipment, and so on) and the delivery of these inputs?

Who is responsible for supplying labor and management?

Who is responsible for keeping records?

Who has access to the land, facilities, and records? When is access allowed?

What are the facility requirements? Do new facilities need to be built? Who provides financing for new facilities and on what terms?

Is there an "exclusivity of use" clause for any facilities?

Does a livestock contract allow the producer to have any other livestock on the farm?

Who is responsible for manure removal and facility cleaning?

Who is responsible for postharvest tillage and other operations?

Is the risk of loss passed from one party to the other in the terms of the contract?

Disputes:

    Does the contract require an alternative dispute resolution method in the event of a dispute?

    Does it require mediation?

    Does it require arbitration?

    Does it specify or limit who may serve as a mediator or arbitrator?

    Where will the mediation or arbitration take place?

    Is the arbitration binding?

    In the event mediation or nonbinding arbitration is unsuccessful, what happens?

    Does the contract specify that a certain state's law governs disputes under the contract?

        How much would it cost to hire a lawyer licensed in that state?

    Does the contract set a venue (i.e., location) for any lawsuit that might be filed?

        How much would it cost to bring a claim in that location?

A Concluding Comment

Signing a contract, especially for the first time, creates a very different business environment for the farmer. Old relationships are not necessarily the same. New risks may be larger than envisioned. If the terms in the contract or its legal consequences are not fully understood, an attorney should be consulted. If the financial and tax consequences of the contract are not fully understood, a lender, tax professional, extension educator, agricultural consultant, or other knowledgeable advisor should be consulted. Another usually good source of information and advice are other producers who have experience with contracts. Talk to them.

Summary Points

- Contracts have been used in farming for a long time.
- Production contracts have increased in popularity and frequency of use in recent years.
- Contracting generally reduces all participants' exposure to risk by combining both market and production functions.
- Farmers enter into contracts for various reasons, including income stability, improved efficiency, market security, and access to capital.
- Processors enter into contracts to control input supplies, improve responses to consumer demand, and expand and diversify operations.

- Production contracts specify the quality and quantity of the commodity to be produced, the quality and quantity of production inputs to be supplied by the contracting firm, the quality and quantity of services to be provided by the grower, and the type, magnitude, and schedule of compensation that the grower will receive.
- The manner in which contracts are constructed affects the legal relationship between the producer and the contractor and, thus, the legal obligations of each party. There are four main types of production contracts.
- With sales contracts or market-specific production contracts, the farmer produces the crop or livestock and agrees to sell the product at harvest to the contractor. The farmer retains ownership until harvest and consequent sale.
- A production-management contract increases the control of the buyer over the production process.
- A production-management contract may be considered a bailment when someone else is entrusted with the possession of property, but has no ownership interest in it.
- Personal service contracts or resource-providing contracts specify that the producer is to provide services, not commodities, to the contractor.
- Evaluating a contract involves answering three basic questions. First, does the contract improve profitability? Second, how does the contract change the risks that the farmer faces? Third, how does the contract change legal relationships and obligations?
- The best way to evaluate profitability is to develop two sets of budgets: one for operating under the contract and one without.
- Avoiding risk is a major reason for contracting, but contracts can introduce new risks.
- Signing a contract can change legal relationships, expectations, and obligations from established, yet informal, relationships.
- A checklist can help a farmer evaluate potential contracts.
- If economic and legal implications of a potential contract are not fully understood, a farmer should seek expert advice to help evaluate the contract.
- The decision to sign a contract is each party's own responsibility.

REVIEW QUESTIONS

1. What are some reasons for a farmer to sign a contract?
2. What are some reasons for a processor or other buyer to sign a contract?
3. What are the four basic requirements for an agreement to be a legally enforceable contract?
4. To be legally enforceable, what types of contracts must be in writing?

5. Even though a contract does not have to be in writing, what are the benefits of putting it in writing?
6. Describe and contrast the four types of production contracts.
7. What are the three basic questions involved in contract evaluation?
8. Using the checklist in this chapter, evaluate the example contract below. (Since budget information is not available, profitability cannot be evaluated completely.)

[NOTE: THE FOLLOWING IS AN EXAMPLE CONTRACT. NAMES ARE FICTITIOUS. IT IS NEITHER A LEGAL INSTRUMENT NOR A MODEL LEGAL INSTRUMENT. LEGAL COUNSEL IS NEEDED TO CREATE A LEGAL INSTRUMENT.]

Feeder Pig Purchase Agreement

THIS AGREEMENT made this 1st day of February, 2003, between Harold Smythe, of Southtown, Anystate, herein after called Producer and Prairie Pigs, Inc., of Northtown, Anystate, herein after called Buyer.

1. Pig Prices: Pigs will be priced at 50% of the meat price on the next Monday close of the Chicago Mercantile five (5) months out with a floor of $25.00 and a ceiling of $35.00 per pig delivered to the Buyer. These pigs will be approximately fourteen (14) to nineteen (19) days old and weigh eight (8) pounds or more. Pigs that weigh less than eight (8) pounds will be priced according to Buyer's discount schedule. Pigs weighing 10 lbs. or more will receive an additional 50 cents per pound.

2. Payment: Buyer agrees to pay the agreed upon price for the pigs upon delivery.

3. Delivery: Producer agrees to sell to Buyer and Buyer agrees to purchase from Producer approximately 250 head of early weaned pigs every four (4) weeks. Producer shall provide groups of approximately 250 head every four (4) weeks. Producer shall notify Buyer of expected delivery dates at least three (3) days prior to delivery.

4. Risk of Loss and Transfer of Ownership: The Producer shall bear all risk of loss while the pigs are in transport due to natural causes, for other death losses cargo Insurance will be provided by the carrier. Ownership of said pigs shall pass at the time of delivery to Buyer.

5. Health: Producer agrees to provide all health permits necessary to qualify the pigs for shipment, and to notify Buyer immediately if there are any production problems that may affect the pigs delivery date, amount of delivery, or quality of pigs delivered.

6. Buyer's Right of Rejection: Buyer reserves the right to reject individual pigs at the time of delivery. The value of any pig in question will be adjusted according to Buyer's discount schedule. Any total discount greater than $250.00 per delivery group must be reported to Producer prior to acceptance of pigs by the Buyer. Buyer must notify Producer of any complaints within twenty-four (24) hours of delivery. The Producer reserves the right, at its sole discretion, to inspect and remove any pigs rejected and subject to discount schedule.

7. Duration of Contract: This agreement shall be effective immediately upon execution by the parties hereto and shall terminate five (5) years following the date of the first delivery of pigs to the Buyer's facility. This agreement, if not terminated in writing at least six (6) months prior to the end of said five (5) year period, will be deemed renewed until either party terminates the Agreement by giving the other party a six (6) month written notice of termination.

8. Notices: That all notices are required to be given pursuant to the Agreement or that are given pursuant to the relationship of the parties shall be in writing and shall be sent by postage prepaid, certified or registered mail to the addresses as given below.

9. Acts of God: Neither party shall be liable for damages due to delay or failure to perform any obligations under the Agreement if such delay or failure result directly or indirectly from circumstances beyond the control of such party. Such circumstances shall include, and shall be limited to, acts of God, acts of war, civil commotions, riots, strikes, lockouts, acts of government in either its sovereign or contractual capacity, perturbation in telecommunications transmission, inability to obtain suitable equipment or components, accident, fire, water damages, flood, earthquake, or other natural catastrophes.

10. Arbitration: The parties agree that should any dispute arise as to the terms or execution of the contact that they will submit the controversy to arbitration in the manner as allowed by Anystate and that they will abide by the arbitrator's decision.

11. Independent Contractor: It is understood and agreed that the parties to the Agreement are independent contractors.

12. Governing Laws: It is agreed that this Agreement shall be governed by, construed, and enforced in accordance with the laws of Anystate.

13. Waiver: The waiver by either party of any breach or violation of this Agreement shall not operate or be construed as a waiver of subsequent breach or violation thereof.

14. Entire understanding: This Agreement constitutes the full understanding of the parties and supersedes any and all prior agreements relating thereto, whether written or oral, that may exist between the parties hereto:

	By: _____
Witness	Producer

	By: _____
Witness	Buyer

NOTE

1. This chapter is intended to provide a basic understanding of the types of contracts and the issues involved in contract evaluation and choice. This chapter is not professional legal advice and should not be acted upon without professional legal advice. Laws and rules vary by state and over time.

13

Staffing and Organization

Managing the workforce (or human resource management) is becoming more important in agriculture as more farms have employees. Not everyone needs, or can take, the same levels of responsibility, authority, and accountability. Not everyone deserves the same level of compensation. That is why we need to talk about managing people. Also, while they do not (or should not) manage each other, the people involved in farms with multiple operators (be they related or unrelated; husband or wife; parents, children, or siblings; on-farm or off-farm) need to understand many of the same principles to ensure good communication and good management of the whole farm. This chapter starts with some basic information on human needs. The bulk of the chapter deals with the process of staffing or human resource management: assessing needs, designing jobs, recruiting, interviewing, hiring, training, motivating, leading, directing, evaluating, and compensating. At the end of the chapter is a section on business organization.

HUMAN AND EMPLOYEE NEEDS

The first fact to remember in human resource management is that employees and all fellow workers are humans. People are different, maybe very different, from each other, but they are still human. Those differences do not need to be avoided; in fact they cannot be avoided, and we probably should not even try to avoid them.

Human Needs

As humans, we have basic needs that need to be met. Maslow (1970) identified seven basic human needs. Those seven needs are listed below in Maslow's order of priority. How they relate to workforce management is also described. For workforce management, meeting the highest priorities is obvious; paying attention to the other needs can create an even better workforce.

1. *Physiological.* We all need air, water, and food. We will instinctively and immediately fight anything or anyone that limits our ability to breathe. Our bodies will tell us we need to drink or eat. Our cultures have set up schedules for meals and breaks for water and food.

   An employee's physiological needs are met by access to air, water, and food at work. By being paid, they can have these at home as well. Work rules need to allow time and space for water and food on the job. Sufficient ventilation is also critical. (These also relate to the second need, safety, but there is a point when the problem moves from an immediate life-threatening situation to a longer term safety concern.) Providing access to retirement programs and disability insurance are also ways to meet physiological needs, although in a long-term sense.

2. *Safety.* We all want to avoid pain and danger although we may have different pain and danger thresholds. We find comfort in being sheltered and not exposed to bad weather or potential harm. An employee needs safety guards, tractor cabs, safe equipment, rest, and protection from the weather. Again, work rules that allow for rest breaks provide for safety by decreasing the chance for accidents due to drowsiness or inattentiveness. Health insurance gives us access to medical care and protection from financial loss that may occur if medical bills are large.

3. *Belonging and Love.* We all need to be part of a group although that need may be expressed in different degrees. Even a "loner" can feel a need to belong to a group, albeit at a distance. We need the security and comfort of a close relationship, someone to talk with, although we have different definitions of "close."

   Belonging is a two-way street. The individual needs a group or a person. From the other direction, the group or other person needs to signal positively that the person does indeed belong to the group or to the relationship.

   An employee needs to feel he or she is part of the business. This may not be expressed directly and may be expressed in negative terms, but at some level, the need to belong is present. An employee needs to have pride in the business, to receive respect from management, to receive some communication about the business. A manager needs to avoid the "them vs. us" atmosphere.

   Interpersonal relationships between employees can range from very good, which creates a productive atmosphere, to very bad, which can be devastating to morale, productivity, and safety. A manager needs to pay attention to how workers relate to each other and how to avoid problems that may occur.

4. *Esteem and Self-esteem.* We need to be accepted by others. At some level,

we want the status and recognition by others that we have done well. We need to be able to look at what we have done and tell ourselves that we are satisfied with that effort.

An employee needs "company"-esteem, which can come from promotions or merits, increases in responsibility, and awards (even little ones). Besides monetary awards and job changes, a manager can signal acceptance publicly in a variety of ways: certificates that signal management's recognition of accomplishments (e.g., 10 pigs/litter; 25,000 pounds of milk per cow); awards of hats or coats for productivity, safety, or other accomplishments; pictures and announcements in a local newspaper or "company" newsletter. Even things that seem small can be very good for self-esteem and desired by employees when done in a public, respectful way.

5. *Self-Actualization.* We need to realize our potential and use it. We want to develop and expand our potential. We want to know what we are good at, use that skill or knowledge, and improve ourselves and our productivity. This skill or knowledge may be work related or it may not. Hobbies or avocations may be the place where this human need is met.

A good interview process can help fit the right person to the right job. This is the first step in allowing an employee to realize and use his or her potential. An employee also needs training to continue to improve skills. Future advancement within the company, or career ladders, can help an employee see how he or she can advance. By setting realistic goals, a manager can help an employee realize accomplishment and the satisfaction of using his or her skills. However, a manager also needs to be cautious and not set unrealistic goals. Unrealistic goals could cause underachievement because an employee cannot see any possibility of accomplishing the goal.

6. *Cognitive Understanding.* We need to know and understand ourselves, our lives, our environment, and how they relate to each other. We need to understand "why we are here." This is the hardest human need to meet. Other needs obviously have a higher priority. We need to find this cognitive understanding only after all our other needs have been met.

For an employee, this cognitive understanding may be easier to accomplish than as a human in the world. An employee needs to know where he or she fits within a company and the company's goals. This can be done through communications, training, and answers to questions like why are we doing this and what is the future direction of the company?

7. *Aesthetic Needs.* Although Maslow says this is hard to see, he does say that, after other needs are met satisfactorily, humans have a need for beauty. For employees this can be a need for clean, nice places to work.

Employee Needs

While Maslow's list of needs provide a very good background for under-standing people and their needs, his list is not the only way to classify work-ers' needs. When focusing directly on workforce management, employees needs can be grouped into four areas: responsibility, authority, accountability, and compensation. These areas will be used to define jobs.

Responsibility

Whether a worker is an employee, a partner, or a child, he or she needs to have the responsibility of doing assigned duties. They need to be responsible for something. These responsibilities may be called duties. Responsibility builds self-esteem, respect, belonging, self-actualization.

Authority

If a worker has the responsibility for a job, he or she needs authority to do the job. The worker needs to decide how to do it and to acquire the resources to do it. Perhaps the job and the resources are well defined (feeding the cattle or driving the combine, for example), but the worker still has to have the author-ity to get and operate the needed equipment. If the person is trained to use the equipment, valuable time can be wasted if the worker is waiting for approval to use an obviously needed piece of equipment. Having authority helps build self-esteem, self-actualization, and career development.

Accountability

If you have the responsibility and authority to do a job, you should also be held accountable for getting the job done. If there is no accountability, you can have anarchy! A manager needs to describe the job and the standards against which the results will be measured. Having accountability and meeting standards help with esteem, self-actualization, career development, respect, and accept-ance by management.

Compensation

Compensation can be more than money. Monetary compensation allows a worker to obtain food, housing, clothes, transportation, and other physiologi-cal needs. A residence on the farm or products from the farm (milk and meat, for example) can also be part of the compensation package. Compensation can also provide health and disability insurance to meet physiological and safety needs. Providing year-end bonuses can allow a worker to participate in the benefits of a good year and a good effort and show the worker that he or she does belong to the company and is not just an employee. Providing a vehicle to a worker not only fills physiological needs but also improves belonging and esteem if given as a reward for length of service or status of job, for example.

Is "all this stuff" about human needs and employee needs really important? *IT IS!* From a survey, Agri Careers of New Hampton, Iowa, found that *over 80%* of the reasons that employees gave for quitting a job could have been prevented (Thomas and Erven 1989). The reasons given for leaving and the percentage of respondents that gave that reason are: 7%—limited time off, 8%—lack of training, 10%—lack of recognition, 13%—lack of achievement, 13%—lack of responsibility, 14%—low salary, and 17%—problems with the boss and the boss's family.

HUMAN RESOURCE MANAGEMENT

In an ongoing operation—as most farms and businesses are—human resource management (managing the workforce) can be described as several steps or tasks including the following nine:

1. Assessing of the present situation
2. Developing tentative job descriptions
3. Matching present employees (or partners and family members) to those tentative job descriptions
4. Developing job descriptions for the remaining tasks
5. Recruiting, interviewing, and hiring of employees who fit those job descriptions
6. Training employees
7. Motivating, leading, and directing employees and colleagues
8. Evaluating performance
9. Compensating employees

For another view of human resource management, these tasks can be classified in terms of the four broad functions of management introduced in chapter 1: planning, organizing, directing, and controlling. By realizing where the different tasks fit in terms of functions, we can perform each more efficiently by knowing which "management hat" we need to be wearing. It also becomes very evident that some tasks of workforce management take place under different management functions. For example, the setting of wages and incentives is done as part of the planning function, but the actual compensation occurs under either directing or controlling (in terms of incentives to obtain the desired performance).

1. Planning: assessing the situation, estimating the number and type of people needed, writing job descriptions, setting employee goals and standards, setting wages and incentives

2. Organizing: interviewing, hiring
3. Directing: training, leading, supervising, motivating, compensating
4. Controlling: evaluating employees and determining incentive pay; taking corrective actions: reassign, retrain, release

Assessing the Situation

Assessing the present situation involves determining the amount of labor available, listing the jobs that need to be done and the labor required for each job, and then comparing the amount available and amount required to estimate any hiring needs. A standard form such as the one in figure 13.1 can be helpful for organizing this information. This form breaks the labor available into operator, partner, family, hired, as well as custom operator. Off-farm work is listed as a need or demand for labor to (1) reflect the increasing frequency of farmers having off-farm employment and (2) emphasize the idea that this choice is an allocation of labor resources very similar to the allocation of land to crops. Custom operators are included to be sure that all labor needs are covered and that nothing is forgotten; for example, harvest labor is still needed even though a custom combine operator will be doing that labor. Labor needs are broken into that needed by the crops and the livestock and the amount that is used on the whole farm and not directly allocated to either crops or livestock. Identifying the amount available and needed by season will help determine whether any additional labor needed should be year-round or seasonal.

Job Descriptions

After determining what and how much labor is needed, jobs need to be designed that include the tasks listed as needing to be done. In job design, a manager considers what tasks need to be done and what combination of tasks will be both complementary with each other and interesting to the worker. Putting all the "good" tasks in one job and the "distasteful" tasks in another job will likely create one job that a worker likes and stays in for a long time and a second job that no one likes and they all leave quickly. For example, the job that includes cleaning the barns should also include some more pleasant tasks. Jobs with distasteful tasks may need other benefits such as extra incentive payments, shorter work hours on distasteful days, and so on.

Part of job design (and thus developing good job descriptions) involves worrying about workers' job satisfaction. If an employee likes a job, he or she will probably perform well, which will improve the overall performance of the business. If an employee does not like the job, performance will suffer and perhaps the employee will leave, both of which will hurt the performance of the business.

	Total hours for year	Distribution of hours			
		Dec. thru March	April May June	July August	Sept. Oct. Nov.
Suggested hours for full-time worker	2400	600	675	450	675
My estimate for full-time worker					
LABOR HOURS AVAILABLE					
Operator (or Partner No. 1)					
Partner No. 2					
Family labor					
Hired labor					
Custom machine operators					
(1) TOTAL LABOR HOURS AVAILABLE					
(2) OFF FARM WORK					
DIRECT LABOR HOURS NEEDED BY CROP					

Crop enterprises	Acres	Total Hr./Acre				
(3) TOTAL LABOR HOURS NEEDED FOR CROPS						

DIRECT LABOR HOURS NEEDED BY ANIMAL ENTERPRISE						
Animal enterprises	No. Units	Total Hr./Unit				
(4) TOTAL LABOR HOURS NEEDED FOR ANIMALS						
(5) Total Hours of Indirect Labor Needed						
(6) TOTAL LABOR HOURS NEEDED (lines 2 + 3 + 4 + 5)						
(7) Additional Labor Hours Required (line 6 minus line 1 IF >0)						
(8) Excess Labor Hours Available (line 1 minus line 6 if >0)						

Figure 13.1. Labor estimate worksheet.

Job satisfaction and job dissatisfaction are on two different scales (Herzberg 2002). They are not the endpoints of the same continuum. Different factors affect these two ideas. So a worker can, at the same time, be both satisfied and dissatisfied with a job. Factors that are intrinsic to the job or the work itself are potential satisfiers. These include:

Achievement—the ability to do a task well, to improve oneself

Recognition—acknowledgement of work well done

The work itself—the job design is interesting and can be seen as useful to the business and society

Responsibility—the ability to have control, to be in charge of at least a small part of the job

Advancement—growth and progress in job status.

Factors that are extrinsic to the job or work itself are potential dissatisfiers. These include:

Company policy and administration—the rules, procedures, directions, dictates from top management

Supervision—poor supervision, bad bosses, high stress, can make a good job unbearable

Pay—low pay, both in absolute and relative terms, can make a worker unhappy

Working conditions—unsafe conditions, poor equipment, poor ventilation, are examples of why workers can be unhappy and thus "short timers" in the job

A job description needs to be developed and written for advertising, interviewing, hiring, and evaluating the employee. The act of writing can help clarify the job and also serve as a communication tool within a multioperator business. A job description should describe the job to be done as well as the authorities given the employee and measures by which the employee will be evaluated. The evaluation measures should be both quantitative (i.e, measurable) and controllable by the employee. The amount of detail will depend on the job and the employee's knowledge. The job description should be clear enough that both potential applicants, current employees, and management understands what is to be done. Besides the title, a complete job description should include:

1. Responsibilities
2. Authorities
3. Evaluation measures
4. Required qualifications (These are minimal levels of training, knowledge, personal characteristics, physical abilities, flexibility in time, flexibility in tasks, and other minimal requirements for the employee to do the job. Required qualifications are used to determine whether an applicant can do the job.)

5. Desired qualifications (These are additional levels of training, knowledge, etc., that would allow an applicant to do the described job even better. Desired qualifications are used to compare and rank applicants.)
6. Supervision (amount and, if applicable, direct supervisor)
7. Time expectations (hours per day and per week, starting and ending times, overtime rules, seasonal differences, etc.)
8. Beginning wage or salary and any bonuses and incentives
9. Benefits (insurance, vacation, time off, etc.)

An example job description for a farm operator assistant is shown in figure 13.2. Examples of quantitative measures for evaluation, potential interview questions, and possible tests of ability needed in the job are included with the description.

Recruiting, Interviewing, and Selecting

Any advertising needs to be done in proper places for job being filled and where appropriate applicants will see it. The advertisement needs to be written clearly with the minimum requirements listed. The application form should be written to show whether applicants meet minimal requirements and to provide enough additional information to help select the proper ones to interview. Interviewing is done with job-specific questions crafted to evaluate the candidates, their knowledge, and their ability to accomplish the job needed. Part of the interview should be a test to determine their skills if these are not already known or obvious to the interviewer. These tests could be written or demonstrations of skills such as tractor driving, computer operation, livestock knowledge, pesticide application procedures, and plant identification. A general interview form is shown in figure 13.3, but job-specific questions and tests need to be crafted and designed in order to complete the form.

Training

Training is needed regardless of previous experience. Orientation, that is, the initial training session, can communicate expectations, process and instill the "team" feeling. If a new employee has worked elsewhere, training in the current job on the current farm can show how the new employer likes the job done. Even little details can be big factors in quality management and/or cost control. Training for review is always good especially for seasonal jobs (e.g., combine operation, pesticide application); it promotes safety. Training can also help in career advancement, personal development, and employee self-esteem. Training doesn't have to be directly related to the job. Nor does it have to be done on the farm or by the employer. Certification of pesticide applica-

This is a job description for a farm operator assistant on a 1,000-acre crop farm. We grow corn, soybeans, alfalfa, and wheat in east central Minnesota. We rent most of our land and are spread out across several counties.

Duties & Responsibilities
  1. Plant and harvest crops
  2. Cultivate row crops
  3. Cut, rake, bale alfalfa
  4. Drive grain truck as needed
  5. Do general maintenance on equipment and tractors
  6. Buy parts and supplies for general maintenance
  7. Notify owner of needed repairs

Authorities
  1. Do general maintenance on schedule
  2. Buy needed parts and supplies within budget provided by owner
  3. Decide which fields need to be cultivated

Accountability
  1. The employee will report to the owner.
  2. Expenditures for parts and supplies have to be appropriate and within budget.
  3. Machinery has to be maintained.
  4. Row crop fields have to be cultivated.

Evaluation Standards
  1. Equipment and tractors properly maintained
  2. Timeliness of cultivation
  3. Level of weed infestations in field
  4. Quality of field work
  5. Purchases within budget
  6. Punctuality and regularity of work time

Minimum Qualifications
  1. Able to drive a tractor in a straight line
  2. Able to lift 80 lbs. from ground to height of pickup bed
  3. Able to do general maintenance on machinery
  4. Able to order and buy parts and supplies
  5. Background knowledge of farming and agriculture

Figure 13.2. Job description for a farm operator assistant. (*Continued on next page.*)

tors is a very good example of training that is almost always done off the farm by certified trainers. Employers should not forget the effect of going to training with their employees; that act shows the importance of the training and also helps give the employee a little more self-esteem because the boss is going to the same training.

Desired Qualifications
1. Ability to operate combines and other machinery
2. Ability to keep maintenance records
3. Willingness to work long hours during planting and harvesting
4. Licensed to drive grain truck
5. Ability to identify weeds

Compensation package:
$8 per hour, time and a half when over 50 hours per week, health benefits for worker, family benefits available at ½ cost, meals while working, pickup truck available

Three quantitative measures for evaluating performance in this job:
A. Machinery maintained on schedule
B. Level of weed control in cultivated fields
C. Punctuality and regularity of work schedule

Five questions you might ask in an interview with an applicant:
A. Describe your experience with driving large machinery.
B. Describe your knowledge of weed control.
C. What would you check before starting and operating a tractor?
D. How deep should you run a cultivator?
E. How would you decide it is time to buy more oil for the machinery?

Three tests that would evaluate the applicant's abilities needed for this job:
A. Drive a tractor in a straight line for 200 yards
B. Lift an 80 lb. bag from the ground to a pickup
C. Change oil and filter on a tractor

Figure 13.2. (*Continued.*)

Motivation, Leading, Directing

There are many aspects of motivation, and many books on the subject are available. Much of motivation is directly related to Maslow's list of basic needs and how those needs can be interpreted for employees. Monetary reward is certainly one motivating factor but not the only one. As an employer or supervisor, providing an employee the knowledge of why the job needs to be done and how the job will be done can help that employee understand where he or she fits in the operation and business. The methods and styles of motivation, leading, and directing will vary with the individual employers and supervisors and, on the personal side of motivation, vary with the employee and how each will respond to various incentives and ideas. Poor supervision can be costly. Intentional and unintentional disrespect of employees by the supervisor and/or the manager can be disastrous in terms of direct costs, poor productivity, and high employee turnover.

Date: _____
Interviewer: _____
Job being interviewed for: _____

Name of Applicant: _____

Address: _____ Phone: ( ) _____

Past work experiences:

_____

Reasons for leaving former job:

_____

Present skills and certificates relevant to this job:

_____

Other questions and information relevant to this job:

_____

Personal goals and aspirations relative to farming:

_____

Why are you applying for **this** job?

_____

Results of tests relevant to this job:

Figure 13.3. Interview form (Adopted from Thomas and Erven 1989). (*Continued on next page.*)

Evaluation

Evaluation is the major source of control of employees. A formal evaluation session is needed at least annually and preferably more frequent, maybe quarterly. But evaluation needs to be done and communicated to the employee when ever it is needed, not just once a year. Both commendations and reprimands are needed. A new employee needs evaluation very often at the begin-

Characteristics (Rate those relevant to this job.)	Low				High
1. Leadership qualities	1	2	3	4	5
2. Ability to work with others	1	2	3	4	5
3. Receptiveness to receiving directions	1	2	3	4	5
4. Motivation to learn	1	2	3	4	5
5. Willingness to perform physical labor	1	2	3	4	5
6. Training and background in:					
Livestock production	1	2	3	4	5
Animal nutrition	1	2	3	4	5
Animal health	1	2	3	4	5
Crop production	1	2	3	4	5
Pesticide application	1	2	3	4	5
Mechanical skills	1	2	3	4	5
Machinery operation	1	2	3	4	5
Vehicle operation	1	2	3	4	5
Management concepts	1	2	3	4	5
Finance	1	2	3	4	5
Marketing	1	2	3	4	5
Ability to manage others	1	2	3	4	5
Ability to compromise	1	2	3	4	5
Ability to identify problems	1	2	3	4	5
Ability to analyze situations	1	2	3	4	5
Ability to make a decision	1	2	3	4	5
Ability to understand directions	1	2	3	4	5
7. Personal goals	1	2	3	4	5
8. Initiative and imagination	1	2	3	4	5
9. Motivation	1	2	3	4	5
10. Determination	1	2	3	4	5
11. Willingness to ask questions	1	2	3	4	5

Comments from references:

Comments by current employees:

Overall rating:

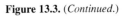

Figure 13.3. (*Continued.*)

ning of employment to be sure both parties know what is being done and whether it is being done as expected. Small, frequent corrections of procedures by close supervision may help improve overall, long-run performance. Good training may decrease the need for close supervision and frequent evaluation. If there are signs of trouble, supervision and evaluation may need to be increased. If possible, evaluation should be very frequent so each employee knows the expectations and standards and whether they are being met or not. An older employee still needs evaluation both when something seems to be wrong and when a manager observes tasks being done well.

Evaluation should be done on the basis of quantitative standards or measures written in the job description, discussed in the job interview, and agreed to at the time of hire. These standards should be appropriate to the job and skill level required. They should be measurable and affected by actions controllable by the employee. Standards should not be vaguely worded such as "exerts good effort." A better standard would be the amount of work done, the efficiency with which work is accomplished, production levels, and so on. Some examples include acres covered per day, fuel use per acre, bushels over area average crop yield, total production, downtime for machinery repairs, pigs weaned per litter, pounds of milk per cow, feed costs per unit of animal production, productivity increases over previous years, and so on.

The standards and measures should not be completely beyond the control of the employee. Other factors should not have a large impact on the measures. For example, if the farrowing house manager has no control over herd genetics, the average ham and loin percentage should not be used as a standard, but the number of pigs born per litter and weaned per litter are very appropriate measures.

Some jobs do not lend themselves well to quantitative measures as the only measure. The products of these jobs have both qualitative and quantitative aspects. Employees in these jobs can still be given some measure of quantitative evaluation by giving them a numerical qualitative score on a frequent basis and then using an average score to provide the employee something quantitative.

For example, one plant nursery's employee evaluation form had every employee rated from zero to ten every day. The employees were evaluated on a moving three-week average of those scores to avoid daily fluctuations. The nursery used this system because the plants or trees needed to be potted, moved, pruned, and cared for in a high-quality manner otherwise sales were hurt. So employees needed to be evaluated on both the physical amount of work and the quality of their work. Since the tasks were very different, a quantified estimate of quality was difficult, if not impossible to develop. So, the subject measure of both quality and quantity was adopted. Crop and livestock farmers may benefit from a similar employee evaluation system if the needed tasks are also very different from one hour, day, or season to the next. This daily evaluation also provides an immediate evaluation of the employee's work and hopefully positive responses by the employee.

At the evaluation session, the basic question is whether the employee met the standards or not. As said earlier, both commendations and reprimands are needed. Commendations reinforce good behavior, show respect for the employee, build self-esteem and company-esteem, and build trust in management by the employee. Reprimands are needed to correct behavior and reinforce management's desired performance standards. The evaluation discussion should include points or ideas on why the standards were met or not met. Constructive criticism may be the best motivation for improvement compared to negative approaches.

Negative evaluations should result in one of four actions. These are the four Rs:

Retrain	Make sure the employee understands what is to be done and how it is to be done.
Reassign	Give the employee new responsibilities. A faulty interview process may have caused a person to be placed in the wrong job.
Reevaluate	Before the last option (firing) is used, reevaluate the person. Remember that the initial evaluation may be faulty. Perhaps the evaluator was having a bad day for reasons other than the employee's performance. Perhaps the manager and the employee just had a disagreement or a fight over a small part of the job. Perhaps the system in which the employee works needs redesign, not the employee.
Release	Firing should be the last option to be considered. Firing is the most expensive option due to the costs of firing and to recruit, interview, hire, and train a new employee. Firing can also create potential legal problems. For these reasons and for just a better base from which to make decisions, keeping a record of quantitative measures of job performance will help establish goals and a basis for releasing employees.

Compensation

Compensation involves the base salary and any incentives and benefits or fringes that the business provides. The overall level of compensation is determined by the local labor market, that is the supply of and demand for labor in the local area. If workers have other businesses as potential employers, farmers will likely have to pay similar wages and benefits as those jobs in order to attract employees. If those businesses provide health insurance for the employee and the employee's family, a common benefit, farmers will find they have to provide that also. Other factors that will affect pay levels and labor availability are the presence of unions, the cost of living in the area, and labor regulations.

Farmers, just like any other business owners, will have to evaluate the costs of labor in their local area versus the benefits of having the work done. If the cost of labor is deemed too expensive for the farm, alternatives to labor could be evaluated—alternatives such as mechanization or computerization to replace labor or to improve the productivity of labor. If productivity is improved, the benefits may then be high enough to compensate labor according to the local market demands.

Incentives

The basic rule of compensation is to reward what you want done. In addition to a base wage or salary, incentives and bonuses may also be desirable. Incentives are usually based on volume of business, productivity, longevity, and profitability. Reaching a goal in business volume, a certain amount of milk produced per month, for example, can be a reason to pay a bonus to those involved in reaching that volume. Other incentives can be for overtime work and for achieving stated levels of productivity and efficiency. These may be based on, for example, acres covered per hour, animals detected in heat or sick, pigs saved per litter, feed conversion rates, milk per cow, crop yields, and so on. Some incentives may be based on the employee's length of employment with the farm to reward loyalty. If the farm has a profitable year, an astute manager will be sure that employees also share in the profit. Thomas and Erven (1989) suggest that normal incentives and bonuses should total 2–5% of the cash wages for semiskilled workers, 4–10% for skilled workers, and 5–40% for supervisory and management employees.

Incentives may be informal. That is, bonuses, rewards, and benefits are awarded when good work and efforts are shown but no sure expectation of receiving the bonus is felt by the employee. However, to be effective in improving and maintaining good performance by employees and for the business, incentives need to have formal structures. Then employees can understand and expect rewards for certain behaviors. Structured, formal incentive programs are most likely to succeed if they have (1) established standards, (2) clearly linked superior performance with pay or a valued reward, and (3) carefully considered what type of performance the incentive stimulates (Billikopf 2000). Effective incentives are designed so the larger the amount earned by the employee, the greater the benefits obtained by the farmer also.

Potential incentives can be evaluated in a partial budget framework before being implemented. For example, the value of increasing the number of pigs weaned per sow per year by 0.5 can be estimated for the farm and compared to what may have to be paid to employees to pay better attention to sow and piglet care and any additional expenses incurred for that care. The amount and cost of overtime needed to plant crops in a timely manner can be compared to either the additional cost of investing in bigger machinery for timely planting or the potential value of yield losses due to not planting in a timely manner.

As an example of evaluating the worth of an incentive program, consider a farmer evaluating whether to pay his milker $1 per cwt. to raise the milk production of three hundred cows from 21,000 to 23,000 lb. per cow by taking better care of the cows. This farmer has to decide if the incentive would be profitable for him and whether it is sufficient to encourage the milker to do the extra work. For the whole herd, the increase of 20 cwt. per cow would mean an increase of 6,000 cwt per year. If this farmer's variable costs for this increase in milk production is $10 per cwt. (which includes the proposed incentive of $1 per cwt to the milker) and milk is worth $13 per cwt., the additional value of the increase would be $3 per cwt. for a total net value of $18,000 per year for the farmer. The farmer does have some ability, it appears, to increase the incentive, but changes in these costs and prices are not figured into this simple analysis. Now the farmer must decide if the $1 per cwt. is enough incentive for the milker to encourage the extra care of the animals. Most likely, the farmer may need to talk to the milker, explain the process, and try the incentive program for a specified test period to see if the milker responds. The milker would, of course, have to understand the rules and that a test period was being used. If the farmer does not ensure that the milker does understand the program or is unable to obtain the desired gains in productivity, the milker could easily become dissatisfied with the job and performance could suffer—the opposite of the desired result.

To be effective in obtaining the desired results intended by incentives, employees must have a fair amount of control over the measures used. Incentives based on crop yield or animal genetics may not be of interest to employees because they cannot control weather or the choice of animals. However, if the incentives are to plant correctly or to observe animals in heat or sick, employees can control those activities. The incentive must also be large enough to create some interest in achieving the goal by employees. However, care must be taken to provide incentives for good work but to design the incentives so that employees do not see a set of rewards that create behavior that is not good for the benefit of the whole farm.

Fringes or benefits usually include health insurance for the employee and the employee's family. Housing, utilities, and food are also common benefits. If the employee has any supervisory duties or works somewhat unusual hours, a farm vehicle may be part of the compensation package. Vacations (with pay) are also becoming more common. A survey of U.S. hog farms showed the most common benefits for paid employees were paid vacation, major medical coverage, paid holidays, workers' compensation, and paid sick leave (table 13.1). Other benefits were also common.

As an employer wrestles with the question of how much to pay, the question first raised in quality management should be considered again: should inputs (labor in this case) be purchased or hired on the basis of the level of cash wages only? No, Deming would say; wages and the entire compensation package (as well as the entire labor management and production process) should be chosen on the basis of lowest product costs for the business, not just the lowest cash cost per hour.

As an example, consider two real, neighboring farmers in California. The first farmer paid a low hourly wage, used poor (and cheap) labor management practices (e.g., a lot of yelling), had low productivity and a high turnover of employees, and, thus, had a high labor cost per unit of product. The second farmer paid a high hourly wage for the area, used good labor management practices, had high productivity and a low turnover of employees, and, thus, had a low labor cost per unit. Which farmer do you think had the better income statement?

Table 13.1. Percentage of Employees on U.S. Hog Farms Indicating the Availability of Benefits, 2000

Benefit	Percent
Medical insurance	71.9
Dental insurance	35.0
Disability insurance	32.8
Life insurance	46.4
Paid vacation	87.4
Paid holidays	69.7
Workers' compensation	67.3
Unemployment insurance	37.3
Paid sick leave	60.8
Pension/retirement plan	48.2
Profit-sharing plan	19.8
Housing	31.3
Paid utilities	17.1
Vehicle	19.7
Processed meat	42.1
Continuing education	24.3
Other	11.1

Source: Hurley et al., 2000.

BUSINESS ORGANIZATION

In this section we will look at three areas. First, the organizational chart, second, the board of advisors or directors for the farm business, and third, how the business is organized from a legal viewpoint: sole proprietorship, partnership, corporation, cooperative, or any of the other variations of those basic legal forms.

The Organizational Chart

Is one person in charge or is a committee? Who is in charge of what? How much responsibility and authority does each person have? For a single-person business, these are easy questions. For a multioperator business, even a family business, these questions can be very involved and contentious. Considerable thought is needed to answer these well and is actually part of the strategic plan of the farm. Good communication on these issues and an understanding by all parties will go a long way in helping any business achieve its strategic and financial objectives and goals. Knowing the answers to these issues can help a multioperator, multiemployee farm work more efficiently and more productively.

The complexity of the business and the ownership, knowledge, and desires of those involved will be important factors in designing an organizational chart. For example, a farmer who runs her cash corn-soybean farms as a sole proprietor with only one employee may not need an extensive chart. Even if the employee is in charge of the equipment and the repair shop, both the farmer and the employee probably know who is in charge of the whole operation.

A farm that has several strategic parts and several people involved will have a more complicated chart. For example, consider two partners with two employees who operate a dairy and crop farm with a seed dealership "on the side." As part of their strategic planning and their staffing plan, these two partners should have discussed who has interest, knowledge, and skills needed in each part of the business. As an example of one possible result of this discussion, their organizational chart could have one partner in charge of the dairy and one in charge of both the crops and the seed dealership (fig. 13.4). This split in responsibility could be due to both interests and the seasonality of work demands: the dairy has a steady work demand throughout the year while the crop and seed dealership has different seasonal work demand patterns. Through talking with their employees and evaluating their skills, the partners have decided that one of the employees can be in charge of milking and the milk house and one can be in charge of the machine shop and equipment main-

Plainview Dairy

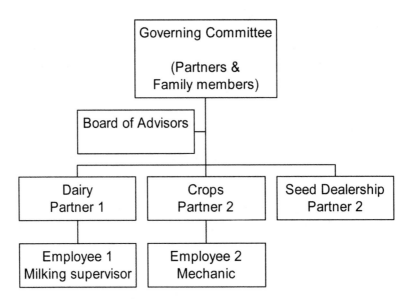

Figure 13.4. Organization chart for a farm.

tenance. This split of responsibility can be part of their motivation and recognition of employees as well as the partners' realization that they can't make all decisions. If their families are involved closely in the business (perhaps as one of the employees), they too should be involved in this discussion. This involvement is shown in the chart as the family being part of the governing committee.

Board of Advisors

Farming is a complicated business and will continue to become even more complicated. A board of advisors can be a valuable management tool for understanding and being prepared for this increasing complexity.[1] The members of a board of advisors could, but don't have to, include the farm's banker, seed and/or fertilizer dealer, a nonfarm business owner, another farmer perhaps, and others who have knowledge of farming, the locality, and the broader economy. Obviously, a farmer would include only those who could contribute management wisdom, are trusted by the farmer, and would keep business information confidential. In many ways, a farmer may already talk with this group of people individually at their businesses and in the coffee shops. The advantage of formally convening a board of advisors is to gain their collective concentration on this specific farm in a business atmosphere (versus the coffee shop). The board can provide a third-person view of the farm as a business without being emotionally or financially involved in the family, the farm history, or the business directly. The board members probably need to be compensated for their time and wisdom but this may be a small cost compared to the better decisions that could be made because of their advice. Other, large businesses have boards of directors and consultants; there is no reason not to use this source of management expertise in farm businesses of any size.

Business Organization

A key issue involved with a farm or ranch operation is the legal form of how the business will be operated. The most common forms of business organization for farms are presented in this section. A summary of the information is found in table 13.2.[2]

Sole Proprietorship

A sole proprietorship is owned by one person solely. Its main advantage is simplicity of formation and governance. If the business operates under the name of the owner (e.g., Joe Smith or Smith Farms), no legal documents need to be drafted. In Minnesota and other states, if another name is used that is not close to the owner's name (e.g., Plainview Farms), a legal form does have to be filed stating the assumed name and the owner. The owner bears personal liability for all debts of the business as well as any other liabilities. That is, when the owner signs as the owner of a sole proprietorship, he or she also signs as an individual person. The owner is limited in the availability of fringe benefits and retirement benefits, as well as the deductibility of health insurance. The sole proprietorship business ceases to exist upon death of the owner.

General Partnerships

A general partnership consists of two or more partners who have joined to operate a business. Each partner is personally liable for all debts incurred by *any* of the other partners in the operation of the business. A general partnership is flexible and fairly easy to form, but forms do have to be filed with the government. A general partnership passes all income and deductions to the partners; a general partnership does not pay income taxes, for example. A general partnership ceases to exist when the partners dissolve the partnership. Partnerships can allow people to pool their financial, management, and other resources to create, hopefully, a better advantage in the marketplace compared to the partners having their own sole proprietorships.

Table 13.2. Comparison of the Major Types of Business Arrangements

	Sole proprietorship	Partnership	Corporation
Nature of entity	Single individual	Association of two or more individuals	Legal entity separate from shareholders
Source of capital	Personal funds or loans	Partners' contributions or loans	Contribution of shareholders for stock, sale of stock, or loans
Management decisions	Proprietor	Agreement of partners	Shareholders elect directors who manage business through officers
Liability	Personally liable	General partner(s) liable for . all partnership obligations Limited partners' liability is limited to their investment	Shareholders not liable for corporate obligations (but individual shareholders or officers of small corporations may be asked to cosign loans)
Limits of business	Proprietor's discretion	Partnership agreement	Articles of incorporation and state law
Life of business	Terminates on death	Agreed term, but terminates on death of partner	Perpetual in most cases
Effect of death	Liquidation	Liquidation or sale to surviving partner	No effect; stock passes by will or inheritance
Transfer of interest	Terminates proprietorship	Dissolves partnership; but new partnership may be formed	No effect on continuity; stock transferable to anyone if not restricted
Income taxes	Business income is combined with other income on individual tax return	Partnership files IRS information report; each partner's share of partnership income is added to her or his individual taxable income	C: corporation files a tax return and pays income tax; salaries to shareholder employees are deductible; shareholders pay tax on dividends received S: corporation files IRS information report; shareholders report their shares of income, operating loss, and capital gains on individual returns
Alternative forms		General and limited partnerships	Regular (C), Tax option (subchapter S), Limited Liability

Limited Partnerships

A limited partnership is similar to a general partnership except that one or more partners are liable only to the extent of their investment. A limited partnership consists of two or more partners who have joined to operate a business with at least one partner personally liable for any and all debts of the partnership and at least one partner limited in liability. The advantage of the limited partnership (over the general partnership) is its ability to attract outside capital from individuals who are willing to put some money at risk but do not want to become personally liable for all debts of the partnership. The disadvantage to the general partner(s) is their unlimited liability for the debts of a larger business.

Sub Chapter-C Corporations

Corporations are separate entities that, once formed, have unlimited life. One or more stockholders own stock in the corporation and their liability is limited to their investment. Shares are easily transferable. Owners are not personally liable for debts of the corporation—unless they also sign as individuals and not just as officers of the corporation. The initial cost of forming a corporation is greater than a partnership or a sole proprietorship; taxation rules and forms are more involved; and the business must be run more formally.

Sub Chapter-C corporations are like the corporations listed on the stock exchanges but they can also be privately held corporations. There is no limit on the number of shares, types of shareholders, or the number of owners of Sub Chapter-C corporations. Sub Chapter-C corporations can be large (like 3M or IBM), or they can be small (like some farm corporations).

One disadvantage of Sub Chapter-C corporations is the double taxation of their profits. Sub Chapter-C corporations pay taxes on corporation income, and the owners pay taxes on any dividends paid to them. Dividends paid to the owners are not deductible by the corporation. Thus, the total tax rate can be quite high. However, owners can be employees of a Sub Chapter-C corporation, and as long as the level of compensation is reasonable compared to the services provided, that compensation is most likely deductible.

This double taxation can be quite vexing if the corporation owns land that appreciates in value. In the past, many small Sub Chapter-C corporations (such as farms) did not own land, but rented land owned personally by the owners of the corporation. The ability to form limited liability companies (discussed later) have eliminated the need for this separation of ownership and operation.

Sub Chapter-C corporations can deduct health insurance costs for employees. They also can utilize many retirement and fringe benefits not available to other business forms.

Sub Chapter-S Corporations

Sub Chapter-S corporations are limited in the number and type of shareholders. They are generally taxed as a partnership so they also have the disadvantage of limited deductibility of fringe benefits, retirement, and health insurance costs. Income, however, can be passed through to the owners—like a partnership.

Sub Chapter-S corporations were designed to help avoid double taxation problems but this was not entirely successful. So land should not be held by Sub Chapter-S corporations either since land appreciation could be double taxed also.

Limited Liability Companies

Limited liability companies have the tax benefits of a partnership (no double taxation) and the limited liability of a corporation. Limited liability companies can hold land without any penalty of double taxation on the appreciation of that land (or any asset) because income is passed through to the owners. They are a fairly new legal form of business organization. Most states now allow this form. The owners of the business are called members and "managers" run the business. The disadvantage of the limited liability company is the complexity and organizational cost. It also can't fully utilize all fringe and retirement benefits.

SUMMARY POINTS

- Human resource management is becoming more important for farm managers.
- Maslow's seven basic human needs are physiological, safety, belonging and love, esteem and self-esteem, self-actualization, cognitive understanding, and aesthetic needs.
- Since employees are human, these basic human needs are important to employees and, thus, managers also.
- An employee also needs responsibility, authority, accountability, and compensation.
- Human resource management involves several steps: assessing the situation; developing tentative job descriptions; matching present employees with tentative job descriptions; developing job descriptions for remaining tasks; recruiting, interviewing, and hiring employees; training employees; motivating, leading, and directing employees; evaluating performance; and compensating employees.

- Assessing the situation involves determining and comparing the type, amount, and timing of available labor and the type, amount, and timing of labor needs.
- A job description needs to include the employee's responsibilities, authorities, evaluation standards, required and desired qualifications, supervision, time expectations, compensation information, and benefits.
- Proper recruiting, interviewing, and selection of employees helps ensure better fit with jobs and, thus, productivity of the employees and profitability of the farm.
- Testing of applicants' skills is a much better method to ascertain their ability to perform a job compared with trusting applicant's and references' statements.
- Training is needed regardless of previous experience.
- Monetary reward is one method for motivating employees but not the only method.
- Appropriate and effective methods for motivating, leading, and directing will vary with both the supervisor and the employee.
- Evaluation is the major source of control of employees and should be done whenever needed.
- The basic question in evaluation is whether the employee met the standards described in the job description or not. Both commendations and reprimands are needed.
- Negative evaluations should result in one of four actions: retrain, reassign, reevaluate, or release.
- Release or firing is the most expensive option and should be the last choice.
- Compensation involves the base salary, incentives, and benefits. It is set according to the local job market and the performance of the employee.
- Formal incentive programs allow an employee to anticipate and work toward specific goals that are desirable for both the employee and the farm.
- Low labor costs are not directly related to low cash wages.
- The complexity of the farm business and the number of people involved increase the need for a formal description of how the farm and the people involved are organized and work together.
- A board of advisors can be a valuable management tool for a manager facing an increasingly complex business environment.
- Three common legal forms of business organization are the sole proprietorship, partnership, and corporation with the sole proprietorship being the most common among farms in the United States.
- Other legal forms include limited partnerships, Sub Chapter-S corporations, and limited liability companies.

- Each legal form has different benefits and requirements and is appropriate in different situations.

REVIEW QUESTIONS

1. What are the seven basic human needs according to Maslow? Describe what these are in terms of any human and also in terms of what an employee needs.
2. Describe the four basic needs of employees.
3. What are employees' main reasons for leaving jobs according to the survey cited in the text? How important is salary in their reasons?
4. Is all that stuff about human needs and employee needs really necessary? Why not just pay an employee well and expect him or her to perform well?
5. Describe the main steps in personnel management.
6. What are the intrinsic job factors that are potential satisfiers for an employee?
7. What are the extrinsic job factors that are potential dissatisfiers for an employee?
8. Why have job descriptions?
9. Why write a job description even for one employee or for a parent-child partnership?
10. What items should be included in a job description?
11. Describe what is involved in recruiting, interviewing, and selecting. Why are these so important?
12. Why should new employees be trained?
13. Why should "old" employees be trained?
14. How often does an employee need to be evaluated?
15. Against what should an employee be evaluated?
16. How can employees be given a quantitative but subjective evaluation?
17. If an employee has a negative evaluation, should he or she be fired immediately?
18. Is firing an employee always the cheapest way to improve performance? Why or why not?
19. What should be included in the basic compensation package?
20. Why should an incentive package be based mostly on those items that an employee can control?
21. If partial budget analysis shows that an incentive plan is profitable for the owner, will the employee respond as the owner wants him or her to respond? Why or why not?
22. Why and when should a farm manager develop an organizational chart?

23. What are the reasons a farm manager may want to have a board of advisors?

24. Describe the main legal forms of business organization. What types of farms would you expect to find using each form?

NOTES

1. The board of advisors is not the same as a board of directors that may be required under some legal forms of business organization.

2. This section is intended to provide a basic understanding of the choices in business organization. It is not professional legal advice and should not be acted upon without professional legal advice. Laws and rules vary by state and over time.

14

The Future Farm Manager

Many trends affect agriculture today: larger and fewer farms, exposure and access to world markets, industrialization of farming, environmental and food safety concerns, changes in government policy, business mergers, contracts and other marketing changes, to name a few. Change has always been part of agriculture. In this chapter, we will look at how some farm managers are responding and how they anticipate responding to these trends.

FARMERS' RESPONSES

To learn directly how farmers are responding to the trends and changes and their ideas for the future, I continually talk to many farmers and others close to farmers such as extension educators, professional farm managers, bankers, and other advisors. I talked with and listened to farmers from a broad range of ages, sizes, and geographies. I heard many of these ideas from farmers of all size farms, not just large farms. Perhaps surprisingly, many but not all needs are the same for different sizes and types.

From these conversations, I have organized the common ideas into four themes or categories: (1) moving down the supply chain, (2) redesigning the production system, (3) availability of new technologies, and (4) broader management skills. Let us look at each of these.

Moving Down the Supply Chain

"Moving down the supply chain" is one of the stronger ideas I heard. The knowledge of and desire to participate farther down the supply chain (or demand chain as its more accurate name may be) is widespread. As one farmer said, "Just producing and selling won't work." The idea and desire to capture more value was expressed by almost all the farmers. Most of them realize that they, as farmers, needed to be producing for the final consumers' needs, not for the farmers' needs, wants, and advantage. They have heard of the concept of creating value for the consumer and they want to capture more of that value for themselves.

As part of this desire to move down the chain (and, as will be discussed later, reduce risk), both *marketing and production contracts* will increase in frequency and importance. Forward contracts, hedging, options, and other marketing tools continue to increase in importance for farm managers. Production contracts will come in two different forms. The first form is the traditional contract. A farmer signs a contract with an elevator or company to produce a product of specified quality and quantity. Seed and vegetable production contracts are common examples of this type of contract. The recent increase in contracting for high-oil corn and similar products are also examples of this type. The potential for new products from genetic engineering (such as nutraceuticals) are another source of actually traditional contracting.

Another form of production contracting that we will see more of in the future is *contracting between neighbors*. While this has always happened, the formality of the contract will increase. Now we often hear of the livestock producer, for example, telling the neighboring crop grower that he wants to buy a certain quantity of corn silage. In the future, the producer and grower may develop a working agreement or contract that specifies which crop variety to plant, how much to plant, and how to value the crop based on the protein, energy, and other nutrient characteristics that create value for the livestock producer and, thus, the crop grower.

Preserving the identity of grains by variety provides the potential to capture greater value. Farmers are certainly aware of the value of and concern about genetically modified (GM) and non-GM varieties. They also know about the potential for niche markets with certain varieties and special crops. (Most are also aware of the small size of niche markets.) Research has shown processors (such as General Mills) that certain varieties possess characteristics that are better both for processing and for the consumers. These companies are aware of the increased value for the consumer and themselves, so they already provide an incentive to farmers to grow and deliver those desired varieties (if they are not mixed with other varieties). Thus, many farmers already participate in the value of *identity preservation*. Identity preservation does create problems for both the farmer and industry. These problems include increased logistics of keeping grains separate, moral risk of mixing different varieties, and so on. In spite of these problems, identity preservation is now required and is being done for many crop and livestock products. It will only increase. Those who do not segregate will not receive any of the potential premiums.

Organic farming usually moves a farmer down the supply chain and is also a new redesign of the production system. Organic products are a small but rapidly increasing area for the food market because consumers are asking for these products. Food retailers are responding faster than other parts of the food supply chain. Many farmers are seeing this as a way to achieve the multiple goals of income, environmental protection, and personal health.

A traditional way to capture more value is to market directly to consumers. A new and yet not so new idea that is being pursued is for farmers to form (new) *marketing cooperatives* rather than trying to contact the consumers as individual farmers. This is not a new idea obviously; we have several large, established marketing cooperatives already in existence: Cenex Harvest States, Land O' Lakes, Farmland, Sunkist, Diamond walnut, and many more. The "newness" is in the recent formation and discussion of small, local cooperatives for the purpose of marketing directly with local consumers or retailers. Slightly larger cooperative efforts have also formed in recent years, such as a group of farmers who own a rather small hog processing firm in southwest Minnesota that recently negotiated an export agreement with Japan.

In addition to cooperative marketing, I also hear farmers talking about cooperative production efforts or *joint ventures with other farmers*. These range from sharing machinery with neighbors to forming companies to build livestock facilities. This joint work is especially common in hog and dairy production but is also seen in specialty products such as fish production. In some of these joint efforts, the organizers saw local resources and developed the idea of how to use these resources to produce a product the consumers wanted.

I should note that while these are cooperative efforts in one definition of the word *cooperative,* the new entities and efforts may or may not be organized and filed legally as a cooperative. These new entities may also be various forms of partnerships and corporations.

Redesigning the Production System

Farmers will always seek more economic efficiency and increased productivity. Farmers have cut costs many times. In recent years, they have begun talking again about how to cut costs. Sometimes this was expressed in terms of needing a lower capitalization level. Others talked about the need to change the production system—perhaps slightly, perhaps radically. Organic production methods were mentioned as were the need for less substitution of capital for labor, increasing productivity through tiling wet areas, use of GIS technology (as discussed later), and the use of more efficient equipment in livestock buildings and for product handling.

This discussion is an appropriate reaction to the need to design a production system that delivers a product at a cost that can compete in the market. Farmers were looking for ways they could respond to the market instead of just wanting the market to supply a price that covers their current costs of production.

Farm managers in the future will continue to be concerned about balancing increased productivity with the risks of higher costs due to investments for raising productivity. I heard some farmers and farm managers talking about the need to lower capitalization on farms. For example, they wanted to purchase fewer inputs (herbicides, for example) and use more on-farm resources (such as, labor and mechanical cultivation). This idea runs counter to the usual trend of increased substitution of capital for labor, and I don't think the new trend will reverse the old substitution in total. The concern for timeliness and for "covering ground" remains. The idea of substituting labor for capital is due to the concern of covering cash costs in the current period of low prices. However, I suspect this current concern affects a current and young generation of farmers and, thus, will carry into the future. Future managers will more seriously consider alternatives that do not rely solely on purchased inputs.

Another idea for redesigning the production process is the expressed desire to share risk between landowners and tenants through flexible cash rents. Landowners may not want to take on the responsibility and risks of grain marketing required by share rental agreements. As cash rents increase, tenants may not want to take all the risks of both production and price. When both the owner and tenant see the need to share risk, new rental agreements will be written that establish a base cash rent and the rules by which the landowner is paid additional cash rent based on that year's yield and, perhaps, price.

Sharing machinery ownership is another old and, I found, newly fashionable way to reduce ownership costs as well as enjoy the potential benefits of better mechanical technology. If there is some geographic dispersion of the farms, they may also enjoy increased efficiency in the use of the machinery in response to differing weather and soil conditions. These shared ownership agreements appeared to be working well between farmers who had similar-sized farms. The agreements were not limited to relatives or legal partnerships. The farmers had agreed beforehand that repairs and maintenance costs would be shared in proportion to the land use regardless of where the machine broke down. They also felt that the decrease in ownership costs due to sharing was far larger than any yield losses due to any decrease in timeliness.

Availability of New Technologies

The precision farming tools of yield monitors, geographical positioning systems (GPS), geographical information systems (GIS), and so on, are increasing in use, to a degree. The farmers could see that they would be using them more in the future, but they also viewed the whole package as too expensive right now. They were very interested in the ability to farm by areas within a field via precision agriculture rather than the whole field. They knew of relatively small areas (bald areas, wet areas, etc.) that were not productive enough but it was currently too expensive in terms of timeliness to change any seeding, fertilizer, or herbicide application as they passed through or drove around those areas.

The farmers did express some concern over the spread of e-commerce and the potential impact on local communities. While lower input prices are desirable, that was not their only concern. They did not want to lose their local service and said they would be willing to pay somewhat higher prices to keep that local company in business. On the personal side of e-commerce, farmers also wanted local businesses to use the internet to improve the local service not to replace it. The local service they wanted the most on the internet was the ability to check prices, order inputs, and leave messages after regular business hours.

These are not the only new technologies in which farmers are interested. Many other ideas were mentioned but not extensively or by large numbers of farmers. As discussed in an earlier section, farmers were interested in other new technologies such as genetically modified crops and animals and organic production methods. They were also interested in new techniques of the old technology paths such as mechanical guidance systems, computer-assisted field work, automated information collection and processing, and so on.

Broader Management Skills

For the future, the farmers themselves could see the need to move beyond production management. Marketing will still be important and still could be improved by many farmers. However, more management time and skills will be needed in risk, strategic, and personnel management.

Risk management was described by farmers as protecting their income and protecting their farm resources. It was and will be more than just the traditional tools of hedging, options, and insurance. As contracts become more prevalent, farmers are becoming more concerned about contract evaluation and negotiation techniques. With more farms having more employees, personnel selection and management become important not just for production management but also for risk management. The wrong employee or an improperly trained employee can quickly change the potential outcome in a negative direction. Job descriptions, selection tests, training, evaluation, and incentive packages will become important tools to the farm manager of the future. Another set of management tools for the future is the set used for controlling the process to better ensure the actual process is proceeding as planned, to be ready to take corrective actions, and to alter plans in response to changes in the situation a farmer faces. Strategic management itself is seen as a risk management tool.

Strategic management involves positioning the farm for the future. It is "big picture" thinking. The farmers I talked to were very concerned about and taking some actions to move and change their farm into one ready for the future. Contracting with companies and neighbors, as mentioned earlier, is one example of strategic change. Cooperating with others for livestock production is another option that has happened and is still being looked upon as having good potential (with memories of bad results too). Another example of strategic change is the hog farmer who could not see a good future in hog farming, so he and his son developed a business of cleaning pits and spreading manure for neighbors. Accomplishing this did not involve just the addition of manure hauling and dropping of hogs, they also had to change from corn and soybeans to alfalfa because the demand for hauling was highest in the spring and fall so conflicted with planting and harvest. Alfalfa hay had a different time demand that did not conflict with the manure hauling business.

Some farmers were looking at their resources with new eyes once they thought of individual parts of the farm as profit centers. For example, the farm shop can be used for many activities in the off-season. One farmer used it as a base for a snow removal business; another as a place to assemble machinery for the local dealer. This is one example of how a resource became even more useful when it was viewed as a "mini-business" or profit center, and the manager said, "How can this resource be better used?"

Farmers are also looking at how to respond to some unknown events in the future. They know they are outnumbered both nationally (less than 2% of the population are farmers) and locally. The local decisions on county boards are being driven more and more by the nonfarming rural population who do not always have the same goals and ideas as farmers. They are discussing and planning how to respond to this situation. They don't have all the solutions, but, as one farmer said, "It's odors now, dust and noise are next." The potential changes in national farm policy are also not known, and farmers are trying to decide how to respond to the directions those changes may take.

Personnel management came up in our discussions in the sense of losing the traditional source of workers since rural children were "going to the city" and the need to understand a new source or workers (and neighbors) from different cultures. However, not all children are going to the city; I also heard that some people want to live in rural areas and work on farms but do not want to own a farm and be responsible for all the decisions and financial risk that owning involves.

Some of the farmers described an open style of management that they found very valuable. Instead of feeling they had to make all the decisions, these farmers were willing to seek out the ideas of employees and fellow farmers. They felt they ended up with more and better ideas than if they had not asked. I am sure the employees had a very different sense of belonging and loyalty to those farmers than they would with a dictatorial farmer (although I have not knowingly talked to such a farmer).

Opportunities for Consultants

These conversations point to some opportunities for consultants and advisors for farmers. Some farmers could convert these opportunities into businesses to complement their farm businesses. First, the farmers themselves realized that they cannot be or do everything that needs to be done. Many, if not all, future farmers will still like the task of producing, but not all will like the business side of farming. Many farmers would probably agree with the one who said to me, "I enjoy being in the field and doing something different every day. I don't like doing paperwork." They were interested in hiring someone to do some of the business work for them while still maintaining ownership control over their farm. Another farmer described this interest as wanting to hire a "CEO" for his farm. He described this person's duties as preparing financial information, finding and evaluating new opportunities and contracts, and working with the various government agencies. The CEO would be responsible to the farmer/owner who would act more as the chair of the board (and still be able to drive the tractor, work with the animals, etc.). While hiring a "CEO" may seem too extreme to some farmers, an opportunity seems apparent. Financial advisors, tax accountants, bookkeepers, and other advisors could use this desire to develop new lines of their businesses to help farmers do the new paperwork in the future.

The traditional jobs for consultants will continue: bookkeeping, crop consultants, veterinary services, and marketing services. Some new ideas I heard included contract identification and evaluation, evaluation of new ventures, and making market connections between geographically separated individuals. The list of potential opportunities will most certainly increase since the complexity of farming is not decreasing.

CONCLUDING COMMENTS

Success is defined in may ways and comes in many forms. Success is not just profit maximization, nor is it just getting bigger. Each farmer and each farm family has its own view of what success is. In this text, we have looked at how to understand the wider industry and economy and how to position the farm within that wider industry and economy to best meet your definition of success. We have also covered a long list of management tools and techniques for you to use in implementing your chosen strategy. However, since the world is a risky place, we also discussed how to make risky decisions to increase the chances of accomplishing your goals and objectives.

SUMMARY POINTS

- Change has always been part of agriculture.
- How farmers are responding to change was grouped into four categories in this chapter.
- For farmers, "moving down the supply chain" means producing closer to the final consumer and capturing more of that value.
- Farmers will always seek more economic efficiency and increased productivity, and they will always be redesigning the production system to meet the current needs and demands of the marketplace.
- New technologies are being developed continually and, as part of their effort to improve efficiency and productivity, farmers will observe and test the latest technologies to see whether they are worth adopting.
- Farmers are and will be expanding their management skills and activities beyond production management. Many have already done so, and most will soon follow.
- The increasing complexity of farming and the demands of the marketplace are providing new opportunities for consultants.
- Success is defined in many ways and comes in many ways.
- Each farmer and farm family needs to decide what its definition is and how its success can be obtained in the future.

REVIEW QUESTIONS

1. Do you see some other ways to adapt that were not mentioned in the four categories discussed in this chapter?
2. Describe some other trends and changes you see affecting agriculture.
3. How do you think farmers can adapt to the trends and changes you just listed?
4. What is your definition of success?

Appendix A

Estimating the Annual Costs of Capital Assets

ESTIMATING AVERAGE ANNUAL MACHINE COSTS

Estimating machinery costs is a complicated, tedious job if done accurately. The procedures, equations, and data for estimating machinery costs are explained and contained in two publications from the American Society of Agricultural Engineers (ASAE 1998, 2001). Lazarus and Selley (2002b) provide a very good description of the procedures and issues encountered when trying to estimate machinery costs. Fortunately, we do not have to go through these procedures every time. Cost estimates are available for typical situations through extension publications (e.g., Lazarus and Selley 2002a). Also, computer programs and spreadsheets that have the equations and data embedded in them are often available through extension services (e.g., Lazarus 2002).

To increase our understanding of how machinery costs are estimated, let us consider a planting operation that uses a 130 HP tractor and a 30 ft. grain drill. We start with the basic price and use information as listed in table A.1.

As a first step, the ending salvage value and the average annual investment are calculated since they are used in subsequent calculations.

The *salvage value (SV)* equals list price times remaining value pct. (RV%). The RV% comes from estimates made by and available on the web from ASAE (1998, 2001). For the tractor, the RV% after ten years is estimated to be 37.1%, and the tractor is estimated to be worth $28,270 after ten years:

$$RV = 76,200 \times 0.371 = \$28,270$$

Table A.1. Basic Price and Use Information for Example Machines

	130 HP tractor	30-ft grain drill
List price	$76,200	$34,600
Purchase price	70,000	30,000
Annual use	450 hr	80 hr
Useful life	10 yr	10 yr

For the grain drill, the RV% after ten years is estimated to be 40.5%, and the grain drill is estimated to be worth $14,013 after ten years:

$$RV = 34,600 \times 0.405 = \$14,013$$

The *average investment* = (purchase price + salvage value) / 2. For the tractor, the average investment over ten years is:

$$(70,000 + 28,270)/2 = \$49,135$$

For the grain drill, the average investment is:

$$(30,000 + 14,013)/2 = \$22,007$$

The next cost to be estimated is the *Annual Capital Recovery Charge (ACRC)*. The annual capital recovery charge must cover (1) purchase cost less salvage value plus interest on the unrecovered amount and (2) interest on salvage value:

$$ACRC = (\text{Purchase Price} - SV) \times CRF + (SV \times \text{Interest Rate})$$

The Capital Recovery Factor (CRF) is an annuity factor equal to $\{i/[1 - (1 + i)^{-n}]\}$. These CRFs can be found in appendix D, table D.4.
For the tractor:

$$
\begin{aligned}
ACRC &= [(70,000 - 28,270) \times CRF\ (10\ \text{yr}, 6\%)] + [28,270 \times 0.06] \\
&= [41,730 \times 0.1359] + 1,696 \\
&= \$7,367\ \text{per year}
\end{aligned}
$$

For the drill: $ACRC = [(30,000 - 14,013) \times 0.1359] + [14,013 \times 0.06]$
$$= \$3,013\ \text{per year.}$$

The *interest expense* for machinery is used to estimate the opportunity cost of capital invested in the machine. Choosing the interest rate is critical to having correct estimates. The chosen rate should be the real interest rate, that is, the common, nominal interest rate adjusted for inflation. The real interest rate is

$$i = \frac{(1+r)}{(1+f)} - 1$$

where i = real interest rate, r = nominal interest rate, and f = inflation rate—all expressed as proportions between 0 and 1.

This formula comes from the relationship that the nominal rate reflects both the real rate and the inflation rate and it is multiplicative relationship not additive. For example, an asset's value should increase by both the real rate and the inflation rate, that is:

$$(1 + r) = (1 + i) \times (1 + f) = (1 + i + f + if)$$

The real interest rate is often approximated as the difference between the nominal rate and the inflation rate:

$$i' = r - f$$

but this approximation is inaccurate by the magnitude of the product "if" as seen in the previous equation. If the real rate and the inflation rate are low, the inaccuracy may be inconsequential, but if the rates are high, the inaccuracy may affect the calculations enough to cause wrong decisions to be made.

Taxes, Insurance, and Housing (TIH) costs are often expressed as percentages of the average investment. Some states do not tax personal property such as farm machinery, so for this example, the percentage is set at 0%. The insurance cost is typically 0.8% of average value with a range of 0.6% to 1.0%. The housing cost is typically 2.0% of average value with a range of 0.5% to 3.0%. For the planting example, let us use these percentages: T = 0.0%, I = 0.8%, and H = 2.0% for a total TIH of 2.8%.

For the tractor, total TIH costs are estimated to be:

49,135 × 0.028 = $1,376 per year

For the grain drill, total TIH costs are estimated to be:

22,007 × 0.028 = $616 per year

Adding ACRC and TIH, we estimate the *total ownership costs* to be $8,743 per year for the tractor and $3,629 for the drill (table A.2).

Ownership cost per hour is calculated by dividing the annual cost by the estimated annual usage. The *total ownership cost per hour* of planting with this tractor and grain drill is estimated to be $64.79/hour ($19.43 + $45.36).

The *fuel cost per hour* = gallons per hour times price per gallon. The consumption rate (gal./hr.) can come from either performance records or engineering equations and is based on engine size (ASAE 1998, 2001). For fuel-use calculations, engine size is expressed as the horsepower measured at the power take off or, simply, PTO HP. (Remember that PTO HP is different from draw bar HP and engine HP.) For example:

gal./hr. = a × PTO HP

For gasoline: a = 0.06.
For diesel fuel: a = 0.044.
For LP: a = 0.072.

Machinery *lubrication cost* is usually estimated to be 10% to 15% of the fuel cost.

For the planting example, the 130HP tractor that uses diesel fuel has a fuel consumption and cost and lubrication cost of:

Fuel use: 130 HP × 0.044 = 5.7 gal/hour.
Fuel cost: 5.7 × $.90/gal = $5.13/hour.
Lubrication cost = 5.13 × 15% = $0.77 per hour.

The grain drill requires no fuel, so how should lubrication costs be estimated? The easiest way is to increase the % of tractor fuel; which is why 15% is used in this example.

Table A.2. Machinery Ownership Costs for Planting Example

	Tractor	Drill
Annual capital recovery cost	7,367	3,013
Taxes, insurance, & housing	1,376	616
Total ownership costs per year	$8,743	$3,629
Estimated annual usage	450 hr	80 hr
Ownership cost per hour	$19.43	$45.36

Machinery *repair and maintenance cost estimates* are based on estimates of the accumulated repair costs over the life of the machine. The engineering equation is:

Accumulated Repair and Maintenance Cost (RMC) = List Price × RMC%

The machine type and the total use (in hours) determine the RMC%. The RMC% has been developed from economic engineering data and is available on the web from ASAE (1998).

Repairs and Maintenance Cost Per Hour =

$$\frac{\text{Accumulated Repair and Maintenance Cost}}{\text{Total Hours of Use Over Useful Life}}$$

For the planting example, the tractor is expected to have a useful life of ten years, estimated use of 450 hours per year, total hours over its life of 4,500 hours, and from ASAE's website, a RMC% of 14.2% of the tractor's list price. Thus, the total accumulated repair cost (over the useful life of the tractor) is estimated to be:

RMC = $76,200 × 14.2% = $10,820

And the average repair cost per hour is estimated to be:

$10,820 / 4,500 hours = $2.40 per hour

The grain drill is expected to have a useful life of ten years, estimated use of 80 hours per year, total hours over its life of 800 hours, and again from ASAE's website, a RMC% of 20.0% of the grain drill's list price. Thus, the drill's total accumulated repair cost (over the useful life of the drill) is estimated to be:

RMC = $34,600 × 20.0% = $6,920

And the average repair cost per hour is estimated to be:

$6,920 / 800 hours = $8.65 per hour

The machine operating costs per hour are estimated to be $8.30 for the tractor; this includes $5.13 for fuel, $0.77 for oil and lubrication, and $2.40 for repairs and maintenance. For the grain drill, there are no fuel, oil, or lubrication costs, so the cost per hour is the $8.65 for repairs. Thus, the total machine operating costs are $16.95 per hour for the planting operation.

For many decisions, the costs per hour need to be converted into *costs per acre*. This can be done by using engineering estimates of machine capacity measured in acres per hour.

Acres per hour = SWE / 8.25
= [(Speed in MPH) × (Width in feet) × Efficiency] / 8.25

The coefficient 8.25 converts the English units of speed and width into machine capacity as measured in acres per hour. Performance data for machine and operation efficiencies are estimated from engineering data and are available on the web from ASAE (1998).

In many calculations, hours per acre is easier to use than acres per hour:

$$\text{Hours per acre} = \frac{1}{\text{Acres per hour}}$$

and machine cost per acre = Cost per hour × Hours per acre.

For the planting example:

Acres per hour = (5 MPH × 30 ft. × 0.70) / 8.25 = 12.73 ac./hr.

Hours per acre = 1 / 12.73 = 0.079

The machinery ownership costs per acre are $5.12 (= $64.79 × 0.079) and the machinery operating costs are $1.34 (= $16.95 × 0.079).

Labor cost is usually budgeted as an operating cost even though the labor may be full-time and thus fixed. The labor required depends on the machinery time adjusted for other work required by the machine and operation:

Labor hours/acre = machine hours/acre × adjustment factor

The adjustment factor varies from 1.0 to 1.2 depending on how much preparation time and infield maintenance time is required by the machinery and the operation plus an allowance for operator comfort. An adjustment factor of 1.2 is used the most.

For the planting example,

Labor hours per acre = 0.079 × 1.2 = 0.095 hours/acre

The labor cost per acre is simply the labor hours per acre multiplied by the wage per hour. For the planting example, the labor cost is

$0.095 hours per acre $\times$ \$8.00 per hour = \$0.76 per acre

Adding these together, we find the estimated machinery cost for this planting operation is \$7.22 per acre (table A.3).

Table A.3. Estimated Costs Per Acre for Planting Example

Machine	Ownership costs	Operating costs
Tractor	\$19.43/hour	\$8.30/hour
Grain drill	45.36	8.65
Total costs per hour	64.79	16.95
Machine hours per acre	.079	.079
Machinery cost per acre	\$5.12	\$1.34
Labor cost per acre		\$0.76
Total cost per acre		\$7.22

ESTIMATING THE ANNUAL COSTS OF BUILDING SERVICES

Since salvage value (SV) is typically zero for buildings, we can simplify the calculation for average investment and several other calculations. For example, usually:

Average investment = (Purchase price + SV) / 2.0

but since SV = 0, we can write:

Average investment = Purchase price $\times$ (1 / 2.0)

This way of writing the relationship allows us to write this and other formulas as multiplicative relationships of the building's purchase price, which will simplify the calculation process.

Straight-line depreciation = (Purchase price − SV) / Useful Life

or, we can see:

Depreciation = Purchase price $\times$ (1 / Useful life)

Interest = Average investment $\times$ Interest rate

or:

= Purchase price $\times$ (Interest rate / 2.0)

Repair & Maintenance = Purchase price × (R&M rate / 2.0). The repair and maintenance (R&M) costs vary with the amount and level of use and the type of building. They vary from 1.0% to 3.5% of purchase price. A typical value is 2.0% for normal use.

Taxes = Purchase price × (Tax rate / 2.0)

Insurance = Purchase price × (Insurance rate / 2.0)

As an example of calculating a building cost, consider a building with a purchase price of $130,000 and a salvage value of $0 after a useful life of 25 years. Using the calculations described above, the cost categories can be expressed as a percent of purchase price (table A.4).

Thus, for this example building, the average annual cost for building services is $13,065 = $130,000 × 0.1005.

Table A.4. Annual Building Costs Estimated as a Percentage of the Purchase Price

	Percentage of purchase price
Depreciation (1/25)	4.00%
Interest (6%)	3.00%
Repairs & maintenance	2.00%
Taxes (1.5%)	0.75%
Insurance (0.6%)	0.30%
Total	10.05%

Appendix B

Supplemental Information for Financial Analysis

OTHER FINANCIAL MEASURES

The financial measures presented in chapter 7 are very common in financial analysis, but they are not the only ones used. Many other measures can be calculated. Some are purely inverses of other ratios. Others express the same information as other measures but in a different way. Here, a few other measures are described because they are common in some areas and institutions, they are historically important, or they are useful for some specific analyses.

The *acid-test ratio* is essentially the same as the current ratio for measuring liquidity but the acid-test ratio includes only those current assets that are cash or nearly cash. All current liabilities are counted, but items such as growing crops or other assets that cannot be easily converted into cash are excluded. It is named the acid-test ratio because of the requirement of counting only cash or near cash current assets.

The *debt service coverage ratio* is another coverage ratio similar to the term debt coverage ratio, but it also includes land rent as a fixed financial obligation. It shows the farm's ability to meet debt servicing obligations, pay family living expenses, and pay tax obligations. The farther above one the debt service coverage ratio is, the more secure the farm is in its ability to service its debt and meet other obligations. A value of less than one shows an inability to meet all obligations. It is defined as:

Debt service coverage ratio =

$$\frac{\text{(net farm income - family living - income taxes + interest payments + land rental)}}{\text{(annual debt service + land rental)}}$$

where annual debt service = interest on all loans + principal on term debt.

The *debt-servicing ratio* is another common ratio in some areas. It measures the farm's principal and interest payments as a proportion of gross farm income. It represents the production required to service a farm's debt. If this ratio is greater than one, the farm has or will have a cash flow and liquidity problem; the farm will not have enough gross cash income to meet debt servicing requirements. A lower ratio suggests a better ability to meet these requirements even if lower than expected income were to occur. It is defined as:

$$\text{Debt-servicing ratio} = \frac{\text{principal} + \text{interest payments}}{\text{gross farm income}}$$

The *debt burden ratio* measures the net cash income as a percentage of the total debt burden of the farm. It shows the size of the total debt burden relative to the annual net cash income of the farm. Pictorially, this ratio can be viewed as measuring the size of the river trying to erode the mountain. Some evidence suggests that a debt burden ratio of 25% or more is healthy. A ratio of less than 15 may signal trouble and the need for further analysis. It is defined as:

$$\text{Debt burden ratio} = \frac{\text{net cash income}}{\text{total liabilities}} \times 100\%$$

The *times-interest earned ratio*, also called the interest coverage ratio, shows the degree to which annual earnings before interest and taxes (EBIT) are sufficient to pay annual interest commitments. Preliminary evidence shows that a ratio of greater than two suggests a financially healthy farm and one that could meet its interest obligations. A value of less than one definitely suggests trouble and a value between one and two suggests potential trouble and the need to evaluate the business further.

$$\text{Times-Interest earned ratio} = \frac{\text{EBIT}}{\text{interest paid}}$$

The *times-earnings ratio* is also used at times. It measures the burden of outstanding debt relative to the farm's earnings after interest (but before taxes). It calculates the number of years needed to repay outstanding debt. It is defined as:

$$\text{Times-Earnings ratio} = \frac{\text{total debt}}{\text{earnings after interest}}$$

The *gross ratio* measures the cost of producing one dollar of production, that is, the efficiency of production. It is defined as:

$$\text{Gross ratio} = \frac{\text{total expenses}}{\text{value of production}}$$

The *expense structure* measures the inflexibility of a farm's expenses by the proportion of those expenses that are fixed. It suggests how well a farm can respond to changes in its economic and physical environments. It is used to analyze a farm's opportunities or problems after an initial analysis has been completed. The expense structure is defined as:

$$\text{Expense structure} = \frac{\text{fixed cash expenses}}{\text{total cash expenses}}$$

Different types of farms will have different standards for the expense structure. A fruit orchard is expected to have a higher expense structure because it will have more capital tied up in fixed assets than an annual crop farm. Farms that have made recent major capital purchases will have a higher expense structure than farms that are nearly debt free. These differences are to be expected; they do not say that one type of farm is worse or better than another type.

The *debt structure* shows the balance between current and total liabilities. It is expressed as the percentage of total debt held as current debt. It shows the proportion of the debt that is due in the next twelve months. This ratio is also called the current debt ratio. It is defined as:

$$\text{Debt structure} = \frac{\text{current liabilities}}{\text{total liabilities}} \times 100\%$$

The *asset structure* shows the balance between current and total assets. It is expressed as the percentage of total assets held as current assets. It is defined as:

$$\text{Asset structure} = \frac{\text{current assets}}{\text{total assets}} \times 100\%$$

Reviewed together, debt and asset structure can help determine potential solutions to some problems, particularly questions regarding the potential for debt restructuring. For instance, suppose a farm has a low debt service coverage ratio and one proposed solution is to restructure the debt to decrease current obligations (at the expense of higher long-term obligations). The debt structure would show whether the farm has the potential to move current debt to long-term debt or if the structure is already relatively low in current debt and restructuring is not a possibility. The asset structure will show whether there is and would be enough balance between the types of assets and debts to permit restructuring as a possible solution.

INVENTORY OF RESOURCES

The farm inventory of resources is a list of all the farm assets a farmer has at a given time with a value established for each item. In the sense that it is a tool for planning, the inventory of resources should include more than the usual list of physical and financial assets. It should also include such items as the available line of credit, managerial abilities, and land and buildings that could be rented or purchased.

The three main reasons for taking an inventory are:

1. the beginning of each fiscal year,
2. when applying for a loan, and
3. when considering or making a major change in the business such as new rentals, partners, and other changes in the business.

Other reasons for having an up-to-date inventory include:

- Providing a basis for determining the extent to which resources are available or lacking
- Showing the condition of machinery, equipment, and improvements
- Providing a basis for setting a value on the farm business
- Aiding in the preparation for a financial statement
- Applying for a loan
- Providing information needed for the preparation of income tax statements, insurance claims, or rental agreements

Steps in Taking an Inventory

The inventory of resources is a large and valuable piece of knowledge for a manager. The job of preparing the inventory can be made easier if it is done in an orderly process. One such process is described in the following steps.[1]

1. Start with the land.
    a. Proceed from field to field until the entire farm is covered.
    b. Identify the fields and building sites on a map.
    c. Note in each field:
        (1) Size of the field, soil type, topography, condition
        (2) Crop planted, if any, and its stage, condition, expected yield
        (3) Improvements such as fences, tiles, irrigation systems, and bridges. (Be sure these improvements are entered in the farmer's depreciation schedule.)
2. Buildings should be listed next after the land.
    a. Note the type of structure, size, use, and condition. (Be sure these are entered in the farmer's depreciation schedule.)
    b. While at each building, record its contents: grain, hay, other feed-

stuffs, livestock, machinery, chemicals, and other supplies. Also, record the information noted below for each type of asset found in the building.

3. Livestock should be listed by kind, age, sex, weight, condition, and location. (Breeding and working livestock should be entered in the depreciation schedule.)

4. Machinery, equipment, implements, and tools should be identified by manufacturer, use, capacity or size, serial or identification number, condition, and location. (These should be entered in the farmer's depreciation schedule.)

5. Operating inputs (seed, feed, fuel, fertilizer, spare parts, etc.) should be listed by type, kind, amount, condition, and location.

6. Harvested crops
    a. Both on-farm and off-farm storage sites
    b. Note the type, amount, condition, and location.
    c. If needed, note the weight of the units (e.g., 60 bags of barley seed @ 80# each).

7. Financial assets
    a. Cash on hand and in liquid accounts
    b. Other capital accounts
    c. Stocks, bonds
    d. Deferred patronage dividends
    e. Accounts receivable
    f. Available line of credit
    g. Other

8. Managerial knowledge, experience, skill, and ability

9. A record of liabilities is needed to determine the financial position of the farm and its ability to obtain additional capital. Liability records include:
    a. Loan Information: date, form, for what, amount, length, rate (and other specifications if it is a nontraditional loan)
    b. Current portions of the principal and interest payments
    c. Accounts payable
    d. Unpaid, accrued interest
    e. Delinquent principal and interest payments
    f. Charge and credit card account information

10. Family records are needed to provide complete information for business analysis and planning. These records include:
    a. Personal expenses
    b. Off-farm income and expenditures
    c. Investments and other assets
    d. Farm products used by family
    e. Personal liabilities

11. Availability of other resources that could be purchased, rented, or hired. This list should cover land, buildings, machinery, labor, and custom work services. This list is very useful to planning the operation and growth of a farm.

Note

1. These inventory steps are adapted from the unpublished notes of Bruce Jensen, Professor Emeritus, College of the Sequoias, Visalia, California, 1985, and Libbin and Catlett (1987).

CHECKING RECORD ACCURACY

Several methods exist to check the accuracy of a set of records besides the visual impression of a neat, precise set of records kept in a timely and orderly manner. These methods or procedures are outlined below.

1. Does cash-in equal cash-out? Does all the cash spent, saved, or otherwise used account for all the sources of cash? A record of personal and nonfarm expenses and income is needed to make this check accurately on farms that do not keep separate personal and business checking accounts.
2. Liabilities check
   a. Ending debt = beginning debt − principal payments + new debt − debt forgiveness
   b. If accounts payable (A/P) are included, add: + beginning A/P − ending A/P to the equation.
   c. Partners and corporations may cause discrepancies if individuals hold some debt and it is not recorded properly in the accounts.
3. Income/net worth check
   a. Does the change in net worth balance with the money left after expenses, debt servicing, and family living?
   b. For farms that do not have separate business and personal accounts, nonfarm income, expenses, and investments need to be included in this calculation.
4. Livestock head count by month
   a Ending count = beginning count + purchases + transfers in − sales − deaths + births − transfers out
   b. This needs to be done each month not all at the end of the year.
   c. The monthly check helps maintain schedules and record sales and purchases, and keeps the manager on top of the operation. It may also help in the crop/feed check.

5. Production records for both crop and livestock
   a. Does the reported production check with the acreage or number of breeding livestock?
   b. Do the yields and production levels seem reasonable?
   c. These records may force recall of other production and resources, such as storage, sales, and other land/animals.
6. Crop/feed check
   a. Do the crop production estimates balance with feed fed, crops bought, crops sold, and crops stored?
   b. The quality of this check depends upon quality of the estimates of crop production, feed fed, and storage amounts.
   c. The crop/feed check may never balance completely due to having to use estimates, but the <u>process</u> of working through the check may stimulate memories and records to provide better records.
7. Average sales price and weight or yield
   a. Does the calculated average yield seem reasonable compared with the reported market average price?

$$\text{Calculated average price} = \frac{\text{Total sales \$}}{\text{Total prod. wt. or volume sold}}$$

   b. Estimate total production by using the market average price and the actual farm sales. How well does this backward estimate compare with the actual production?

$$\text{Estimated total production} = \frac{\text{Total sales}}{\text{Mkt. ave. price}}$$

   c. Estimate average yield or production per animal from total farm sales, average market prices, and the number of acres or animals. How well do these estimates compare with actual yields and productivities?

$$\text{Estimated yield} = \frac{\text{Total sales}}{\text{Market ave. price} * \text{acres}}$$

$$\text{Estimated production} = \frac{\text{Total sales}}{\text{Market ave. price} * \text{head}}$$

   d. These backward estimates are useful for discovering missing information or other mistakes. Since market averages are used, some small deviations between the estimates and the actual numbers will exist. These small deviations are to be expected; we are interested in finding large deviations that suggest major problems.

Appendix C

Estimating Subjective Probabilities

Probabilities of certain events often are needed to evaluate a decision fully. The rules for estimating probabilities, how to choose intervals for probability estimation, and some methods for calculating probability are explained in this appendix.

Probabilities are classified into three types based on the information source and how they are calculated. These types are:

1. Empirical Probabilities

    These probabilities are based on historical and/or experimental data and not on personal views of the future. Usually they are preferred to subjective probabilities. However, when empirical probabilities are applied to future events, they become subjective probabilities because the assumption that historical data can predict the future is a subjective assumption.

2. Deductive Probabilities

    These probabilities can be deduced by the information or circumstances surrounding the event(s) under consideration. Without doing extensive genetic research, we deduce the probability of having either male or female offspring is .5 for each.

3. Subjective Probabilities

    Often we have no or insufficient data to estimate probabilities empirically and we cannot deduce what the probabilities of certain events are. For these occasions, we need to have methods for developing a set of probabilities based on how we view the future. These methods involve subjective, not objective, views of the future and, thus, result in what is called subjective probabilities.

RULES FOR ESTIMATING PROBABILITIES

Three basic rules need to be followed when estimating probabilities. These are simple rules, but must be followed or any resulting analysis is worthless.

The first rule says the probability of any event must be 0, 1, or any number between 0 and 1. An event with a probability of 0 is certain not to occur. An event with a probability of 1 is certain to occur. No event can have a probability lower than 0 or a probability higher than 1. Mathematically, this is written:

$$0 \leq P(O_i) \leq 1$$

where O_i = occurrence of the ith event, and $P(O_i)$ = probability of the occurrence of the ith event.

The second rule requires that all possible events are included such that the list is collectively exhaustive of possible outcomes. If this rule is met, the sum of the probabilities of all possible events will be equal to 1:

$$\sum_{i=1}^{M} P(O_i) = 1$$

For example, next week the cattle price may go up, it may go down, or it may remain the same as this week. These are all the possible events that may happen to the cattle price, so the sum of the probabilities of these three events must be 1. If the sum is less than 1 or greater than 1, we have done something wrong in the way that we estimated the probabilities of the individual events. In a more complicated situation, the sum may be less than 1 because we have failed to include all possible events; the solution is to include all possible events and then re-add the probabilities to see if this second rule is satisfied.

The third rule is that all events are mutually exclusive. That is, the events do not include parts of another event. For example, two events which are not mutually exclusive are: a cattle price increase of five cents per pound and a positive price increase; the first event is included in the second event. Mathematically,

$$P(O_i \text{ or } O_j) = P(O_i) + P(O_j), \text{ for all i and j where } i \neq j$$

CHOOSING DATA INTERVALS FOR PROBABILITY ESTIMATION

The intervals chosen for any probability estimation process are critical to the accuracy of the work, the complexity of the analysis, and the time required to complete the analysis. Important factors to consider are the relevant range of the entire distribution, the number of intervals, the width of the intervals, and the midpoint of the intervals.

The relevant range for a set of prices, weather data, and yields, includes the values most likely to occur. It does not necessarily include all of the observed values. If we consider a set of soybean prices that ranged from $2 to $11 per bushel over a long period, we may decide that the relevant range for this year's analysis is $3 to $9 per bushel. All values do not have to be included just because they have occurred in the past.

Observations that are "outliers" (that is, quite a distance from the main group of observations) do not have to be included in the relevant range. The reasons for these outliers need to be considered, but adding them in the range will add more time and complexity to the analysis. For instance, suppose some very high soybean prices were due to an export embargo in a past year, those high prices may be excluded from the relevant range of prices for the current year if we do not expect to see another embargo this year. However, the number of observations has to be large before a value can be treated as a "true outlier" rather than part of a normal range.

The number and width of the intervals chosen are related. The necessary complexity of the analysis and the time required to do the analysis are components in deciding the width and number of intervals used. For instance, soybean prices could be divided into 10-cent intervals between $3 and $9 per bushel, but that would add unnecessary complexity and a large amount of computational time. A more reasonable interval may be 50 cents. This would result in 12 intervals between $3 and $9 per bushel.

Also, all intervals do not have to be equal in width. A soybean price interval of 50 cents may be correct in the $4 to $8 per bushel range and an interval of $1 may be correct from $3 to $4 and from $8 to $9 per bushel. This would result in ten intervals between $3 and $9 per bushel.

Another factor used in determining the final set of intervals is the choice of the midpoint for each interval. Midpoints are used to represent the interval as a single number. For most intervals, the midpoint will be the average of the interval limits. That is, for the interval $5 to $6, the midpoint would be $5.50. For some intervals, such as the lowest and highest of the range, a midpoint may be the value most expected in that interval, but not the mathematical midpoint. If the highest interval for the soybean price is from $8 to $11, a midpoint of $9 may be selected over $9.50 because $9 is more representative of the prices that may occur in that high price interval.

Since midpoints affect the understandability and computational ease of a set of prices or yields, intervals may be set so that "rounded" midpoints may be chosen. For instance, starting a 25-cent interval on a half cent can make a price distribution easier to use and understand. For example, a midpoint of $5.25 for an interval of $5.125 to $5.375 may be easier to use and understand than a midpoint of $5.375 for the interval $5.25 to $5.50.

Usually, interval boundaries are adjusted to avoid overlapping boundaries. Instead of having intervals such as $5.00 to $5.50 and $5.50 to $6.00 and wondering where to put a price of $5.50, the intervals could be redefined as $5.01 to $5.50 and $5.51 to $6.00. The midpoints can still be $5.25 and $5.75, respectively. For better readability, lower boundaries may be rounded numbers (for example, 6.00 rather than 5.99 or 6.01).

Let us return to the soybean price example and choose a final set of intervals. After considering the information needed and the desire to have some understandable midpoints, a set of ten intervals is chosen (table C.1). In this set, the final intervals have uneven widths. Also, not every midpoint is the mathematical midpoint of the interval. Furthermore, it is not the only set of intervals that could be developed; other information, situations, and years may require another set of intervals.

In summary, the basic rules in choosing intervals are accuracy, understandability, and ease of calculations. These rules may compete with each other. Thus, to decide which rules are most important in a specific situation, we need to decide how much accuracy is required, what set of numbers is available, what we understand, and how we will be doing the calculations. The final choice depends upon the situation and the user.

ESTIMATING SUBJECTIVE PROBABILITIES

While we want to use all the objective information available, sometimes we do not have good, objective information. This lack of data may be due to the absence of information, the cost of obtaining the information compared with the potential benefits, or insufficient time to find the information before a decision must be made.

Decisions still need to be made even though good, objective information is not available. So, for risky decisions, we still need to be able to develop probabilities for analysis. Since these probabilities are based on subjective views and not objective data, they are called subjective probabilities.

Table C.1. Soybean Price Intervals
and Midpoints

Soybean Price Interval ($/bushel)	Midpoint ($/bushel)
3.00–4.75	4.00
4.76–5.25	5.00
5.26–5.75	5.50
5.76–6.25	6.00
6.26–6.75	6.50
6.76–7.25	7.00
7.26–7.75	7.50
7.76–8.25	8.00
8.26–9.25	8.75
9.26–11.00	10.00

Several methods for developing subjective probabilities are methods described in this section: (1) direct estimation, (2) cumulative probabilities, (3) conviction weights, (4) sparse data methods, and (5) triangular distribution.

The *choice of method* depends upon how much information is available and upon personal preference. When some information is available, it should be used even if subjective information is needed to complete the estimation. For example, if the amount of historical information is not sufficient to calculate good empirical probabilities, the sparse data method is preferred. The choice of method also depends upon how well the manager knows and likes each method. With these methods, we should remember that the resulting probabilities can be adjusted to reflect better perceptions and to include new information.

Direct Estimation

The direct estimation method is just what its name says. Once the intervals are chosen, the manager starts assigning probabilities to each interval according to his perceptions of the marketplace, weather, and other conditions that may affect the distribution under consideration. There is no intermediate step; probabilities between and including 0 and 1 are specified until they are satisfactory and the sum is equal to 1.0. An example of direct estimation would be someone specifying the probabilities for the soybean price intervals in table C1 without consulting historical price frequencies.

The use of a pencil or a computer spreadsheet is recommended for this method since many changes may be made to the probabilities. Unless a person is very familiar with estimating probabilities and knows the events being considered, the conviction weight method is recommended over the direct estimation method.

Cumulative Probabilities

This method also involves estimating directly, but in a different way than the direct estimation method. The cumulative approach starts with estimating the probability that a price or yield will be equal to or less than a certain level. This method is called "cumulative" because, for example, the probability of the soybean price being equal to or below $5 is equal to the probability of the price being equal to or below $4 plus the probability of the price being between $4 and $5.

As with direct estimation, there are no preliminary steps other than specifying the intervals. The manager looks at the set of intervals, considers the market and other factors in the situation, and estimates the probability that the soybean price, for example, will be in the lowest interval. Then the manager estimates the probability of the price being in the lowest two intervals, the lowest three intervals, and so on until there is a probability of 1.00 of the price being in the highest or lower interval. No rule says that the cumulative probability cannot reach 1.00 before the highest interval.

Once the cumulative probabilities have been estimated, the probability for each interval is calculated. This is done by subtracting the cumulative probability for the next lower interval from the cumulative probability for each interval.

As an example, let us consider the probabilities of various soybean yields. A farmer considers his/her past yields, the county yields, and experimental yields and develops a set of intervals (table C.2).

Using the cumulative probabilities method, the farmer specifies the cumulative probabilities for each interval. For example, the farmer decides the yield may be in the first interval 5% of the time. Then he decides that the yield will be in the second interval or the lower 12% of the time. By the time the interval 37.5 to 42.4 bushels is considered, the farmer thinks the typical yield is included so the cumulative probability should be more than 0.5. So, for that interval, the cumulative probability is set at 0.72. The cumulative probabilities are estimated for each interval from lowest to highest.

The probability for each interval is calculated by subtracting the cumulative probability of the next lower interval from the cumulative probability of the interval under consideration. For instance, the probability of the soybean yield being in the interval 32.5 to 37.4 is .19: this is the cumulative probability that the yield will be less than 37.5 bushels (that is, 0.44) minus the cumulative probability that the yield will be less than 32.5 bushels (that is, .25 from the next lower interval).

Conviction Weights

In this method, the estimation of probabilities depends upon how strongly we feel that one event will happen relative to another event happening. Specifi-

Table C.2. Soybean Yield Probabilities

Yield interval (bu./acre)	Midpoint (bu.)	Estimated cumulative probability	Interval probability
15.0–22.4	19	.05	.05
22.5–27.4	25	.12	.12 - .05 = .07
27.5–32.4	30	.25	.25 - .12 = .13
32.5–37.4	35	.44	.44 - .25 = .19
37.5–42.4	40	.72	.72 - .44 = .28
42.5–47.4	45	.82	.82 - .72 = .10
47.5–52.4	50	.90	.90 - .82 = .08
52.5–57.4	55	.95	.95 - .90 = .05
57.5–62.4	60	.98	.98 - .95 = .03
62.5–67.4	65	1.00	1.00 - .98 = .02
		TOTAL:	1.00

cally, we weight our conviction that an event will happen within a certain interval relative to all other intervals. These conviction weights are then used to estimate the probabilities of those events happening.

The use of conviction weights is useful for two reasons. First, they may be easier to use than estimating probabilities directly. Conviction weights are usually between 0 and 100; these may be easier to understand than numbers between 0 and 1. Second, the weights may disguise the process and, thus, keep some of our biases out of the resulting probabilities.

After the desired set of intervals is developed, we start by choosing the interval in which the price, for example, will most likely be. This interval is given a conviction weight of 100. The next step is to specify weights for the remaining intervals based on our conviction that the price will occur within that interval rather than the initial interval. More than one interval can have a conviction weight of 100, if we feel that they all have an equal chance of containing the final price or yield.

The conviction weight method has two nice features. Both of them decrease calculation time. First, conviction weights of more than 100 are allowable. This is convenient for those occasions when, after starting, we decide an interval has a greater chance than the one we chose initially. We can give this new interval a conviction weight greater than 100 and not have to adjust the other conviction weights. Second, when estimating our conviction weights, we do not have to be concerned with the rules of estimating probabilities; those are taken care of in the next step.

After specifying and adjusting our conviction weights to our satisfaction, the sum of the conviction weights is calculated. The probability of the final event (price, for example) being in a certain interval is that interval's conviction weight divided by the sum of all the conviction weights. (Just as if the conviction weights were frequencies and not weights.)

As an example of the conviction weights method, let us estimate the probabilities of a range of yields for corn in Minnesota. The range and intervals to be considered are specified in table C.3.

Since the county average yield has been about 120 bushels per acre, the yield interval 115.1 to 125 bushels is chosen as the interval in which the yield is most likely to be found. That interval is given a conviction weight of 100. The rest of the intervals are given weights based on our conviction of the actual yield being in that interval compared to the first conviction weight of 100.

In this example, the conviction weights for all the intervals add up to 455. This sum is used to calculate the probability of the actual yield being in each interval. For example, the first interval (75 to 85 bu.) has a conviction weight of 15; the probability for that interval is .03 (15/455). The probability for the most likely interval (115.1 to 125 bu.) is .22 (100/455). The probabilities for the rest of the intervals are calculated in the same way. If rounding has caused

Table C.3. Corn Yield Probabilities

Yield interval (bu./acre)	Midpoint (bu.)	Conviction	
		Weight	Probability
75–85	80	15	15/455 = .03
85.1–95	90	35	35/455 = .08
95.1–105	100	65	65/455 = .14
105.1–115	110	85	85/455 = .19
115.1–125	120	100	100/455 = .22
125.1–135	130	80	80/455 = .18
135.1–145	140	50	50/455 = .11
145.1–155	150	25	25/455 = .05
	TOTAL:	455	1.00

the sum to be slightly under or over 1.0, the individual probabilities may need to be adjusted so that the sum is equal to 1.0.

Sparse Data

Many times data are available but not enough to estimate empirical probabilities satisfactorily. Often, this "sparse data" situation occurs with yields for a specific farm. The situation is not hopeless. The data are still useful to help guide a subjective approach to estimating probabilities.

The procedures for a sparse data situation are similar to the cumulative probability method. After the intervals are set, the available data are used to develop a preliminary set of probabilities. This preliminary set is refined to the final set of probabilities.

The crux of the sparse data method is the ordering of the available data from smallest to largest, and then dividing the cumulative probability over the range of the data. The division is made according to the ranking of the available data. The probability that the actual yield will be equal to or less than an observed yield is set at that yield's rank divided by the sum of the number of observations plus one. Mathematically, the probability that the sum will be equal to or less than an observed yield is expressed in the following equation:

Cumulative probability = $k/(n+1)$

where k is the rank of the specific yield in question, and n is the total number of observed yields.

For example, we have five soybean yields from two years and three similar fields: 39, 31, 45, 34, and 42. The ordering is done as shown in table C.4. The preliminary cumulative probabilities are estimated from the five observed yields. The final set of probabilities is developed on the basis of the preliminary probabilities and our subjective views of what else may affect the actual distribution of yields.

In this example, the five observed yields are used as a guide to develop the final set of probabilities. The preliminary probabilities are refined by our knowledge of the circumstances surrounding these yields and where they are in relation to the interval boundaries. For instance, observed yield of 34 is the second lowest and, thus, causes a preliminary estimate of .33 for the interval 32.5 to 37.4 bushels. Since 34 is in the lower side of that interval, the final cumulative probability is increased to .40 to account for higher yields that may occur within that interval. Some of the other probabilities are modified by this same reasoning. The probabilities for the upper and lower intervals are developed by subjective views of yield potential.

If two or more observed yields are in the same interval, the preliminary cumulative probability is calculated from the highest rank in that interval. In this example, the third and fourth observed yields (out of 5) were in the same interval. So the preliminary cumulative probability for this interval is 0.67 [4/(5 + 1)].

The reliability of the probabilities suggested by these methods depends upon our view of the situation surrounding the soybean yields. The cumulative probabilities estimated directly (table C2) are slightly different. Do the observed yields carry enough information to change our views? Only your opinion in the specific situation can answer that question.

Triangular Distribution

The triangular distribution method involves specifying three events: the lowest, most likely, and highest yields or prices that may occur. These three events

Table C.4. Soybean Yield Probabilities, Revisited

Yield interval (bu./acre)	Observed yields (bu.)	Prelim. cum. probability	Refined cum. probability	Interval probability
15.0–22.4			.03	.03
22.5–27.4			.10	.07
27.5–32.4	31	1/(5 + 1) = .17	.20	.10
32.5–37.4	34	2/(5 + 1) = .33	.40	.20
37.5–42.4	39,42	4/(5 + 1) = .67	.67	.27
42.5–47.4	45	5/(5 + 1) = .83	.83	.16
47.5–52.4			.92	.09
52.5–57.4			.98	.06
57.5–62.4			1.00	.02
62.5–67.4			1.00	.00

are used to estimate the probabilities that the actual yield or price will be in each interval.

The range between the lowest and highest values should encompass almost all of the yields or prices that may occur, but does not have to include every possible yield that may occur in even the wildest situation. For instance, the lowest corn yield possible is zero, but if we ignored this utter failure, the lowest likely yield may be 75 bushels compared with a most likely yield of 115 bushels and a highest yield of 155 bushels per acre.

The triangular distribution method starts from the assumption that the probability is .5 that the actual value will be on either side of the most likely value. This assumption is softened by two features of this method. First, the most likely yield or price does not have to be the midpoint between the lowest and highest values. Second, the number of intervals above and below the most likely value does not have to be equal. However, to use these procedures, the most likely value also has to be one of the interval boundaries.

Once the intervals have been developed and the lowest, most likely, and highest values have been specified, the probabilities are calculated in three steps.

1. The interval boundaries are given values between zero and one. For the intervals below the most likely yield or price, the values are increments between zero and 0.5. For the intervals above the most likely yield or price, the values are increments between .5 and one. The increments for either side of the most likely value are calculated by dividing 0.5 by the number of intervals on that side of the most likely value. For instance, if four intervals are on the lower side, the increment is 0.125 (0.5/4) and the values for the interval boundaries are 0.0, 0.125, 0.25, 0.375, and 0.5.
2. The interval boundary values are transformed into subjective probabilities that the event will be greater than that boundary. For example, the values of 0.125 and 0.25 are changed into the probability that the yield will be greater than the yields associated with those interval boundaries. (This is the reverse of the cumulative probability, which was the probability that the actual event will be less than a certain level.) This is done by different methods depending on where the boundary lies in relation to the most likely value.
   a. At the lowest boundary, which is also the least likely value, the probability that the yield or price will be greater than that value is 1.0 by definition.
   b. For the interval boundaries less than the most likely value, the equation for the probability of the yield or price being greater than the boundary is:

Probability $= 1 - (x^2/0.5)$

where x is the value between zero and 0.5 for each interval boundary.

   c. For the interval boundary at the most likely value, the probability is 0.5 that the yield or price will be greater than the most likely value.

   d. For the intervals greater than the most likely value, the equation for the probability of the yield or price being greater than the boundary is:

$$\text{Probability} = (1 - x)^2 / 0.5$$

   where x is the value between 0.5 and one for each interval boundary.

   e. For the highest boundary, the probability of a value greater than that boundary is 0.0.

3. The probability that the actual price or yield will be in a specific interval is the difference between the probability that the value will be greater than the lower boundary and the probability that it will be greater than the upper boundary.

For an example, let us reconsider the corn yields for the farmer in southwestern Minnesota. The intervals are the same as in table C3. The most likely yield is now 115 bushels per acre. The lowest and highest yields are 75 and 155 bushels, respectively. The three steps and the resulting probabilities are shown in table C.5.

The triangular distribution method assumes what is called a "normal" distribution in statistics. If the data are not a normal distribution and the triangular distribution is used, the resulting probabilities may not be accurate. However, having probabilities that may be slightly inaccurate is better than not having any information, so this method can be used if better information is not available and it is understood. Since these are subjective probabilities, they could be refined, after initial calculation, to reflect non-normality. However, this may require prior experience with calculating probabilities.

SUMMARY POINTS

- Probabilities are classified as empirical, deductive, and subjective based on the information sources and how they are calculated.
- Empirical probabilities are based on historical and/or experimental data and not on personal views of the future.
- When empirical probabilities are applied to future events, they become subjective probabilities because the assumption that historical data can predict the future is a subjective assumption.
- Deductive probabilities can be deduced from the information or circumstances surrounding the event(s) under consideration.
- Subjective probabilities are used when we have insufficient data to estimate probabilities empirically and we cannot deduce what the probabilities of

Table C.5. Corn Yield Probabilities, Revisited.

Steps 1 and 2:

Interval boundary (bu./acre)	Boundary value (bu.)	Subjective probability
75	0.000	1.000
85	.125	$1 - (.125^2/0.5) = .969$
95	.250	.875
105	.375	.719
115	.500	.500
125	.625	$(1 - .625)^2/0.5 = .281$
135	.750	.125
145	.875	.031
155	1.000	0.000

Step 3:

Yield interval (bu./acre)	Midpoint (bu.)	Interval probability
75.0–85	80	$1.000 - .969 = .031$
85.1–95	90	$.969 - .875 = .094$
95.1–105	100	$.875 - .719 = .156$
105.1–115	110	$.719 - .500 = .219$
115.1–125	120	$.500 - .281 = .219$
125.1–135	130	$.281 - .125 = .156$
135.1–145	140	$.125 - .031 = .094$
145.1–155	150	$.031 - .000 = .031$
	TOTAL:	1.000

certain events are. Subjective probabilities are developed based on our personal views of the future.
- Three basic rules should be followed when estimating probabilities. First the probability of any event must be 0, 1, or any number in between. Second, any list of possible events should be collectively exhaustive of all possible events. Third, all events are mutually exhaustive.
- Intervals need to be chosen appropriately for accurate estimation of probabilities. The relevant range needs to be included and boundaries should be chosen to ease the estimation process. The number and width of the intervals need to balance the need for detailed information and the complexity of the calculations. Interval widths and boundaries can be chosen to provide midpoints that are understandable and easy to use in subsequent calculations.

- Subjective probabilities can be estimated in five ways: direct estimation, cumulative probabilities, conviction weights, sparse data method, and triangular distribution.
- The choice of method depends upon how much information is available and upon personal preference.
- Conviction weights can be used when we have some knowledge and understanding of the events but have very little data.
- The sparse data method is used when we have some empirical data but not enough to perform statistical analysis.
- The triangular distribution method can be used when we have a basic idea of the minimum, maximum, and most likely levels and expect the values to be distributed approximately normal.

REVIEW QUESTIONS

1. How are historical or empirical probabilities different from subjective probabilities?
2. Estimate the interval probabilities for soybean prices six months from now using the price intervals listed in table C.6 and your own conviction weights based on your understanding of current market conditions and trends. (If market conditions have changed drastically since this worksheet was developed, don't hesitate to change the price intervals.)

Table C.6. Estimating Soybean Price Probabilities Using Conviction Weights

Soybean price intervals ($/bushel)	Midpoint	Your conviction weight	Interval probability
4.25–4.74	4.50		
4.75–5.24	5.00		
5.25–5.74	5.50		
5.75–6.24	6.00		
6.25–6.74	6.50		
	TOTALS		1.00

3. Ralph and Marge Podanski are evaluating whether they should keep spring wheat in their corn and soybean rotation. They have grown wheat for seven years and are not entirely sure that it is profitable enough for them. They want to have a better idea of the possible distributions of yields, but they need some help. Using the sparse data method, help the Podanskis calculate the probabilities of different yield intervals. They have started by setting up the table below using 6-bushel yield intervals. Their yields for the

past seven years are: 42, 51, 37, 40, 41, 32, and 44 bushels per acre. Based on their experience, they realize that the highest probable yield is 60 bushels per acre and in all but the worst possible conditions, their lowest possible yield is 30 bushels per acre.

Table C.7. Estimating Wheat Yield Probabilities Using the Sparse Data Method

Yield interval (bu/ac)	Yield midpoint	Actual yields	Preliminary cumulative probability	Refined cumulative probability	Interval probabilities
30.1–35	32.5				
35.1–40	37.5				
40.1–45	42.5				
45.1–50	47.5				
50.1–55	52.5				
55.1–60	57.5				
					Total = 1.00

4. Mike Thompson is considering adding spring wheat to his list of potential crops. Since he does not have any yield history, he cannot even use the sparse data method. After talking with his neighbors and some crop consultants and advisors, he decides his most likely spring wheat yield is 45 bushels per acre; the lowest possible yield is 25; and his highest possible yield is 65. Using the triangular distribution method, estimate the probabilities of the actual yield being within each 5-bushel interval between 25 and 65 bushels per acre. Some calculations are already done.

Table C.8. Estimating Wheat Yield Probabilities Using the Triangular Distribution Method

Steps 1 and 2:

Interval boundary (bu/ac)	Boundary value		Subjective probability
25	0.0		1.000
30	0.125	$1 - (0.125^2 / 0.5) =$	0.969
35			
40			
45	0.5		0.500
50	0.625	$(1 - 0.625)^2 / 0.5 =$	0.281
55			
60			
65	1.0		0.000

Step 3:

Yield Interval (bu/ac)	Midpoint		Interval probability
25–29.9	27.5	$1.000 - 0.969 =$	0.031
30–34.9			
35–39.9			
40–44.9			
45–49.9			
50–54.9			
55–59.9			
60–65			

Appendix D

Interest Tables

The following tables contain the time value of money factors explained in Chapter 9, "Investment Analysis." They are used in many decisions and calculations including loan amoritization (chapter 8), investment analysis (chapter 9), land valuation (chapter 10), and cost recovery for budgeting (Appendix A).

Table D.1. Amount of 1 at Compound Interest

	2%	3%	4%	5%	6%	$V^f_n = (1 + i)^n$ 7%	8%	9%	10%	11%	12%
1	1.0200	1.0300	1.0400	1.0500	1.0600	1.0700	1.0800	1.0900	1.1000	1.1100	1.1200
2	1.0404	1.0609	1.0816	1.1025	1.1236	1.1449	1.1664	1.1881	1.2100	1.2321	1.2544
3	1.0612	1.0927	1.1249	1.1576	1.1910	1.2250	1.2597	1.2950	1.3310	1.3676	1.4049
4	1.0824	1.1255	1.1699	1.2155	1.2625	1.3108	1.3605	1.4116	1.4641	1.5181	1.5735
5	1.1041	1.1593	1.2167	1.2763	1.3382	1.4026	1.4693	1.5386	1.6105	1.6851	1.7623
6	1.1262	1.1941	1.2653	1.3401	1.4185	1.5007	1.5869	1.6771	1.7716	1.8704	1.9738
7	1.1487	1.2299	1.3159	1.4071	1.5036	1.6058	1.7138	1.8280	1.9487	2.0762	2.2107
8	1.1717	1.2668	1.3686	1.4775	1.5938	1.7182	1.8509	1.9926	2.1436	2.3045	2.4760
9	1.1951	1.3048	1.4233	1.5513	1.6895	1.8385	1.9990	2.1719	2.3579	2.5580	2.7731
10	1.2190	1.3439	1.4802	1.6289	1.7908	1.9672	2.1589	2.3674	2.5937	2.8394	3.1058
11	1.2434	1.3842	1.5395	1.7103	1.8983	2.1049	2.3316	2.5804	2.8531	3.1518	3.4785
12	1.2682	1.4258	1.6010	1.7959	2.0122	2.2522	2.5182	2.8127	3.1384	3.4985	3.8960
13	1.2936	1.4685	1.6651	1.8856	2.1329	2.4098	2.7196	3.0658	3.4523	3.8833	4.3635
14	1.3195	1.5126	1.7317	1.9799	2.2609	2.5785	2.9372	3.3417	3.7975	4.3104	4.8871
15	1.3459	1.5580	1.8009	2.0789	2.3966	2.7590	3.1722	3.6425	4.1772	4.7846	5.4736
16	1.3728	1.6047	1.8730	2.1829	2.5404	2.9522	3.4259	3.9703	4.5950	5.3109	6.1304
17	1.4002	1.6528	1.9479	2.2920	2.6928	3.1588	3.7000	4.3276	5.0545	5.8951	6.8660
18	1.4282	1.7024	2.0258	2.4066	2.8543	3.3799	3.9960	4.7171	5.5599	6.5436	7.6900
19	1.4568	1.7535	2.1068	2.5270	3.0256	3.6165	4.3157	5.1417	6.1159	7.2633	8.6128
20	1.4859	1.8061	2.1911	2.6533	3.2071	3.8697	4.6610	5.6044	6.7275	8.0623	9.6463
21	1.5157	1.8603	2.2788	2.7860	3.3996	4.1406	5.0338	6.1088	7.4002	8.9492	10.8038
22	1.5460	1.9161	2.3699	2.9253	3.6035	4.4304	5.4365	6.6586	8.1403	9.9336	12.1003
23	1.5769	1.9736	2.4647	3.0715	3.8197	4.7405	5.8715	7.2579	8.9543	11.0263	13.5523
24	1.6084	2.0328	2.5633	3.2251	4.0489	5.0724	6.3412	7.9111	9.8497	12.2392	15.1786
25	1.6406	2.0938	2.6658	3.3864	4.2919	5.4274	6.8485	8.6231	10.8347	13.5855	17.0001
26	1.6734	2.1566	2.7725	3.5557	4.5494	5.8074	7.3964	9.3992	11.9182	15.0799	19.0401
27	1.7069	2.2213	2.8834	3.7335	4.8223	6.2139	7.9881	10.2451	13.1100	16.7386	21.3249
28	1.7410	2.2879	2.9987	3.9201	5.1117	6.6488	8.6271	11.1671	14.4210	18.5799	23.8839
29	1.7758	2.3566	3.1187	4.1161	5.4184	7.1143	9.3173	12.1722	15.8631	20.6237	26.7499
30	1.8114	2.4273	3.2434	4.3219	5.7435	7.6123	10.0627	13.2677	17.4494	22.8923	29.9599
31	1.8476	2.5001	3.3731	4.5380	6.0881	8.1451	10.8677	14.4618	19.1943	25.4104	33.5551
32	1.8845	2.5751	3.5081	4.7649	6.4534	8.7153	11.7371	15.7633	21.1138	28.2056	37.5817
33	1.9222	2.6523	3.6484	5.0032	6.8406	9.3253	12.6760	17.1820	23.2252	31.3082	42.0915
34	1.9607	2.7319	3.7943	5.2533	7.2510	9.9781	13.6901	18.7284	25.5477	34.7521	47.1425
35	1.9999	2.8139	3.9461	5.5160	7.6861	10.6766	14.7853	20.4140	28.1024	38.5749	52.7996
36	2.0399	2.8983	4.1039	5.7918	8.1473	11.4239	15.9682	22.2512	30.9127	42.8181	59.1356
37	2.0807	2.9852	4.2681	6.0814	8.6361	12.2236	17.2456	24.2538	34.0039	47.5281	66.2318
38	2.1223	3.0748	4.4388	6.3855	9.1543	13.0793	18.6253	26.4367	37.4043	52.7562	74.1797
39	2.1647	3.1670	4.6164	6.7048	9.7035	13.9948	20.1153	28.8160	41.1448	58.5593	83.0812
40	2.2080	3.2620	4.8010	7.0400	10.2857	14.9745	21.7245	31.4094	45.2593	65.0009	93.0510

Table D.2. Present Value of 1 at Compound Interest

	2%	3%	4%	5%	6%	$V^P_n = 1/(1 + i)^n$ 7%	8%	9%	10%	11%	12%
1	0.9804	0.9709	0.9615	0.9524	0.9434	0.9346	0.9259	0.9174	0.9091	0.9009	0.8929
2	0.9612	0.9426	0.9246	0.9070	0.8900	0.8734	0.8573	0.8417	0.8264	0.8116	0.7972
3	0.9423	0.9151	0.8890	0.8638	0.8396	0.8163	0.7938	0.7722	0.7513	0.7312	0.7118
4	0.9238	0.8885	0.8548	0.8227	0.7921	0.7629	0.7350	0.7084	0.6830	0.6587	0.6355
5	0.9057	0.8626	0.8219	0.7835	0.7473	0.7130	0.6806	0.6499	0.6209	0.5935	0.5674
6	0.8880	0.8375	0.7903	0.7462	0.7050	0.6663	0.6302	0.5963	0.5645	0.5346	0.5066
7	0.8706	0.8131	0.7599	0.7107	0.6651	0.6227	0.5835	0.5470	0.5132	0.4817	0.4523
8	0.8535	0.7894	0.7307	0.6768	0.6274	0.5820	0.5403	0.5019	0.4665	0.4339	0.4039
9	0.8368	0.7664	0.7026	0.6446	0.5919	0.5439	0.5002	0.4604	0.4241	0.3909	0.3606
10	0.8203	0.7441	0.6756	0.6139	0.5584	0.5083	0.4632	0.4224	0.3855	0.3522	0.3220
11	0.8043	0.7224	0.6496	0.5847	0.5268	0.4751	0.4289	0.3875	0.3505	0.3173	0.2875
12	0.7885	0.7014	0.6246	0.5568	0.4970	0.4440	0.3971	0.3555	0.3186	0.2858	0.2567
13	0.7730	0.6810	0.6006	0.5303	0.4688	0.4150	0.3677	0.3262	0.2897	0.2575	0.2292
14	0.7579	0.6611	0.5775	0.5051	0.4423	0.3878	0.3405	0.2992	0.2633	0.2320	0.2046
15	0.7430	0.6419	0.5553	0.4810	0.4173	0.3624	0.3152	0.2745	0.2394	0.2090	0.1827
16	0.7284	0.6232	0.5339	0.4581	0.3936	0.3387	0.2919	0.2519	0.2176	0.1883	0.1631
17	0.7142	0.6050	0.5134	0.4363	0.3714	0.3166	0.2703	0.2311	0.1978	0.1696	0.1456
18	0.7002	0.5874	0.4936	0.4155	0.3503	0.2959	0.2502	0.2120	0.1799	0.1528	0.1300
19	0.6864	0.5703	0.4746	0.3957	0.3305	0.2765	0.2317	0.1945	0.1635	0.1377	0.1161
20	0.6730	0.5537	0.4564	0.3769	0.3118	0.2584	0.2145	0.1784	0.1486	0.1240	0.1037
21	0.6598	0.5375	0.4388	0.3589	0.2942	0.2415	0.1987	0.1637	0.1351	0.1117	0.0926
22	0.6468	0.5219	0.4220	0.3418	0.2775	0.2257	0.1839	0.1502	0.1228	0.1007	0.0826
23	0.6342	0.5067	0.4057	0.3256	0.2618	0.2109	0.1703	0.1378	0.1117	0.0907	0.0738
24	0.6217	0.4919	0.3901	0.3101	0.2470	0.1971	0.1577	0.1264	0.1015	0.0817	0.0659
25	0.6095	0.4776	0.3751	0.2953	0.2330	0.1842	0.1460	0.1160	0.0923	0.0736	0.0588
26	0.5976	0.4637	0.3607	0.2812	0.2198	0.1722	0.1352	0.1064	0.0839	0.0663	0.0525
27	0.5859	0.4502	0.3468	0.2678	0.2074	0.1609	0.1252	0.0976	0.0763	0.0597	0.0469
28	0.5744	0.4371	0.3335	0.2551	0.1956	0.1504	0.1159	0.0895	0.0693	0.0538	0.0419
29	0.5631	0.4243	0.3207	0.2429	0.1846	0.1406	0.1073	0.0822	0.0630	0.0485	0.0374
30	0.5521	0.4120	0.3083	0.2314	0.1741	0.1314	0.0994	0.0754	0.0573	0.0437	0.0334
31	0.5412	0.4000	0.2965	0.2204	0.1643	0.1228	0.0920	0.0691	0.0521	0.0394	0.0298
32	0.5306	0.3883	0.2851	0.2099	0.1550	0.1147	0.0852	0.0634	0.0474	0.0355	0.0266
33	0.5202	0.3770	0.2741	0.1999	0.1462	0.1072	0.0789	0.0582	0.0431	0.0319	0.0238
34	0.5100	0.3660	0.2636	0.1904	0.1379	0.1002	0.0730	0.0534	0.0391	0.0288	0.0212
35	0.5000	0.3554	0.2534	0.1813	0.1301	0.0937	0.0676	0.0490	0.0356	0.0259	0.0189
36	0.4902	0.3450	0.2437	0.1727	0.1227	0.0875	0.0626	0.0449	0.0323	0.0234	0.0169
37	0.4806	0.3350	0.2343	0.1644	0.1158	0.0818	0.0580	0.0412	0.0294	0.0210	0.0151
38	0.4712	0.3252	0.2253	0.1566	0.1092	0.0765	0.0537	0.0378	0.0267	0.0190	0.0135
39	0.4619	0.3158	0.2166	0.1491	0.1031	0.0715	0.0497	0.0347	0.0243	0.0171	0.0120
40	0.4529	0.3066	0.2083	0.1420	0.0972	0.0668	0.0460	0.0318	0.0221	0.0154	0.0107

Table D.3. Annuity Factors (Present Value of 1 Per Annum at Compound Interest)

	2%	3%	4%	5%	$a = [1 - (1 + i)^{-n}]/i$						
					6%	7%	8%	9%	10%	11%	12%
1	0.9804	0.9709	0.9615	0.9524	0.9434	0.9346	0.9259	0.9174	0.9091	0.9009	0.8929
2	1.9416	1.9135	1.8861	1.8594	1.8334	1.8080	1.7833	1.7591	1.7355	1.7125	1.6901
3	2.8839	2.8286	2.7751	2.7232	2.6730	2.6243	2.5771	2.5313	2.4869	2.4437	2.4018
4	3.8077	3.7171	3.6299	3.5460	3.4651	3.3872	3.3121	3.2397	3.1699	3.1024	3.0373
5	4.7135	4.5797	4.4518	4.3295	4.2124	4.1002	3.9927	3.8897	3.7908	3.6959	3.6048
6	5.6014	5.4172	5.2421	5.0757	4.9173	4.7665	4.6229	4.4859	4.3553	4.2305	4.1114
7	6.4720	6.2303	6.0021	5.7864	5.5824	5.3893	5.2064	5.0330	4.8684	4.7122	4.5638
8	7.3255	7.0197	6.7327	6.4632	6.2098	5.9713	5.7466	5.5348	5.3349	5.1461	4.9676
9	8.1622	7.7861	7.4353	7.1078	6.8017	6.5152	6.2469	5.9952	5.7590	5.5370	5.3282
10	8.9826	8.5302	8.1109	7.7217	7.3601	7.0236	6.7101	6.4177	6.1446	5.8892	5.6502
11	9.7868	9.2526	8.7605	8.3064	7.8869	7.4987	7.1390	6.8052	6.4951	6.2065	5.9377
12	10.5753	9.9540	9.3851	8.8633	8.3838	7.9427	7.5361	7.1607	6.8137	6.4924	6.1944
13	11.3484	10.6350	9.9856	9.3936	8.8527	8.3577	7.9038	7.4869	7.1034	6.7499	6.4235
14	12.1062	11.2961	10.5631	9.8986	9.2950	8.7455	8.2442	7.7862	7.3667	6.9819	6.6282
15	12.8493	11.9379	11.1184	10.3797	9.7122	9.1079	8.5595	8.0607	7.6061	7.1909	6.8109
16	13.5777	12.5611	11.6523	10.8378	10.1059	9.4466	8.8514	8.3126	7.8237	7.3792	6.9740
17	14.2919	13.1661	12.1657	11.2741	10.4773	9.7632	9.1216	8.5436	8.0216	7.5488	7.1196
18	14.9920	13.7535	12.6593	11.6896	10.8276	10.0591	9.3719	8.7556	8.2014	7.7016	7.2497
19	15.6785	14.3238	13.1339	12.0853	11.1581	10.3356	9.6036	8.9501	8.3649	7.8393	7.3658
20	16.3514	14.8775	13.5903	12.4622	11.4699	10.5940	9.8181	9.1285	8.5136	7.9633	7.4694
21	17.0112	15.4150	14.0292	12.8212	11.7641	10.8355	10.0168	9.2922	8.6487	8.0751	7.5620
22	17.6580	15.9369	14.4511	13.1630	12.0416	11.0612	10.2007	9.4424	8.7715	8.1757	7.6446
23	18.2922	16.4436	14.8568	13.4886	12.3034	11.2722	10.3711	9.5802	8.8832	8.2664	7.7184
24	18.9139	16.9355	15.2470	13.7986	12.5504	11.4693	10.5288	9.7066	8.9847	8.3481	7.7843
25	19.5235	17.4131	15.6221	14.0939	12.7834	11.6536	10.6748	9.8226	9.0770	8.4217	7.8431
26	20.1210	17.8768	15.9828	14.3752	13.0032	11.8258	10.8100	9.9290	9.1609	8.4881	7.8957
27	20.7069	18.3270	16.3296	14.6430	13.2105	11.9867	10.9352	10.0266	9.2372	8.5478	7.9426
28	21.2813	18.7641	16.6631	14.8981	13.4062	12.1371	11.0511	10.1161	9.3066	8.6016	7.9844
29	21.8444	19.1885	16.9837	15.1411	13.5907	12.2777	11.1584	10.1983	9.3696	8.6501	8.0218
30	22.3965	19.6004	17.2920	15.3725	13.7648	12.4090	11.2578	10.2737	9.4269	8.6938	8.0552
31	22.9377	20.0004	17.5885	15.5928	13.9291	12.5318	11.3498	10.3428	9.4790	8.7331	8.0850
32	23.4683	20.3888	17.8736	15.8027	14.0840	12.6466	11.4350	10.4062	9.5264	8.7686	8.1116
33	23.9886	20.7658	18.1476	16.0025	14.2302	12.7538	11.5139	10.4644	9.5694	8.8005	8.1354
34	24.4986	21.1318	18.4112	16.1929	14.3681	12.8540	11.5869	10.5178	9.6086	8.8293	8.1566
35	24.9986	21.4872	18.6646	16.3742	14.4982	12.9477	11.6546	10.5668	9.6442	8.8552	8.1755
36	25.4888	21.8323	18.9083	16.5469	14.6210	13.0352	11.7172	10.6118	9.6765	8.8786	8.1924
37	25.9695	22.1672	19.1426	16.7113	14.7368	13.1170	11.7752	10.6530	9.7059	8.8996	8.2075
38	26.4406	22.4925	19.3679	16.8679	14.8460	13.1935	11.8289	10.6908	9.7327	8.9186	8.2210
39	26.9026	22.8082	19.5845	17.0170	14.9491	13.2649	11.8786	10.7255	9.7570	8.9357	8.2330
40	27.3555	23.1148	19.7928	17.1591	15.0463	13.3317	11.9246	10.7574	9.7791	8.9511	8.2438

Table D.4. Capital Recovery or Amortization Factors (Annuity Whose Present Value at Compound Interest Is 1)

	2%	3%	4%	5%	6%	7%	8%	9%	10%	11%	12%
					$1/a = i/[1 - (1 + i)^{-n}]$						
1	1.0200	1.0300	1.0400	1.0500	1.0600	1.0700	1.0800	1.0900	1.1000	1.1100	1.1200
2	0.5150	0.5226	0.5302	0.5378	0.5454	0.5531	0.5608	0.5685	0.5762	0.5839	0.5917
3	0.3468	0.3535	0.3603	0.3672	0.3741	0.3811	0.3880	0.3951	0.4021	0.4092	0.4163
4	0.2626	0.2690	0.2755	0.2820	0.2886	0.2952	0.3019	0.3087	0.3155	0.3223	0.3292
5	0.2122	0.2184	0.2246	0.2310	0.2374	0.2439	0.2505	0.2571	0.2638	0.2706	0.2774
6	0.1785	0.1846	0.1908	0.1970	0.2034	0.2098	0.2163	0.2229	0.2296	0.2364	0.2432
7	0.1545	0.1605	0.1666	0.1728	0.1791	0.1856	0.1921	0.1987	0.2054	0.2122	0.2191
8	0.1365	0.1425	0.1485	0.1547	0.1610	0.1675	0.1740	0.1807	0.1874	0.1943	0.2013
9	0.1225	0.1284	0.1345	0.1407	0.1470	0.1535	0.1601	0.1668	0.1736	0.1806	0.1877
10	0.1113	0.1172	0.1233	0.1295	0.1359	0.1424	0.1490	0.1558	0.1627	0.1698	0.1770
11	0.1022	0.1081	0.1141	0.1204	0.1268	0.1334	0.1401	0.1469	0.1540	0.1611	0.1684
12	0.0946	0.1005	0.1066	0.1128	0.1193	0.1259	0.1327	0.1397	0.1468	0.1540	0.1614
13	0.0881	0.0940	0.1001	0.1065	0.1130	0.1197	0.1265	0.1336	0.1408	0.1482	0.1557
14	0.0826	0.0885	0.0947	0.1010	0.1076	0.1143	0.1213	0.1284	0.1357	0.1432	0.1509
15	0.0778	0.0838	0.0899	0.0963	0.1030	0.1098	0.1168	0.1241	0.1315	0.1391	0.1468
16	0.0737	0.0796	0.0858	0.0923	0.0990	0.1059	0.1130	0.1203	0.1278	0.1355	0.1434
17	0.0700	0.0760	0.0822	0.0887	0.0954	0.1024	0.1096	0.1170	0.1247	0.1325	0.1405
18	0.0667	0.0727	0.0790	0.0855	0.0924	0.0994	0.1067	0.1142	0.1219	0.1298	0.1379
19	0.0638	0.0698	0.0761	0.0827	0.0896	0.0968	0.1041	0.1117	0.1195	0.1276	0.1358
20	0.0612	0.0672	0.0736	0.0802	0.0872	0.0944	0.1019	0.1095	0.1175	0.1256	0.1339
21	0.0588	0.0649	0.0713	0.0780	0.0850	0.0923	0.0998	0.1076	0.1156	0.1238	0.1322
22	0.0566	0.0627	0.0692	0.0760	0.0830	0.0904	0.0980	0.1059	0.1140	0.1223	0.1308
23	0.0547	0.0608	0.0673	0.0741	0.0813	0.0887	0.0964	0.1044	0.1126	0.1210	0.1296
24	0.0529	0.0590	0.0656	0.0725	0.0797	0.0872	0.0950	0.1030	0.1113	0.1198	0.1285
25	0.0512	0.0574	0.0640	0.0710	0.0782	0.0858	0.0937	0.1018	0.1102	0.1187	0.1275
26	0.0497	0.0559	0.0626	0.0696	0.0769	0.0846	0.0925	0.1007	0.1092	0.1178	0.1267
27	0.0483	0.0546	0.0612	0.0683	0.0757	0.0834	0.0914	0.0997	0.1083	0.1170	0.1259
28	0.0470	0.0533	0.0600	0.0671	0.0746	0.0824	0.0905	0.0989	0.1075	0.1163	0.1252
29	0.0458	0.0521	0.0589	0.0660	0.0736	0.0814	0.0896	0.0981	0.1067	0.1156	0.1247
30	0.0446	0.0510	0.0578	0.0651	0.0726	0.0806	0.0888	0.0973	0.1061	0.1150	0.1241
31	0.0436	0.0500	0.0569	0.0641	0.0718	0.0798	0.0881	0.0967	0.1055	0.1145	0.1237
32	0.0426	0.0490	0.0559	0.0633	0.0710	0.0791	0.0875	0.0961	0.1050	0.1140	0.1233
33	0.0417	0.0482	0.0551	0.0625	0.0703	0.0784	0.0869	0.0956	0.1045	0.1136	0.1229
34	0.0408	0.0473	0.0543	0.0618	0.0696	0.0778	0.0863	0.0951	0.1041	0.1133	0.1226
35	0.0400	0.0465	0.0536	0.0611	0.0690	0.0772	0.0858	0.0946	0.1037	0.1129	0.1223
36	0.0392	0.0458	0.0529	0.0604	0.0684	0.0767	0.0853	0.0942	0.1033	0.1126	0.1221
37	0.0385	0.0451	0.0522	0.0598	0.0679	0.0762	0.0849	0.0939	0.1030	0.1124	0.1218
38	0.0378	0.0445	0.0516	0.0593	0.0674	0.0758	0.0845	0.0935	0.1027	0.1121	0.1216
39	0.0372	0.0438	0.0511	0.0588	0.0669	0.0754	0.0842	0.0932	0.1025	0.1119	0.1215
40	0.0366	0.0433	0.0505	0.0583	0.0665	0.0750	0.0839	0.0930	0.1023	0.1117	0.1213

Glossary

Abilities: availability, reliability, and maintainability.

Accountability: being answerable for the completion of certain tasks.

Accounting values: values of assets, incomes, costs determined by standard rules of accounting.

Actual cash flow: the cash flow that happened in reality. An actual cash flow is made after the time period is over and once the prices, quantities, and other cash flows are known with certainty.

Amortization: allocating a loan (or other value) over a specified number of periods and, at the same time, accounting for any interest that may also accrue.

Annuity: a series of uniform, periodic payments received (or paid) for either a fixed number of periods or in perpetuity.

Authority: the right and power to command, use, access, and so on.

Average income or expense value: values per unit (acre, bushel, head, for example).

Bailment: a contract in which someone else is entrusted with the possession of property, but has no ownership interest in it.

Balance sheet: shows the value of assets, amount of liabilities, and a farm's net worth at a certain point in time.

Benchmarking: identifying the best and then comparing costs and physical efficiencies between farms.

Budget: projection of future income and expenses; used for planning.

Business life cycle: the stages that a business goes through from its beginning entry or establishment, through growth and survival, to exit or disinvestment.

Business model: the combination of strategy, products, structure, organization, and location that a company uses to sustain itself—that is, generate revenue and profit.

Business organization: both (1) how a business is organized in terms of how owners, managers, and workers relate to each other and (2) the legal form of organization (i.e., sole proprietorship, partnership, corporation, or cooperative).

Business plan: a structured statement of a business' strategic plan, marketing plan, production and operations plan, financial plan and statements, and organization and staffing plan.

Capital: simply put, capital is money. Capital is one of the three main broad inputs for business; the other two are land and labor.

Capital asset: an asset that is not expected to be used up during production and thus has a multiyear useful life.

Cash cost or income: a cost or expense that involves an actual cash transfer.

Cash flow budget: a statement of cash inflows and cash outflows.

Cash flow deviations: differences between the projected cash flow and the actual cash flow.

Cash flow management: managing the flow of cash in and out of a business and analyzing when cash is needed and available for paying farm expenses, loan payments, and living expenses, for example.

Cash flow statement: shows the annual flow and timing of cash coming into and out of a business.

Cash rent: a fixed amount of money paid by the tenant to the landowner for the use of the land and(or) buildings.

Cause and effect diagrams: a diagram that shows a problem or opportunity and the potential causes for the problem or opportunity.

Check sheets: forms that merely require a check made in the appropriate column or spot to indicate the frequency of certain events that relate to quality or process characteristics.

Compensation: the payment or reimbursement to a worker for performing the tasks assigned.

Competitive advantage: having a profit rate higher than the industry average.

Competitive forces: the forces that affect the level of competition within an industry.

Compounding: the mathematical process of calculating the future value of a known, present value. Interest is calculated on both the original amount and any accumulated interest.

Contract: an agreement between two or more persons or parties to do or not to do something.

Contractee: usually thought of, in common agricultural use, as the person or party who does the work of producing the product or providing the service to the buyer, that is, the company or processor. (Contractee is not always commonly used. In other usage, a contractor is the one who does the work. So care must be taken to understand who or what the word refers to in the specific situation involved.) *See* **Contractor**.

Contractor: usually thought of as the buyer in common agricultural use, that is, the company or processor who contracts with the seller, typically a farmer, to grow, raise, or provide a specific product or service. (In other usage, a contractor is the one who does the work. So care must be taken to understand who or what the word refers to in the specific situation involved.) *See* **Contractee.**

Controlling: comparing actual results with goals and objectives and taking corrective actions, if needed.

Conviction weights: weights or scores based on our personal, subjective opinion as to whether a certain event will occur.

Cooperative: an enterprise collectively owned and operated for mutual benefit.

Corporation: a group of persons granted a charter to form a separate entity having its own rights, privileges, and liabilities; a corporation is usually formed to operate a business.

Cost basis: the value or price assigned to an asset by subtracting previously taken depreciation from its original basis or value.

Cost center: an enterprise that incurs costs, such as, land ownership.

Costs of quality: costs that are associated with not meeting the customer's requirements.

Crafting strategy: the managerial process of deciding how to achieve the targeted results within the farm's physical and economic environment and its prospects for the future.

Cumulative probability: the increasing probability that a certain event (price, yield, weight, and so on) will occur as the range of possible occurrences increases. The cumulative probability ranges from 0 to 1.

Data: raw, discrete unprocessed information gathered from original sources.

Debt capital: capital from a liability or other financial obligation on which interest and other fees have to be paid.

Decision criteria: a set of rules that can be used to evaluate the information available and make a decision even though the end results are not known with certainty.

Decision making: the process of gathering information, analyzing alternatives, and choosing the best alternative.

Diagnostic analysis: identifying problems behind the symptom and potential solutions.

Direct costs: costs that are used directly by a specific enterprise.

Directing: coordinating resources, directing and scheduling activities, managing personnel.

Discounting: the mathematical process of calculating the present value of a known, future value. by subtracting potential interest to estimate a present value. Discounting is the opposite of compounding.

Dispatching: deciding which jobs or tasks need to be done next.

Dominant economic traits: those economic traits or characteristics that drive and affect an industry and the firms within that industry.

Economic engineering: using data from manufacturers, university reports, and other sources to prepare a budget (instead of allocating whole-farm records).

Economic environment: all the external factors that affect the economic decisions and results on a farm. These factors are grouped into four areas: resources, markets, institutions, and technology.

Economic profitability: an investment's ability to return a reasonable profit on the initial investment. It is evaluated in one of three ways: the payback period, net present value, and internal rate of return.

Economic values: values that take into account cash values, noncash values, opportunity costs, and other methods not included in standard rules of accounting.

Employee: a person hired to do certain tasks for a business or organization.

Employee needs: responsibility, authority, accountability, and compensation.

Enterprise: a common name for any activity, such as, corn, dairy, or machinery.

Enterprise budget: a statement of what is expected if particular production practices are used to produce a specified amount of product.

Enterprise selection: the process of choosing enterprises or products for a farm.

Entry or establishment stage: starting and establishing the business on a stable keel.

Equity capital: capital from the owner(s) of the business as well as partners and other investors.

Exit or disinvestment: planning for and reaching retirement through sale and transfer of resources and responsibilities.

Expense: a charge that is, or is expected to be, made in return for receiving a product or service.

External analysis: looking outside the firm; studying the forces operating in the general economy and in the industry that the firm belongs to.

Fail-safe plans: a process and product design tool that looks for ways to eliminate the possibility of problems or mistakes occurring.

Financial analysis: evaluating a farm's financial health, that is, its financial position and performance, on the basis of its profitability, solvency, liquidity, repayment capacity, and financial efficiency.

Financial condition and performance: an evaluation of a farm's profitability, solvency, liquidity, repayment capacity, and efficiencies.

Financial control: the process of determining and implementing the necessary actions to make certain that financial plans are transferred into desired results.

Financial efficiency: the ability to use financial resources and expenses well to produce profit.

Financial feasibility: an investment's own ability to generate sufficient cash flow to cover any debt incurred to make the investment.

Financial management: the process of obtaining, using, and controlling capital—both cash and credit.

Financial objectives: targets established for the farm's financial performance.

Financial performance: the financial results of decisions over time.

Financial position: the financial resources controlled by a farm and the claims against those resources.

Financial risk: financial risk has four components: (1) the cost and availability of debt capital, (2) the ability to meet cash flow needs in a timely manner, (3) the ability to maintain and grow equity, and (4) the increasing chance of losing equity by larger levels of borrowing against the same net worth.

Financial statements: reports that organize a farm's financial information three ways—the income statement, the balance sheet, and the cash flow statement.

Fixed cost: a cost that occurs no matter what or how much is produced.

Functions of management: planning, organizing, directing, and controlling. (Compared to the functions of business: production, marketing, finance, and personnel.)

Gantt chart: a chart showing (1) when jobs are to be done, (2) who is doing which jobs, and (3) which equipment is being used on which jobs, by whom and on which days.

GIGO: garbage in, garbage out.

Gross margin: gross income minus variable costs.

Growth and survival stage: improving and expanding the resource base and income generation while protecting the business from the risks in its economic environment.

Human resource risk: disruption in the business due to death, divorce, injury, illness, poor management, improper operation and application of production and marketing procedures, poor hiring decisions, improper training, and so on.

Income: a value that is, or expected to be, received in return for providing a product or service.

Income capitalization: the value of real estate estimated as the present value of the future stream of income due to the productive capability of that real estate.

Income statement: reports income versus expenses for a specific period of time.

Information: processed, interpreted data.

Initial analysis: a first-time analysis of a farm with the assumption of having no previous knowledge of the farm.

Input supply management: identifying the quantity, timing, and source of the inputs needed to meet production output plans.

Internal analysis: looking inside the firm; evaluating a farm's strengths, weaknesses, competitive capabilities, and past and potential condition and performance.

Internal rate of return (IRR): the discount rate that sets the net present value of the investment to zero.

Interval probability: the probability or chance that a certain event (such as the actual price or yield being within a certain interval) will occur. An interval probability will range from 0 to 1. The sum of all the interval probabilities for a certain event will equal 0.

ISO 9000 standards: a set of guidelines (developed by the International Organization for Standards) on how to set up and operate a management system to ensure that products conform to the customer's requirements.

Job design: developing jobs by combining tasks that are complementary with each other and interesting to the worker.

Legal risk: unknown and unanticipated events due to business structure and tax and estate planning, contractual arrangements, tort liability, and statutory compliance, including environmental issues.

Liquidity: the ability of the firm to cover debt during the next 12 months from short-term assets, measured at a certain point in time.

Listed costs: a term used in published budgets to signal that the list of costs may not include every cost found on a specific farm.

Long-range objectives: the results to be achieved either within the next three to five years or else on an ongoing basis year after year.

Long-run: a planning horizon that is more than one year and probably many more years.

Low-cost strategy: being the low-cost producer.

Macro environment: the four dimensions of macroeconomics, social, demographic and the political and legal environments.

Management: making and implementing decisions that allocate limited resources in ways to achieve as best as possible an organization's goals.

Management by exception: setting rules on what size of deviations need management's attention. These rules can be in terms of absolute deviations or percentage deviations.

Marginal input cost: the cost of an additional unit of input.

Marginal product: the product received due to an additional unit of input.

Marginal return or cost: the additional return or cost resulting from an additional unit of output.

Market niche strategy: producing a specialty product for a specific market.

Market pull: make what one can sell.

Market-specific production contract: a contract (also referred to as a sales contract) in which the farmer agrees to produce a specific crop or livestock and to sell the product at harvest to the contractor.

Market value: the value or price assigned to an asset by the marketplace.

Marketing risk: not knowing what prices will be. Unanticipated forces, such as weather or government action, can lead to dramatic changes in crop and livestock prices.

Maslow's seven basic human needs: physiological, safety, belonging and love, esteem and self-esteem, self-actualization, cognitive understanding, and aesthetics.

Maximin: this decision criteria chooses the action that, after identifying the minimum return of each possible action, has the largest minimum return. In other words, this decision criteria chooses the maximum of the minimums.

Minimax: this decision criteria chooses the action that, after identifying the maximum regret of each possible action, has the smallest maximum regret. In other words, this decision criteria chooses the minimum of the maximums.

Mission: a firm's definition of its current business directions and goals; it indicates what a farm is trying to do for its customers.

Models of decision making: different descriptions of how decisions are made (for example, linear, matrix, "garbage can," and so on).

Moral risk: devious and less-than-truthful behavior by individuals and other companies as well as corrupt and criminal behavior.

Net present value (NPV): the sum of the present values of future after-tax net cash flows minus the initial investment.

Noncash income and expense: a real income or cost that does not involve a transfer of cash, such as, depreciation and intrafarm transfers between enterprises (corn to livestock, for example).

Opportunity cost: the net value that could be received from the best alternative use of a resource.

Organizing: acquiring and organizing the necessary resources to carry out the business plan.

Overhead costs: costs that are hard to assign directly to a particular enterprise.

Partial budget: estimate of the net effects of only what changes in the business.

Partnership: two or more partners who have joined to operate a business.

Payback period: the number of years required to recover the initial cost of the investment. The payback period can be estimated using either discounted or undiscounted future after-tax net returns.

Payoff matrix: a table of potential returns or payoffs that could be obtained if certain actions are taken and certain events occur.

Personal service contract: a contract (also referred to as a resource-providing contract) that specifies that the producer is to provide services, not commodities, to the contractor.

Planning: determining the intended strategy and course of action for the business.

Political risk: disruption in plans due to changing policies—both governmental and institutional policies (such as lending policies at a bank).

Porter's five forces: five forces that shape the competitiveness within an industry. The five forces are risk of entry by potential competitors, rivalry among established farms, bargaining power of buyers, bargaining power of suppliers, and substitute products.

Process: a series of actions designed to bring about a particular result.

Process control: the procedures used to monitor production during production for compliance with the original plan and development of corrective actions designed to bring the process back into compliance.

Process design: specifying the inputs, actions, methods, jobs, machines, and steps to be used in the production process.

Process improvement: understanding the current process better and looking for potential ways to improve both the process and the product.

Process map: a description of a method or process of accomplishing a task.

Process quality: (1) how well a process produces a product within specification given by the consumer and (2) how well the process performs for the producer.

Product development: choosing the specific characteristics of the items that could be produced.

Product quality: how well a product meets the specifications given by the customer; its "fitness of use."

Production contract: an agreement between two or more persons or parties to produce a specific product or provide a specific service.

Production-management contract: a contract used when the seller (i.e., farmer or producer), through production decisions, can affect the value of the product to the buyer (say a vegetable or meat processor or seed company) or when the seller, through marketing decisions, can affect the value of the product to the buyer.

Production risk: not knowing what actual production levels will be. The major sources of production risk are weather, pests, diseases, technology, genetics, machinery efficiency and reliability, and the quality of inputs.

Profit center: an enterprise that brings in income, such as, soybeans or hogs.

Profitability: the ability to produce a profit over a period of time.

Projected cash flow: an estimate of what the cash flow may be for a future period of time. A projected cash flow or cash flow budget is made with information available before the time period under consideration.

Quality: meeting or exceeding customer requirements.

Quality control: controlling the already selected production process for the products already chosen and designed. Two major parts of quality control are process control and process improvement.

Quality management: a holistic view of the entire production process from initial design through input purchases and actual production to the service supplied after production.

Quality of conformance: how well the production process has done in producing a product that meets the specifications of the design.

Quality of design: how well the product design meets consumer needs and wants.

Real estate: landed property; usually referring to the land plus buildings and other improvements such as drainage tile, waterways, fences, and so on. Also includes all mineral rights and other natural resources.

Regret matrix: a table of potential regrets that, if a certain event happens having chosen a certain action instead of having chosen any other alternative action. A regret matrix is calculated from a payoff matrix.

Repayment capacity: the ability to cover cash outflow from cash inflows over a period of time.

Resource-providing contract: a contract (also referred to as a personal service contract) that specifies that the producer is to provide services, not commodities, to the contractor.

Responsibility(ies): the duty(ies) or task(s) that a worker is obligated or expected to do.

Risk: we know all the possible outcomes and the objective probability of each outcome occurring.

Safety-first rule: this decision criteria first eliminates all possible actions that violate some safety-first rule (such as, "no losses") and then chooses from the remaining actions the action that has the largest expected return.

Sales contract: a contract (also referred to as a market-specific production contract) in which the farmer agrees to produce a specific crop or livestock and to sell the product at harvest to the contractor.

Scenarios: descriptions of different views or possibilities of what the future may be like.

Scheduling: deciding what activities are to be done, when they will be done, who will do these activities, and what equipment will be needed.

Sequencing: specifying the exact order of operations or jobs.

Service after delivery: the warranty, repair, and replacement after the product has been sold.

Share rent: a percentage of the physical yield that the landowner receives in return for allowing the tenant to use the land. The value of the physical yield received by the landowner will vary with the price of the product. The landowner usually pays some of the production costs as well as receives part of the yield.

Short-range objectives: the organization's near-term performance targets; the amount of short-term improvement signals how fast management is trying to achieve the long-range objectives.

Short-run: a planning horizon of one year or less.

Sole proprietorship: a business owned by one owner.

Solvency: the ability to pay off all debts at a certain point in time.

Sparse data method: estimating probabilities using only a few observations.

Stakeholders: the people, businesses and institutions that have a claim or interest in the farm.

Strategic control: monitoring a farm's strategic performance and its external environment and making adjustments in the strategy and its implementation as needed to accomplish the farm's objectives and its vision.

Strategic objectives: targets established for strengthening the farm's overall position and competitive vitality.

Strategic plan: a statement outlining an organization's mission and future direction, near-term and long-term performance targets, and strategy.

Strategic planning: the process of identifying stakeholders, objectives, vision, and mission; developing internal and external analysis; crafting strategy.

Strategy: the pattern of actions managers employ to achieve organizational objectives.

Strategy implementation: the full range of managerial activities associated with putting the chosen strategy into place, supervising its pursuit, and achieving the targeted results.

Structural change: changes in the makeup of an industry: the number, size, and geographical location of buyers, suppliers, processors, producers, and all other firms as well as the physical, legal, and social network of connections between the firms in an industry.

Sunk costs: variable costs that have become fixed once they have been used or committed to use.

SWOT analysis: developing and analyzing a firm's strengths, weaknesses, opportunities, and threats.

Technology push: sell what one can make.

Time value of money: the value that money has by being used over time.

Total income or expense: income or expense summed over an entire farm or enterprise.

TQM: Total Quality Management is a management philosophy that strives to involve everyone in a continual effort to improve quality and achieve customer satisfaction.

Trend charts: a running plot of measured quality characteristics: stored grain moisture, pigs born per litter, milk per cow per day, acres per day, bacteria or somatic cell counts, and so on.

Triangular distribution: estimating probabilities using only three values: most likely, lowest, and highest.

Uncertainty: we know some (or maybe all) of the possible outcomes, but we cannot quantify the probabilities.

Value of additional product: the value of the marginal product.

Variable cost: a cost that changes as the volume of business changes.

Variable or flexible cash rent: The landowner receives only cash from the payment, but the amount paid each year can vary due to changes in the physical crop yield, the productivity of the animals, or the market price. The entire payment may be variable or there may be a fixed base cash rent plus a variable portion.

Vision: the picture of what the stakeholders want the farm to look like in the future; a view of an organization's future direction and business course; a guiding concept for what the organization is trying to do and to become.

Whole-farm: a combination of enterprises for a farm.

Whole-farm budget: a summary of the major physical and financial features of the entire farm.

Worker: a person who works for a business or organization who may or may not be paid a wage or salary. "Worker" is used as a generic term that includes employees, owners, partners, and family members.

Workforce: the people who work for a business, farm, or any organization.

Workforce management: planning, organizing, directing, and controlling the workers in a business or organization.

Bibliography

Abell, D. F. 1980. *Defining the Business: The Starting Point of Strategic Planning.* Englewood Cliffs, NJ: Prentice-Hall.

American Society of Agricultural Engineers. 1998. *Agricultural Machinery Management Data,* ASAE D497.4 January 1998, available at http://asae.frymulti.com/request2.asp?JID=2&AID=2524&CID=s2000&v=&i=&T=2, accessed December 17, 2002.

American Society of Agricultural Engineers. 2001. *Agricultural Machinery Management,* ASAE EP496. 2 December 2001, available at http://asae.frymulti.com/request2.asp?JID=2&AID=2523&CID=s2000&T=2, accessed December 17, 2002.

Billikopf, G. E. 2000. *Labor Management in Ag: Cultivating Personnel Productivity.* Agricultural Extension, Stanislaus County, University of California. http://www.cnr.berkeley.edu/ucce50/ag-labor/, accessed October 22, 2002.

Boehlje, M. D., and V. R. Eidman. 1984. *Farm Management.* New York: John Wiley & Sons.

Center for Dairy Profitability. 2003. *Agricultural Budget Calculation Software Version 7.1.* University of Wisconsin, http://cdp.wisc.edu/, accessed February 19, 2003.

Cow-calf Production, Partially Improved Pasture, Texas Coastal Bend Area (District 11). 2002. Texas Agricultural Extension Service, Texas A & M University.

Duffy, M., and D. Smith. 2002. Estimated Costs of Crop Production in Iowa—2002,@ FM 1712 Rev. Jan. 2002. University Extension, Iowa State University, Ames.

Duncan, R. A., P. A. Verdegaal, B. A. Holtz, K. A. Klonsky, and R. L. De Moura. Sample Costs to Establish an Almond Orchard and Produce Almonds: San Joaquin Valley North. 2002. Flood Irrigation, AM-VS-02-1. University of California Cooperative Extension.

Farm Financial Standards Council. 1997. *Financial Guidelines for Agricultural Producers II.* Naperville, IL: Farm Financial Standards Council. Revised 1997.

FDA. 2001. *HACCP: A State-of-the-Art Approach to Food Safety.* October, 2001. Http://www.fda.gov/opacom/backgrounders/haccp.html, accessed December 16, 2002.

Ferreira, W. N. 2001. *BUDSYS: A New Tool for Farm Enterprise Analysis.* EER 195, Department of Agricultural and Applied Economics, Clemson University, available at http://cherokee.agecon.clemson.edu/budgets/BudSys.htm, accessed February 19, 2003.

Food Safety and Inspection Service, USDA, 1999. *Guidebook for the Preparation of HACCP Plans.* HACCP-1, September.

Futrell, G., ed. 1982. *Marketing for Farmers.* St. Louis, MO: Doane-Western.

421

Gerhardson, B. 1999. *A Guide to Agricultural Production Contracting in Minnesota*. St. Paul: Minnesota Department of Agriculture.

Hawkins, R., R. Craven, K. Klair, R. Loppnow, D. Nordquist, and W. Richardson. 1993. *FINPACK User's Manual*. St. Paul, MN, Center for Farm Financial Management, Department of Agricultural and Applied Economics, University of Minnesota.

Herzberg, F. 2002. One more time: How do you motivate employees? In Best of HBR on motivation, product number 2772. *Harvard Business Review*, 17–32.

Hill, C. W. L., and G. R. Jones. 1998. *Strategic Management Theory: An Integrated Approach*. 4th ed. Boston: Houghton Mifflin Company.

Hurley, T., J. Kliebenstein, P. Orazem, and D. Miller. 2000. *Employee Management, National Hog Farmer*. Minneapolis, MN: Intertec Publishing Company.

International Organization for Standardization. 2000. *The ISO 9000 Family*. http://www.ISO.ch/iso/en/iso9000-14000/iso9000/selection_use/iso9000family.html, accessed October 25, 2001.

Jensen, B. 1985. Unpublished Notes. College of the Sequoias, Visalia, California.

Kay, R. D., and W. M. Edwards. 1999. *Farm Management*. 4th ed. New York: McGraw-Hill, Inc.

Kunkel, P. L., and S. T. Larison. 1998a. *Agricultural Production Contracts*. FO-7302-GO, University of Minnesota Extension Service.

Kunkel, P. L., and S. T. Larison. 1998b. *Contracts, Notes and Guarantees*. FS-2590-GO, University of Minnesota Extension Service.

Laughlin, D. H., and S. R. Spurlock. 2001. *Mississippi State Budget Generator version 5.5*. Department of Agricultural Economics, Mississippi State University, http://www.agecon.msstate.edu/laughlin/msbg.php, accessed February 19, 2003.

Lazarus, W. 1996. *Minnesota Farm Machinery Economic Cost Estimates for 1996*. FO-6696-B, Minnesota Extension Service, University of Minnesota.

Lazarus, W. F. 2002. MACHDATA.XLS, St. Paul, MN: University of Minnesota, Department of Applied Economics, 2002. [Contact Lazarus at wlazarus@umn.edu for more information.]

Lazarus, W. F., and R. A. Selley. 2002a. *Farm Machinery Economic Cost Estimates for 2002*, FO-6696, St. Paul, MN: University of Minnesota, Minnesota Extension Service, September 4, 2002. Available at: http://www.apec.umn.edu/faculty/wlazarus/mf2002.pdf, accessed December 17, 2002.

Lazarus, W. F., and R. A. Selley. 2002b. Suggested Procedures for Estimating Farm Machinery Costs. St. Paul, MN: University of Minnesota, Department of Applied Economics. [Staff Paper P02-16] Electronically available only at: http://agecon.lib.umn.edu/mn/p02-16.pdf, accessed December 14, 2002.

LeGrande, E. 1963. The development of a factory simulation system using actual operating data. *Management Technology* 3(1, May): 17.

Libbin, J. D., and L. B. Catlett. 1987. *Farm and Ranch Financial Records*. New York: Macmillan Publishing Co.

Maslow, A. H. 1970. *Motivation and Personality*. 2d ed. New York: Harper & Row.

Mintzberg, H., D. Raisingham, and A. Théorêt. 1976. The structure of "unstructured" decision processes. *Administrative Science Quarterly* 21(1976): 246–75.

Molz, R. 1988. *Steps to Strategic Management: A Guide for Entrepreneurs*. Plano, Texas: Wordware Publishing, Inc.

Nordquist, D., and B. Lazarus. 1997. Farm labor laws and regulations for Minnesota, 1997. FM 303, Rev. 12/1997. Department of Applied Economics, University of Minnesota, St. Paul.

Öhlmér, B., K. Olson, and B. Brehmer. 1998. Understanding farmers' decision making processes and improving managerial assistance. *Agricultural Economics*, 18(1998):273–290.

Osburn, D. D., and K. C. Schneeberger. 1983. *Modern Agricultural Management: A Systems Approach to Farming*. 2d ed. Reston, VA: Reston Publishing Company, Inc.

Porter, M. 1980. *Competitive Strategy: Techniques for Analyzing Industries and Competitors*. New York: Free Press, 1980.

Porter, M. 1985. *Competitive Advantage*. New York: Free Press, 1985.

Production Contracts Task Force. 1996a. *Grain Production Contract Checklist*. Office of the Attorney General, Iowa Department of Justice, accessed December 6, 2002, through Iowa State University Extension, http://www.exnet.iastate.edu/Pages/grain/tools/graincontract.html.

Production Contracts Task Force. 1996b. *Livestock Production Contract Checklist*. Office of the Attorney General, Iowa Department of Justice, accessed December 6, 2002, through Iowa State University Extension, URL: http://www.exnet.iastate.edu/Pages/grain/tools/livecontract.html.

Schroeder, R. G. 2000. *Operations Management*. New York: Irwin McGraw-Hill, Inc.

Starner, D. 2001. *MinnCERT*. University of Minnesota, College of Veterinary Medicine. Available at http://www.cum.umn.edu/anhlth_foodsafety/MinnCERT.html, accessed February 18, 2003.

Stevenson, W. J. 2002. *Operations Management*. 7th ed. Boston: McGraw-Hill Irwin, 2002.

Swensen, A., and R. Haugen. 2001. *Projected 2002 Crop Budgets North Central North Dakota*. Farm Management Planning Guide, NDSU Extension Service. North Dakota State University, Fargo.

Swoboda, R. 2001 (February). ISO 9002: What can it do for you? *Wallaces Farmer*, 12–13.

Thomas, K. H., and B. L. Erven. 1989. *Farm Personnel Management*. North Central Regional Extension Publication 329-1989, AG-BU-3613, 1989.

Thompson, A. A., and A. J. Strickland. 1992. *Strategy Formulation and Implementation: Tasks of the General Manager*. 5th ed. Homewood, IL: Irwin.

Thompson, A. A., Jr., and A. J. Strickland, III. 2003. *Strategic Management: Concepts and Cases*. 13th ed. Homewood, IL: Irwin.

Weness, E. 2000. Personal Communication. Minnesota Extension Service, University of Minnesota, Worthington.

Weness, E. 2002a. *Soybean prices*. Minnesota Extension Service, University of Minnesota, updated 10/3/02. URL: http://swroc.coafes.unp.edu/SWFM/Files/graphs/swma_charts/soybeans.htm.

Weness, E. 2002b. *Price Probability*. Minnesota Extension Service, University of Minnesota, updated 6/11/02. URL: http://swroc.coafes.umn.edu/SWFM/Files/fin/price_probability.htm, accessed December 12, 2002.

Weness, E., and B. Anderson. 2001. *A 2002 Crop Budget for Southwest Minnesota*. Minnesota Extension Service, University of Minnesota.

West, T. D., T. J. Vyn, and G. S. Steinhardt. 2002. *Feasibility of One-Pass Tillage Systems for Corn and Soybeans*. AGRY 02-01, Agronomy Department, Purdue University.

Willis, R. E. 1987. *A Guide to Forecasting for Planners and Managers*. Englewood Cliffs, NJ: Prentice-Hall, Inc.

Index